D1187662

SIMPLIFIED DESIGN OF BUILDING STRUCTURES

SIMPLIFIED DESIGN OF BUILDING STRUCTURES

JAMES AMBROSE

Professor of Architecture
University of Southern California
Los Angeles, California

SECOND EDITION

A Wiley-Interscience Publication

JOHN WILEY & SONS

New York / Chichester / Brisbane / Toronto / Singapore

Library of Congress Cataloging in Publication Data

Ambrose, James E.
Simplified design of building structures.

"A Wiley-Interscience publication."
Bibliography: p.
Includes index.
1. Structural design. 2. Architectural design.
3. Buildings. I. Title.

TA658.A5 1986 721 85-22676
ISBN 0-471-80929-2

Printed in the United States of America

10 9 8 7 6 5 4 3 2

Preface

The basic purpose and intended usage for this book remain essentially the same as described in the preface to the first edition. Since that book was published, I have written several others, including new editions of four of the books written by the late Harry Parker. From that experience, as well as the responses I have received from those who have used the first edition, I developed several ideas for the improvement of this book.

The first edition had three example buildings. I have added a fourth building (Building Three in this edition) to further extend the range of building types and permit some examples of additional types of construction.

In keeping with the style developed in my recent books, I have provided SI unit values in the numerical work done along with the traditional units. This issue is still in transition, and most of the basic references are still in traditional units, so that it is much easier to use material from these references in computations done in traditional units.

Most of the material from the first edition has been retained, although some reduction has been made to eliminate repetitive presentations and to make somewhat fuller use of design aids where these are commonly utilized by professional designers. The graphic presentation of the building construction as an aid to visualization of problems has been retained and extended.

Although the book is not intended to be freestanding with regard to reference material, considerable information is presented in the appendixes, which consist mostly of materials adapted or directly reproduced from various references. I am grateful to the International Conference of Building Officials, publishers of the *Uniform Building Code,* and to the American Institute of Steel Construction, the Masonry Institute of America, and John Wiley & Sons for permission to use these materials.

The legions of students who have suffered through my courses in structures over the years have served nobly as testing grounds for my work. If there is any merit in this material, it has been painfully forged by the give and take of my classroom contacts.

I am grateful to the editors and the production staff at John Wiley & Sons for their competent and thorough work in shaping my rough product into something that is presentable.

Finally, I am, as always, grateful to my family for their patience and tolerance. Because I do almost all of my writing at home, I could not possibly have completed this effort without their help and understanding.

James Ambrose

Westlake Village, California
January 1986

Preface to the First Edition

This book is intended to fill a gap that has existed in the technical literature in the area of structural design for buildings. While the subject has usually been well covered with regard to its many topics in an incremental way, there have been relatively few books written to explain the overall process of designing a building structure; beginning with the architect's design drawings and ending with a set of structural plans and details. That is, of course, precisely what the structural designer does in the majority of building design cases, and yet the process has seldom been illustrated. The few attempts to do so have usually consisted of examples of the designer's calculations, with little explanation of the general process or of the relations between the architectural and structural design, and with a minimum of graphic illustration.

The work here consists of the illustration of the design of the structural systems for three relatively ordinary buildings: a two-story residence, a one-story commercial building, and a six-story office building. For each building the presentation begins with a set of architectural design drawings such as would normally be developed early in the building design process. This is followed by the development of an example structural system, with sample calculations for typical elements of the system and some discussion of the alternates and options possible for various situations. For the second and third buildings two separate structures are designed with different materials. Completing the illustration in each case is a set of typical structural plans and details.

In order to keep the work within the range of those with less than a complete training in structural engineering, calculations have been limited to simplified and approximate methods as usually presented in books written for architecture students and others with less than a thorough background in calculus and engineering physics. This means that the work is slightly below acceptable professional standards in some cases, although it is generally sufficient to obtain approximate designs that are useful for cost estimating, for development of the architectural details, and for gaining a general sense of the needs of the structure. The reader with a more rigorous training in engineering may easily pursue the analysis and design calculations to a higher level of accuracy, but will usually find that the end results are not substantially changed.

While this book is essentially intended for self-study, or for use in teaching in architecture or technical school programs, the lack of similar illustrative material should also make it of considerable value to engineering students and engineers in training. In fact, anyone who is interested in the general problem of designing structures for buildings, and who has not actually done it much, should benefit from reading this work.

Two decisions had to be made in developing this material. The first had to do with the selection of the references to be used. These were deliberately chosen to be ones that were generally available as well as being usable by the less than experienced reader. The second decision had to do with the use of English units (feet, pounds, etc.) instead of international units (metrics), which are steadily becoming more widely used in engineering work. Since the references selected all use English units, the decision was a pragmatic one—to reduce confusion for the reader. The necessity for conversion from one system to the other will simply be a way of life for designers in the coming years.

This work has been developed from my experience over some 30 years of involvement in building design, as a student, teacher, writer, and professional designer. Much is owed to the teachers, students, critics, and professional colleagues whose reactions and help have molded that experience and tempered it. I am grateful to the International Conference of Building Officials, the American Institute of Steel Construction, and the Concrete Masonry Association of California for permission to draw extensively from materials in their publications. Reading of the text drafts by my colleagues, Harold Hauf and Dimitry Vergun, provided invaluable assistance and encouragement. Finally, I am indebted to my family for their patience and indulgence and especially to my wife, Peggy, for her faith and her important assistance.

JAMES AMBROSE

Los Angeles, California
March 1979

Contents

SIMPLIFIED DESIGN OF BUILDING STRUCTURES

Introduction

Designing structures for buildings involves the consideration of a wide range of factors. Building structural designers must not only understand structural behavior and how to provide for it adequately, but must also be knowledgeable about building construction materials and processes, building codes and standards, and the economics of building. In addition, because the structure is merely a subsystem in the whole building, they must have some understanding of the problem of designing the whole building. Structures should not only be logical in their own right, but should also relate well to the functional purposes of the building and to the other subsystems for power, lighting, plumbing, heating, and so on.

Formal education in structural design is usually focused heavily on learning the procedures for structural analysis and the techniques and problems of designing individual structural elements and systems in various materials. The whole problem of designing a structure for a building is not well documented, and learning it usually takes place primarily on the job in professional offices. Although this means of learning is valuable in some ways, it does not provide a good general understanding because it is usually limited to the highly specific situations of each design problem.

The principal purpose of the examples in this book is to illustrate the problems and processes of designing whole structural systems for buildings. The procedure used in the examples is to first present a general building design as a given condition, which is followed by the illustration of the selection and design of the various typical elements of the structural system. The buildings shown are not particularly intended as examples of good architectural design but merely as illustrations of common structural design situations.

Although most of the calculations shown are in reasonably complete form, it is assumed that the reader has previously mastered the fundamentals of analysis and design of simple structures. The word "simplified" implies some limit to the complexity of the work, and the general image for this limit is the level of complexity dealt with in the series of books originally written by the late Harry Parker that bear titles beginning with the word simplified. The first six books in the list of references for this work are from that series and should be considered as the basic references for the structural calculations in this work.

Two topics, however, that are not generally developed in the books by Professor Parker are design for lateral forces of wind and earthquakes and the design of building foundations. The references for these topics used in this book are from two books that I wrote in recent years. *Simplified Building Design for Wind and Earthquake Forces* (Ref. 12) develops the general problems of designing for the effects of these forces in a style that is intended to complement the Parker series. *Simplified Design of Building Foundations* (Ref. 11) is simi-

larly conceived and develops the subjects of soil behaviors and the design of ordinary foundation elements of both plain and reinforced concrete.

In many relatively simple structures most structural design problems can be "solved" by the use of tabulated materials from codes, handbooks, and manufacturer's flyers. Where this is possible from readily available sources, the examples show such use. Usually, however, longhand calculations are shown for the purpose of explaining the problems more thoroughly.

For sake of brevity the structural calculations shown are not complete but are limited to the typical elements of the systems. To complete the illustrations, however, the framing plans and other drawings are usually shown in reasonably complete detail.

Construction detailing of structures and of buildings in general is subject to considerable variation, effected by the judgment of individual designers as well as by regional conditions and practices. Although detailing of the construction in the examples has been developed from the recommendations of various codes, industry standards, and other sources, it is not the purpose of this book to serve as a guide for building construction detailing. Details shown are for the purpose of giving complete illustrations and should not therefore be considered as recommended standards.

Although the procedure in the examples is to begin the structural design after the general building design has been predetermined in considerable detail, it is much better practice to involve structural considerations in the earliest design work. Because it is not possible to illustrate this process without a complete presentation of the whole architectural design process, the examples should be accepted with this limitation in mind. It is assumed that there are good reasons for the situations shown in the examples, although it is pointed out occasionally how some changes in material use, in plan layouts, or in other details might result in improvement of the structure.

Because there are several model building codes and hundreds of local codes in use throughout the United States, it is difficult to deal generally with building code requirements. It is not possible, however, to show building design examples without the use of some code criteria. Because of its reasonable thoroughness, we have chosen to use the *Uniform Building Code* (Ref. 7) as a general reference for the work here. Fortunately, except for regional variations of snow, wind, and earthquake problems, structural design criteria are reasonably consistent among most building codes. The reader is cautioned, however, to use the legally enforceable code for any actual design work.

References for structural design information in general tend to be dated, and their use varies regionally. Anyone using this work as a guide for actual design problems should take care to be sure that the references are currently accepted by legally enforceable codes and regulatory agencies. The references used are listed after Chapter 23 and note should be made of their dates.

To keep the work in this book within the scope of those not fully trained in structural engineering use has been made of simplified analysis and design techniques. The reader is encouraged to not accept this simplified approach entirely, but to pursue the mastery of more exact and thorough methods where they are significant to the work. It is hoped that the learning of these simplified techniques will serve as an initial stage in an ongoing development of competency in structural design.

In most professional design firms structural computations are now com-

monly done with computers, particularly when the work is complex or repetitive. Anyone aspiring to participation in professional design work is advised to acquire the background and experience necessary to the application of computer-aided techniques. The computational work in this book is simple and can be performed easily with a pocket calculator. The reader who has not already done so is advised to obtain one. The "pocket slide rule" type with eight-digit capacity is quite sufficient.

Structural computations can for the most part be rounded off. Because accuracy beyond the third place is seldom significant, this is the level used in this work. In some examples more accuracy is carried in early stages of the computation to ensure the desired degree in the final answer. All the work in this book, however, was performed on an eight-digit pocket calculator.

At the time of preparation of this edition, the building industry in the United States is still in a state of confused transi-

TABLE 1. Units of Measurement: U.S. System

Name of Unit	Abbreviation	Use
Length		
Foot	ft	Large dimensions, building plans, beam spans
Inch	in.	Small dimensions, size of member cross sections
Area		
Square feet	ft^2	Large areas
Square inches	in.^{2}	Small areas, properties of cross sections
Volume		
Cubic feet	ft^3	Large volumes, quantities of materials
Cubic inches	in.^{3}	Small volumes
Force, mass		
Pound	lb	Specific weight, force, load
Kip	k	1000 lb
Pounds per foot	lb/ft	Linear load (as on a beam)
Kips per foot	k/ft	Linear load (as on a beam)
Pounds per square foot	lb/ft^2, psf	Distributed load on a surface
Kips per square foot	k/ft^2, ksf	Distributed load on a surface
Pounds per cubic foot	lb/ft^3, pcf	Relative density, weight
Moment		
foot-pounds	ft-lb	Rotational or bending moment
inch-pounds	in.-lb	Rotational or bending moment
kip-feet	k-ft	Rotational or bending moment
kip-inches	k-in.	Rotational or bending moment
Stress		
Pounds per square foot	lb/ft^2, psf	Soil pressure
Pounds per square inch	lb/in.^{2}, psi	Stresses in structures
Kips per square foot	k/ft^2, ksf	Soil pressure
Kips per square inch	k/in.^{2}, ksi	Stresses in structures
Temperature		
Degree Fahrenheit	°F	Temperature

tion from the use of English units (feet, pounds, etc.) to the new metric-based system referred to as the SI units (for Système International). Although a complete phase-over to SI units seems inevitable, at the time of this writing the construction materials and products suppliers in the United States are still resisting it. Consequently, the AISC Manual and most building codes and other widely used references are still in the old units. (The old system is now more appropriately called the U.S. system because England no longer uses it!) Although it results in some degree of clumsiness in the work, we have chosen to give the data and computations in this book in both units as much as is practicable. The technique is generally to perform the work in U.S. units and immediately follow it with the equivalent work in SI units enclosed in brackets [thus] for separation and identity.

Table 1 lists the standard units of measurement in the U.S. system with the abbreviations used in this work and a description of the type of use in structural work. In similar form Table 2 gives the corresponding units in the SI system. The conversion units used in shifting from one system to the other are given in Table 3.

TABLE 2. Units of Measurement: SI System

Name of Unit	Abbreviation	Use
Length		
Meter	m	Large dimensions, building plans, beam spans
Millimeter	mm	Small dimensions, size of member cross sections
Area		
Square meters	m^2	Large areas
Square millimeters	mm^2	Small areas, properties of cross sections
Volume		
Cubic meters	m^3	Large volumes
Cubic millimeters	mm^3	Small volumes
Mass		
Kilogram	kg	Mass of materials (equivalent to weight in U.S. system)
Kilograms per cubic meter	kg/m^3	Density
Force (load on structures)		
Newton	N	Force or load
Kilonewton	kN	1000 newtons
Stress		
Pascal	Pa	Stress or pressure (one pascal = one N/m^2)
Kilopascal	kPa	1000 pascals
Megapascal	MPa	1,000,000 pascals
Gigapascal	GPa	1,000,000,000 pascals
Temperature		
Degree Celsius	°C	Temperature

TABLE 3. Factors for Conversion of Units

To Convert from U.S. Units to SI Units Multiply by	U.S. Unit	SI Unit	To Convert from SI Units to U.S. Units Multiply by
25.4	in	mm	0.03937
0.3048	ft	m	3.281
645.2	in^2	mm^2	1.550×10^{-3}
16.39×10^3	in^3	mm^3	61.02×10^{-6}
416.2×10^3	in^4	mm^4	2.403×10^{-6}
0.09290	ft^2	m^2	10.76
0.02832	ft^3	m^3	35.31
0.4536	lb (mass)	kg	2.205
4.448	lb (force)	N	0.2248
4.448	kip (force)	kN	0.2248
1.356	ft-lb (moment)	N-m	0.7376
1.356	kip-ft (moment)	kN-m	0.7376
1.488	lb/ft (mass)	kg/m	0.6720
14.59	lb/ft (load)	N/m	0.06853
14.59	kips/ft (load)	kN/m	0.06853
6.895	psi (stress)	kPa	0.1450
6.895	ksi (stress)	MPa	0.1450
0.04788	psf (load or pressure)	kPa	20.93
47.88	ksf (load or pressure)	kPa	0.02093
0.566 × (°F − 32)	°F	°C	(1.8 × °C) + 32

The following shorthand symbols are frequently used:

Symbol	Reading
$>$	Is greater than
$<$	Is less than
$\geq$	Equal to or greater than
$\leq$	Equal to or less than
6′	Six feet
6″	Six inches
Σ	The sum of
ΔL	Change in L

Design of even the simplest building structures is not entirely an automatic process. Although the work in this book may appear to use some reasonably logical processes, judgment and compromise are ever-present parts of the design process. With all the facts in hand and with the ability to intelligently interpret them, the structural designer will, it is hoped, proceed logically. In the best of real situations, however, lack of time, of clear information, of experience with problems of a similar nature, and of numerous other factors can leave the designer in something short of an ideal decision-making situation. We have tried to make the examples in this book as "real" as possible in order to present true design conditions. Designing real buildings, however, is both a little more mysterious and a lot more fun than it appears here.

PART ONE

BUILDING ONE

Building One is a two-story, two-family, residential building. The construction to be used is ordinary light wood framing for all the portions of the building above grade and poured concrete for the basement walls, floor, and footings. In professional practice a complete set of structural calculations is seldom done for such a building, because the majority of structural elements are selected from code requirements, handbook tables, or manufacturer's recommendations. Even when local building regulatory agencies require a set of calculations, they are generally limited to special structural elements such as long span beams, special foundations, and unusual lateral bracing. The analysis and design work shown in this example is therefore presented for the purpose of explaining the structure and not to illustrate what must be typically done to obtain a building permit.

CHAPTER ONE

Design of the Wood Frame Structure

1.1. THE BUILDING

The form of the building and the general details of the construction are shown in Figures 1.1 and 1.2. The construction materials and details for such a building vary considerably because of the wide range of climate conditions, local building code requirements, and the practices of local builders. There is often little if any actual structural design work done for such a building because of the common, repetitive nature of the construction. The explanations that follow are included to illustrate the nature of the structure, although most of the elements of the system may be obtained from various tabulations in codes or handbooks.

We have for the most part used the *Uniform Building Code* (UBC) as a reference for design criteria, although this code is not really intended to cover single-family or two-family housing. As in all cases, local codes should be consulted for any actual design work.

1.2. THE STRUCTURAL SYSTEM

The illustrations in Figures 1.2 and 1.3 show the structural system for the building. The materials to be used for the construction are:

Joists, rafters, and studs: No. 2 Douglas Fir–Larch.

Beams and posts: No. 1 Douglas Fir–Larch (4× and wider).

Roof deck, floor deck, and exterior wall sheathing: Douglas Fir plywood, structural grade.

Structural steel: A36, F_y = 36 ksi [248 MPa].

Concrete: stone aggregate, f'_c = 3 ksi [20.7 MPa].

Some of the criteria used for the structural design are:

Floor live load: 40 psf [1.92 kPa].

Roof live load: 30 psf (snow) [1.44 kPa].

Wind: 20 psf on vertical surfaces [0.96 kPa].

Soil: 2 ksf maximum [96 kPa].

1.3. DESIGN OF THE ROOF STRUCTURE

The roof structure consists of structural plywood panels nailed to closely spaced

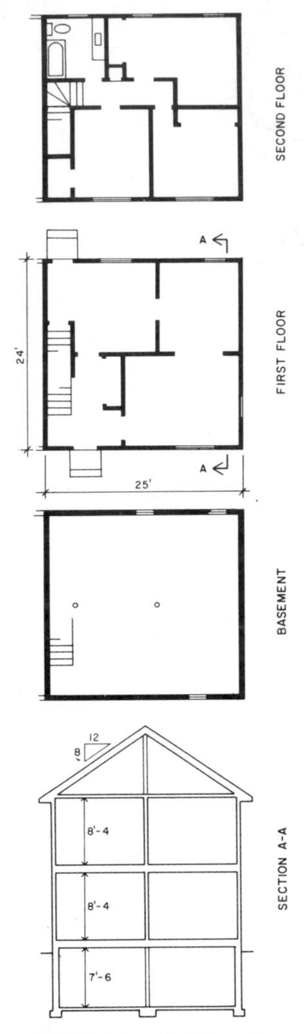

FIGURE 1.1. Building One.

rafters. The rafters span from the exterior bearing walls to the ridge member that is supported by the interior bearing wall. The rafter geometry is as described in Figure 1.4.

Plywood Roof Sheathing

The required thickness for the plywood depends on the grade of the plywood, the roofing materials, the roof live load, and the wind. Logical spacings for the rafters (some even increment of the plywood panel length of 8 ft) are 12, 16, 19.2, 24, 32, and 48 in. For this situation the most common spacings are 16 and 24 in. We will select a 24-in. spacing for which UBC Table 25-S-1 (see Appendix B) requires a minimum of $\frac{3}{8}$ in., 24/0 plywood with the edges blocked, and the plywood face grain perpendicular to the rafters. This selection is, of course, subject to modification when the function of the roof deck as a horizontal diaphragm is dealt with later in the design of the lateral load resistive system.

Rafters

A complete design of the rafter would include considerations for bending stress and deflection due to the various combinations of dead, live, and wind loads. The wind load is assumed to act normal to the roof surface, but the gravity loads would produce bending only as vectors in the direction perpendicular to the rafter span (see Fig. 1.4.). For the relatively low slope rafter and the low value of wind force, it is common to ignore the wind and to design the rafter on the basis of the horizontal span. Tabulations of allowable load/span conditions for ordinary rafter sizes and spacings may be found in various references. One such table is UBC Table 25-U-R-14, which is reproduced in Appendix B. Inspection of the UBC table produces the following:

For the No. 2 Douglas Fir–Larch rafters, from UBC Table 25-A-1, F_b = 1450 psi [10 MPa] for repetitive use and E = 1,700,000 psi [11.7 GPa], where F_b is the allowable maximum bending stress and E is the modulus of elasticity. Then, from UBC Table 25-U-R-14, for the 12-ft span and the 24-in. rafter spacings, select a 2 × 8 rafter.

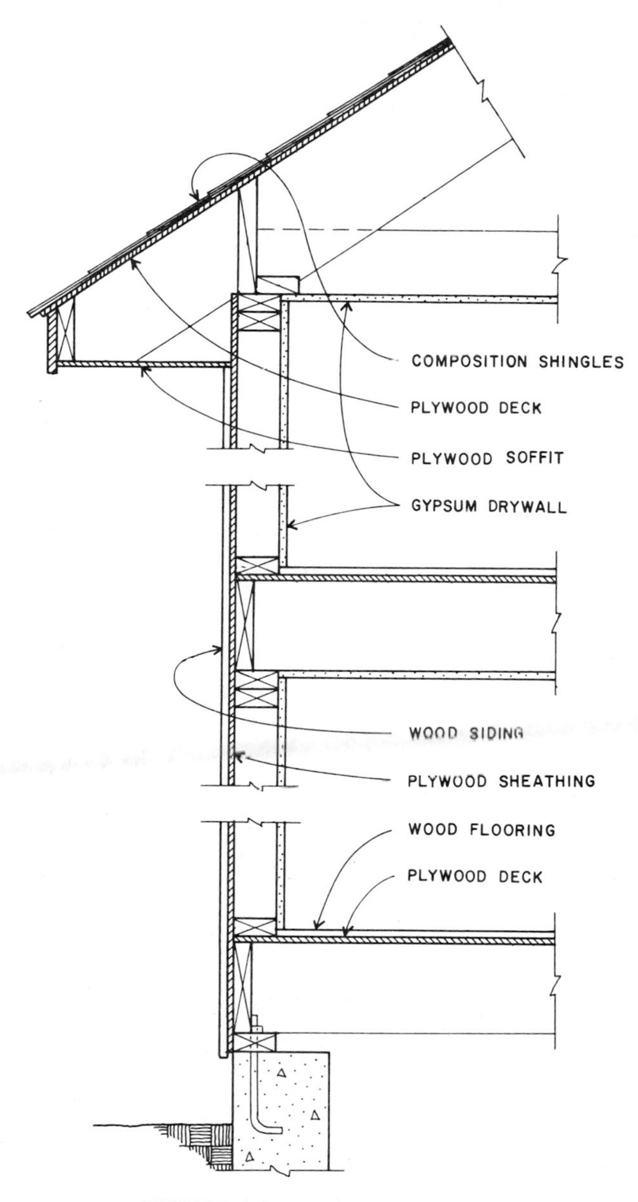

FIGURE 1.2. Typical construction.

Ceiling Joists

Design of the ceiling joists is somewhat arbitrary. Some of the considerations are:

1. Deflection should not be such as to cause visible sag. The straightness of the lumber is probably actually more critical in this regard. "Visible" sag is hard to put a number on.

2. If the crawl space is accessible, it should be assumed that someone may enter it or store materials in it. An old rule is to design for a maximum live load

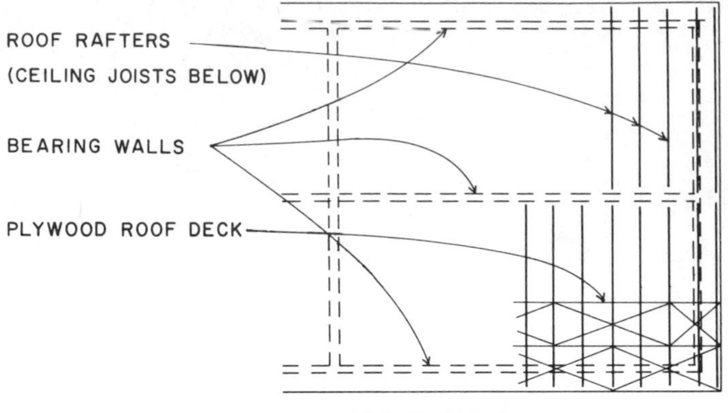

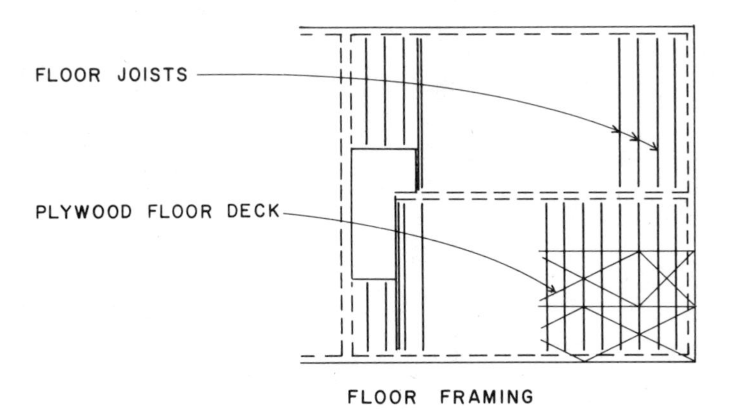

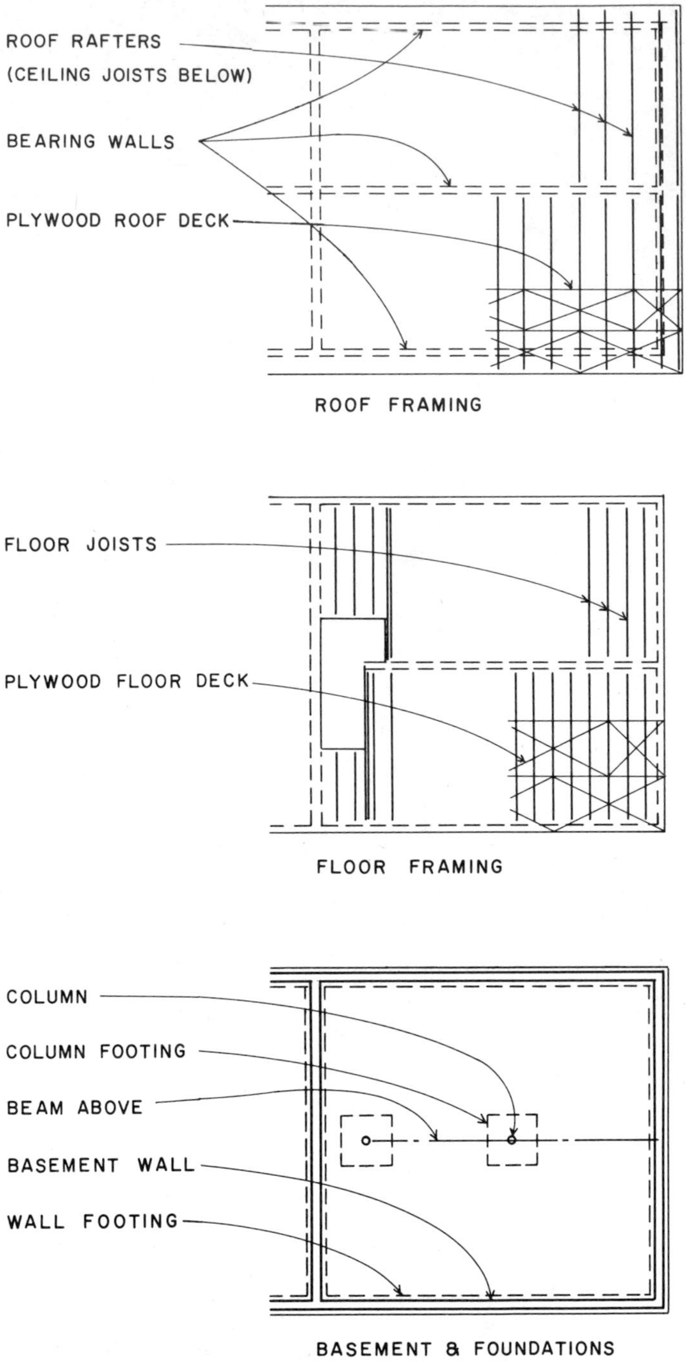

COLUMN
COLUMN FOOTING
BEAM ABOVE
BASEMENT WALL
WALL FOOTING

BASEMENT & FOUNDATIONS

FIGURE 1.3. Structural plans.

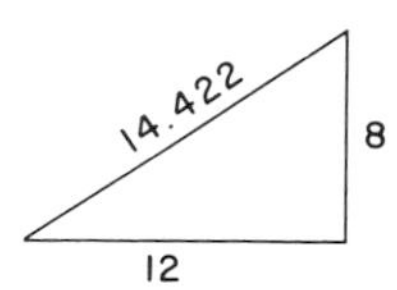

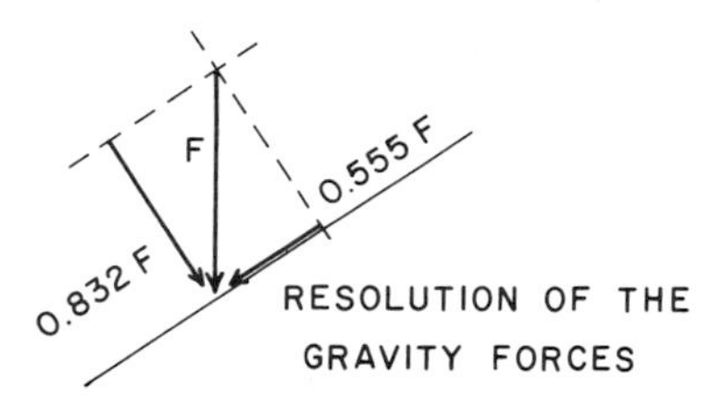

FIGURE 1.4. Resolution of the roof gravity loads.

deflection of $L/360$ in order to prevent cracking of the ceiling, especially if it is plastered. An arbitrary uniform or concentrated load may be used. The UBC Table 23-B requires a uniform load of 10 psf. The UBC Table 25-U-J-6 (see Appendix B) indicates that a 2 × 4 at 16-in. spacing is barely adequate. (Note that the clear span from the outside to inside walls is approximately 11.5 ft.) The UBC tabulated loading is also based on a deflection of $L/240$ under the live load. A slightly more conservative design would be to allow for a maximum deflection of $L/360$ under the weight of a single person (assumed at 200 lb) at the center of the span. Thus

$$\text{maximum } M = \frac{PL}{4} = \frac{200(12)}{4}$$

$$= 600 \text{ ft-lb } [814 \text{ N-m}]$$

$$\text{required } S = \frac{M}{F_b} = \frac{600(12)}{1250}$$

$$= 5.76 \text{ in.}^3 \; [94.4 \times 10^3 \text{ mm}^3]$$

maximum permitted deflection

$$= \frac{L}{360} = \frac{144}{360} = 0.4 \text{ in. } [10 \text{ mm}]$$

$$\text{required } I = \frac{PL^3}{48E\Delta}$$

$$= \frac{(200)(144)^3}{(48)(1{,}700{,}000)(0.4)}$$

$$= 18.29 \text{ in.}^4 \; [7.6 \times 10^6 \text{ mm}^4]$$

A 2 × 6, with S = 7.563 in.^{3} and I = 20.797 in.^{4}, will satisfy these criteria.

Another potential structural function for the ceiling joists is to serve to tie the tops of the walls against the outward thrust of the sloping rafters (see Fig. 1.5). This is not the case in this building because the central bearing wall supports the inside ends of the rafters.

A final consideration for the rafters is the required detailing for the ceiling surface material. In this example the construction might be simpler if the rafters and ceiling joists were at the same spacing. The 24-in. spacing is somewhat high for drywall ceilings, however. Thus the whole interactive relationship of the roof deck, rafters, ceiling joists, and ceiling surface must be considered.

Assuming the construction detailing problems to be solvable, we will settle for ceiling joists of 2 × 6 at 16 in.

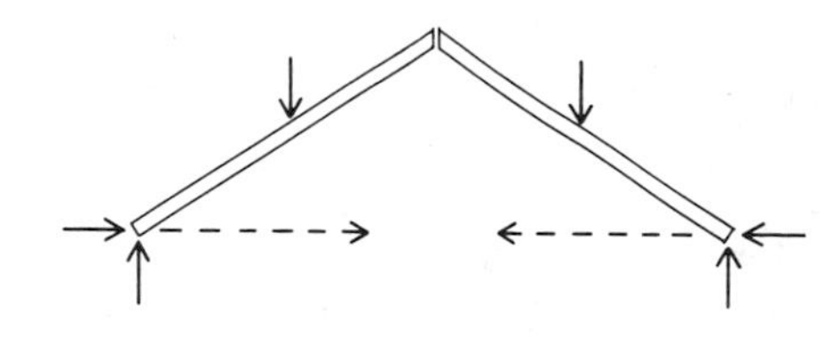

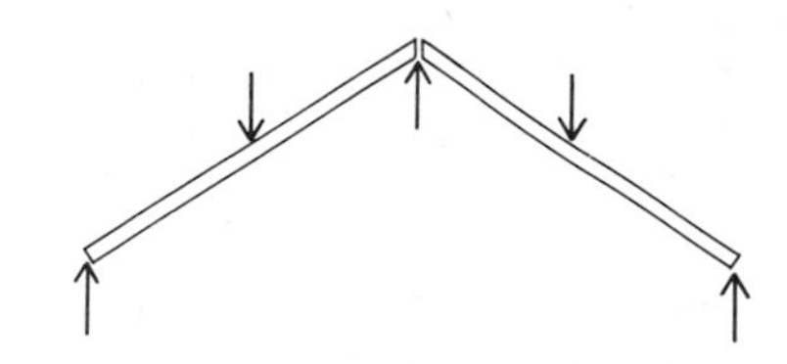

FIGURE 1.5. Stability of the sloped rafters.

1.4. DESIGN OF THE FLOOR STRUCTURE

The typical floor construction will consist of 2× joists at 16-in. centers. The structural floor deck will consist of plywood sheets. These are sometimes available with tongue-and-groove edges; otherwise the edges perpendicular to the joists will be supported by 2× wood blocking. Joists should be doubled at openings, such as for the stair, and under partitions parallel to the joists.

Floor Deck

Selection of the plywood grade and thickness depends on the type of flooring used and on the direction of the plywood face grain with respect to the joists. The face grain is normally in the 8-ft direction of the sheets and the sheets are slightly stronger and stiffer in that direction. However, placing them with the 8-ft direction perpendicular to the joists requires more blocking (every 4 ft).

On the basis of UBC Table 25-S-1 (see Appendix B) the minimum plywood would be $\frac{1}{2}$ in., C-D grade or better, index 32/16 with the face grain perpendicular to the joists.

Floor Joists

The usual practice for the joists would be to size the joist for the maximum span condition and use this size throughout the floor for all joists, headers, and blocking. This provides a level underside for the attachment of the ceiling and allows the top plates of all the stud bearing walls to be at a common height.

For the typical 12-ft span joist the design loading is

Dead load:
Finish floor = 3 psf (hardwood strip)
$\frac{1}{2}$ in. plywood = 1.4
$\frac{5}{8}$ in. drywall = 2.5
Joists (average)
= 2.6

Total dead load
= 9.5 psf [0.45 kPa]

Live load:
40 psf or $(\frac{16}{12})(40)$
= 53.3 lb/ft of joist [778 N/m]

Total load:
49.5 psf or $(\frac{16}{12})(49.5)$
= 66 lb/ft of joist [963 N/m]

We now proceed to determine the three section properties required with A for shear, S for bending, and I for deflection:

$$\text{maximum } M = \frac{wL^2}{8} = \frac{66(12)^2}{8}$$

$$= 1188 \text{ ft-lb } [1.61 \text{ kN-m}]$$

$$\text{required } S = \frac{M}{F_b} = \frac{1188(12)}{1450}$$

$$= 9.83 \text{ in.}^3 \ [161 \times 10^3 \text{ mm}^3]$$

Assuming a 2 × 8, for critical shear distance from the end,

$$\text{maximum } V = w\left\{\left(\frac{L}{2}\right) - d\right\}$$

$$= 66(6 - 0.67)$$

$$= 352 \text{ lb } [1.57 \text{ kN}]$$

required A

$$= \left(\frac{3}{2}\right)\left(\frac{V}{F_v}\right) = \left(\frac{3}{2}\right)\left(\frac{352}{95}\right)$$

$$= 5.56 \text{ in.}^2 \ [3.59 \times 10^3 \text{ mm}^3]$$

allowable live load deflection

$$= \frac{L}{360} = \frac{144}{360} = 0.4 \text{ in. } [10 \text{ mm}]$$

required I

$$= \frac{5WL^3}{384E\Delta}$$

$$= \frac{5(53.3 \times 12)(144)^3}{384(1{,}700{,}000)(0.4)}$$

$$= 36.57 \text{ in.}^4 \ [15.2 \times 10^6 \text{ mm}^4]$$

These requirements are sufficiently met by a 2 × 8 with S = 13.14 in.^{3}, A = 10.875 in.^{2}, and I = 47.63 in.^{4}

For comparison, it may be noted that the UBC Table 25-U-J-1 (see Appendix B) yields 2 × 8 joists at 16-in. centers.

Floor Beam at First Floor

This beam carries the inside end of the first-floor joists and also supports the stud bearing wall that carries the second-floor joists, the ceiling joists, and the rafters. The exact dimensions of the beam depend somewhat on the required construction details. For simplicity we assume the beam to have two equal spans of 11 ft each. The two-span beam will be supported by a post at the center, a post at the stair, and by the basement wall at the other end.

Because the beam supports considerable total roof and floor area, some reduction of the live load is appropriate. The UBC 2306 permits a reduction of 0.08% of the live load for each square foot supported in excess of 150 ft^2. On this basis, using the 11-ft beam span and the 12-ft joist and rafter spans, the reduction permitted is

$$R = r(A - 150)$$
$$= 0.08(396 - 150) = 19.68$$

say 20%, where R = reduction (in percent) permitted for live load; r = reduction factor = 0.08; and A = total loaded area supported by the member.

The code limits the total permitted reduction, but this value does not exceed the limits.

Computations for the beam loading are shown in Table 1.1. The shear, moment, and deflection diagrams for the fully loaded two-span beam are shown in Figure 1.6. Assuming a solid timber beam of Douglas Fir–Larch, No. 1 grade, the allowable stresses from UBC Table 25-A-1 (see Appendix B) are

F_b = 1300 psi $\quad F_v$ = 85 psi $\quad E$ = 1,600,000 psi

[F_b = 8.96 MPa $\quad F_v$ = 0.59 MPa $\quad E$ = 11.03 GPa]

TABLE 1.1. Loads on the Beam

		Load: lb/ft of Beam		
Source	Determination	LL	DL	LL + DL
Roof[a]	LL = 30 psf × 12-ft span	360		
	DL = 7 psf × 14.42-ft rafter		101	461
Second floor ceiling	DL = 5 psf × 12-ft span		60	60
Second floor	LL = (0.80)(40) psf × 12-ft span	384		
	DL = 9.5 psf × 12-ft span		114	498
First floor	LL = (0.80)(40) psf × 12-ft span	384		
	DL = 7 psf × 12-ft span		84	468
Wall	First and second floor, $\frac{5}{8}$-in. gyp			
	DL = 10 psf × 16-ft total height		160	160
	Stub wall to rafters			
	DL = 5 psf × 8-ft height		40	40
Beam	Assume weight		25	25
Total design loads		1128 plf [16.5 kN/m]		1712 plf [25.0 kN/m]

[a] Note that although we have included the rafter span area in determining the live load reduction factor, we have not reduced the roof live load because the value of 30 psf normally indicates snow load that is usually not reduced.

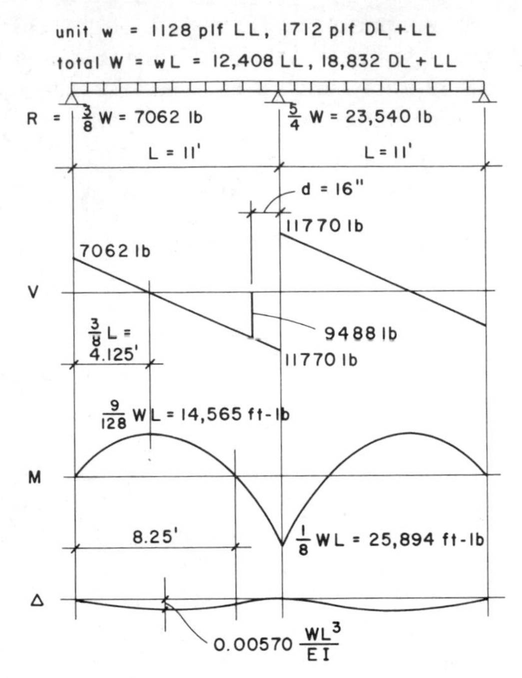

FIGURE 1.6. Analysis of the first floor beam.

Using these stresses and the calculated critical values for V, M, and deflection, we determine the required section properties:

required A

$$= \left(\frac{3}{2}\right)\left(\frac{V}{F_v}\right) = \left(\frac{3}{2}\right)\left(\frac{9488}{85}\right)$$

$$= 167 \text{ in.}^2 \; [108 \times 10^3 \text{ mm}^2]$$

required S

$$= \frac{M}{F_b} = \frac{25{,}894(12)}{1300}$$

$$= 239 \text{ in.}^3 \; [3.92 \times 10^6 \text{ mm}^3]$$

maximum live load deflection

$$= \frac{132}{360} = 0.367 \text{ in. } [9.3 \text{ mm}]$$

required I

$$= \frac{0.00570(WL^3)}{E\Delta}$$

$$= \frac{0.00570(1128 \times 11)(132)^3}{(1{,}600{,}000)(0.367)}$$

$$= 277 \text{ in.}^4 \; [115 \times 10^6 \text{ mm}^4]$$

The potential solid timber choices are shown in Table 1.2. All are somewhat massive and would involve intrusion on the headroom in the basement. It would be wise to consider a glue-laminated beam which has considerably higher shear and bending stresses allowable. From UBC Table 25-C-1, a Douglas Fir–Larch beam of 24F grade has allowable bending stress of 2400 psi and allowable shear of 175 psi. This reduces the A and S property requirements to

$$A = 167\left(\frac{85}{175}\right)$$

$$= 81.1 \text{ in.}^2 \; [52.3 \times 10^3 \text{ mm}^2]$$

$$S = 239\left(\frac{1300}{2400}\right)$$

$$= 129 \text{ in.}^3 \; [2.11 \times 10^6 \text{ mm}^3]$$

This would require a $6\frac{3}{4} \times 13.5$ or a $8\frac{3}{4} \times 10.5$ section.

It would also be possible to use a steel beam. This would entail the use of a 2× nailer on the top of the steel beam, adding slightly to its depth.

We will select the $8\frac{3}{4} \times 10.5$ beam. Because the bending stress is still not critical for this section, it could drop in grade to a 20F.

TABLE 1.2. Choices for the Beam

Required Properties	Optional Choices and Their Properties		
	8 × 24	10 × 20	12 × 16
A = 167 in.^{2}	176	185	178
S = 239 in.^{3}	690	602	460
I = 264 in.^{4}	8111	5870	3568

1.5. DESIGN OF THE WALLS AND COLUMNS

The vertical load elements consist of the stud walls above grade and the concrete walls and steel columns below grade, For ordinary situations the details of the wall construction will usually be covered by the specifications of the building code that apply to this type of building.

Basement Column

The column in the center of the beam carries the larger load from the beam (see Fig. 1.6). The column at the stair carries loads from the stair framing, including the weight of the walls above, so that the loads are probably close enough to require the same size column and footing. For brevity we will design only the center column.

From the beam analysis the column load is approximately 26 kips. For calculation we assume the unsupported height to be 7 ft from the bottom of the beam to the top of the basement floor slab.

From UBC Table 25-A-1 (see Appendix B) the allowable stress and modulus of elasticity are

$$F_c = 1000 \text{ psi}$$

$$E = 1{,}600{,}000 \text{ psi [6.9 MPa, 11.03 GPa]}$$

Assuming a nominal 4× member with least d of 3.5 in.,

$$\frac{L}{d} = \frac{84}{3.5} = 24$$

Then the allowable compression stress is determined as follows:

$$K = 0.671 \sqrt{\frac{E}{F_c}}$$

$$= 0.671 \sqrt{\frac{1{,}600{,}000}{1000}} = 26.84$$

$$F'_c = F_c \left\{1 - \frac{1}{3}\left(\frac{L/d}{K}\right)^4\right\}$$

$$= 1000 \left\{1 - \frac{1}{3}\left(\frac{24}{26.84}\right)^4\right\}$$

$$= 787 \text{ psi [5.43 MPa]}$$

The required area for a 4× member is thus

$$A = \frac{P}{F'_c} = \frac{26{,}000}{787}$$

$$= 33.04 \text{ in.}^2 \text{ [}21.3 \times 10^3 \text{ mm}^2\text{]}$$

This would require a 4 × 12 with A = 39.375 in.^{2}, which hardly seems reasonable. We try a larger d, if d = 5.5 in.,

$$\frac{L}{d} = \frac{84}{5.5} = 15.3$$

$$F'_c = 965 \text{ psi [6.65 MPa]}$$

and the required area for a 6× member is

$$A = \frac{26{,}000}{965} = 26.9 \text{ in.}^2$$

This would permit the use of a 6 × 6 with A = 30.25 in.^{2}.

If a steel post is desired, the usual choice would be a round pipe column. This can be selected from the AISC Manual (Ref. 10) using the column allowable load tables in section 3. For this load and height, with F_y of 36 ksi for the steel, a $2\frac{1}{2}$-in. standard steel pipe is adequate.

Although either column may be used, the steel column provides for slightly better details at the footing and the beam bearing and thus we have shown it in the construction details.

Stud Bearing Walls

The UBC 2517 has numerous requirements and limits for stud wall construction. For this height a 2 × 4 stud is permitted with a maximum spacing of 24 in. Normal procedure would be to check the 2 × 4 studs at 16-in. centers for the heaviest loading condition. If they are not adequate, we would increase them for that wall, work backward to find the heaviest wall loading for which they are adequate, and then use them for all the rest of the walls.

From the beam load tabulation in Table 1.1 we may observe that the first floor stud wall over the beam carries the beam load less the first floor joists and the beam. This is a load of 1219 lb/ft from the tabulation. At 16-in. centers, one stud carries a load of

$$P = 1219\left(\frac{16}{12}\right) = 1625 \text{ lb } [7.23 \text{ kN}]$$

With Douglas Fir–Larch No. 2 studs, we determine from UBC Table 25-A-1 (see Appendix B):

$F_c = 1000$ psi

$E = 1{,}700{,}000$ psi [6.9 MPa; 11.7 GPa]

For the individual stud/column the critical d dimension for buckling will be 3.5 in. because the wall surfacing serves to brace the studs on the weaker 1.5-in. axis. The allowable load on the 2 × 4 with an unsupported height of 8 ft 4 in. is therefore

$$\frac{L}{d} = \frac{100}{3.5} = 28.6$$

$$F'_c = \frac{(0.3)(1{,}700{,}000)}{(28.6)^2}$$

$$= 623.5 \text{ psi } [4.30 \text{ MPa}]$$

$$\text{allowable } P = (F'_c)(\text{area of } 2 \times 4)$$

$$= (623.5)(5.25)$$

$$= 3273 \text{ lb } [14.6 \text{ kN}]$$

Because this is twice the required load and the heaviest loaded wall, we may safely use the 2 × 4 studs at 16-in. centers for all the walls.

Basement Wall

Depending on local codes and practices, these walls may be of solid poured concrete or of concrete masonry units. We will show the design for poured concrete walls. The UBC Table 29-A requires a minimum 8-in. wall thickness for either type of construction.

It is not uncommon for these walls to be built with little or even no reinforcement. We recommend a minimum of reinforcement to be a continuous horizontal No. 4 bar at the top and bottom of the walls. For the best construction it is also recommended that the minimum vertical and horizontal temperature and shrinkage reinforcement be provided as recommended by the ACI Code (Ref. 9).

The typical exterior basement wall is under a combination of loads due to the vertical loads from the construction above and the horizontal pressure from the soil. This results in a combination of axial compression plus bending. For the concrete the critical stress condition will be the net tension stress, which will be the greatest when the axial compression is the least. We therefore look for the outside wall with the least load from the building above. In our example this will be the end walls because the rafters and floor joists are parallel to them.

Figure 1.7 shows the assumed loading for the end wall. Since the bending will produce a maximum moment at approxi-

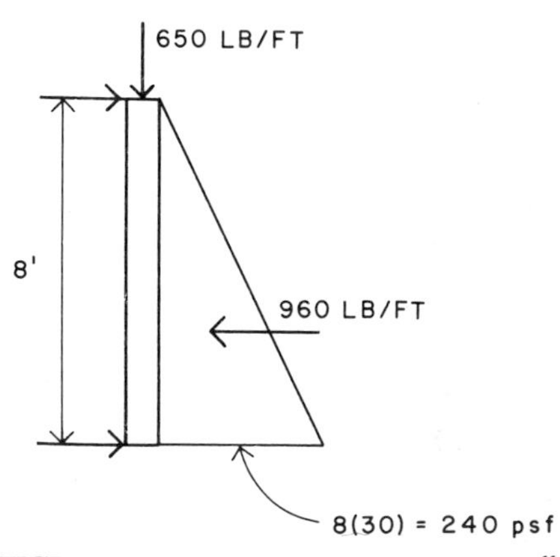

FIGURE 1.7. Forces on the basement wall.

mately midheight of the wall, we have used the axial compression load at midheight for the determination of the maximum tension stress. On this basis the load is:

Stud wall, approximately 20 ft at 10 psf
= 200 lb/ft
4 ft of basement wall at 100 psf
= 400
Portion of roof and floor, say
= 50
Total load at midheight
= 650 lb/ft [9.48 kN/m]

The soil pressure is taken as equal to the pressure in an equivalent fluid with density of 30 pcf. This is the minimum pressure that is usually required by codes, and may be higher if the soil type or the groundwater conditions are more severe.

We now proceed to find the net tension stress due to the combined loading using a 12-in.-wide strip of wall. The maximum moment due to the triangular distributed loading can be found from the beam diagrams in section 2 of the AISC Manual (Ref. 10) or from other handbooks:

$$\text{maximum } M = 0.1283WL = 0.1283(960)(8) = 985 \text{ ft-lb}$$

The section modulus for the 8 × 12-in. wall portion is

$$S = \frac{bd^2}{6} = \frac{(12)(8)^2}{6} = 128 \text{ in.}^3$$

The maximum bending stress is then

$$F_b = \frac{M}{S} = \frac{985(12)}{128} = 92.3 \text{ psi}$$

The compressive stress due to the gravity load is

$$F_c = \frac{P}{A} = \frac{650}{96} = 6.8 \text{ psi}$$

The net tension stress is therefore

$$F_t = 92.3 - 6.8 = 85.5 \text{ psi}$$

For the concrete with F'_c of 3000 psi this is slightly less than 3% of the ultimate compressive strength, which is not usually considered critical. The wall is therefore theoretically adequate without vertical reinforcing.

Note: It would obviously be logical to use a different strip width with SI units.

1.6. DESIGN OF THE FOUNDATIONS

For this construction a strip footing would normally be provided under the basement walls to serve as a foundation as well as a platform for the construction of the wall. For the latter purpose it would normally be made a few inches wider than the wall. For the 8-in. wall we would usually use a minimum 14-in. wide footing for this purpose. The UBC Table 29-A requires a minimum 15-in.-wide by 7-in.-thick footing for the two-story building.

The heaviest loaded walls are the front and rear walls that carry the ends of the rafters and floor joists. The tabulation of the load for this wall is shown in Table 1.3. With the allowable soil pressure of 2000 psf [96 KPa] the 15-in. [380-mm] wide footing will carry a load of

$$w = 2000\left(\frac{15}{12}\right) = 2500 \text{ lb/ft [36 kN/m]}$$

The minimum footing is therefore adequate for the heaviest wall load.

Column Footing

From the column design calculations the center column will place a total load of approximately 27 kips [120 kN] on the footing. If we deduct from the allowable soil pressure for the weight of a 10-in.

TABLE 1.3. Load on the Front Wall Footing

	Loads: lb/ft of Wall Length		
Load Source	DL	LL	Total
Rafters and second floor ceiling	86	180	266
Floor joists (with 100% LL)	99	480	579
Stud wall: 20 ft at 12 psf	240		240
Basement wall: 8 ft at 100 psf	800		800
Footing (estimate)	110		110
Total load on footing	1335 [19.5 kN/m]	660 [9.6 kN/m]	1995 [29.1 kN/m]

[250-mm] thick footing, the required area for this load will be

$$A = \frac{27{,}000}{1875} = 14.4 \text{ ft}^2 \text{ [1.34 m}^2\text{]}$$

A 3-ft 10-in. [1.17-m] square footing will provide A = 14.7 ft^2 [1.37 m^2].

Although calculations can be performed for the footing, there are tabulated designs in various handbooks from which the footing width and thickness and the reinforcing can be determined once the total load and allowable soil pressure are known.

CHAPTER TWO

Design for Wind

There are various problems to be considered in design for wind force on the building. The following discussion deals separately with the issues relating to the three principal building elements involved in wind resistance: the roof, the floors, and the walls.

2.1. THE ROOF

The roof must resist inward and outward pressures. The effect of the inward pressure, as additive to the gravity loads, was treated in the design of the rafters. With the relatively light construction and roofing the upward wind pressure is often larger than the roof dead load in these buildings. Minimum code requirements for the attachment of the rafters to the stud walls will provide some anchorage, but it is usually best to provide more positive anchorage by the use of nailed sheet metal connectors, such as those shown in the construction details (see Fig. 4.2).

The roof also acts as a horizontal diaphragm (even though it is sloped) that transfers the wind force to the vertical bracing elements; in this case, to the shear walls. Wind force on the long side of the building applies force as a uniform load to the edge of the roof. The load area for this pressure, as shown in Figure 2.1, is one half the second-story height plus the roof height. The load per foot of roof edge is thus

$$w = (20 \text{ psf}) \left[\left(\frac{9}{2} \right) + 8 \right]$$

$$= 250 \text{ lb/ft } [3.65 \text{ kN/m}]$$

The roof spans to transfer this load to the two end walls and the center dividing wall, as shown in Figure 2.1. Considering these as two simple spans, the load delivered to the end walls is thus

$$V = (250) \left(\frac{25}{2} \right) = 3125 \text{ lb } [13.9 \text{ kN}]$$

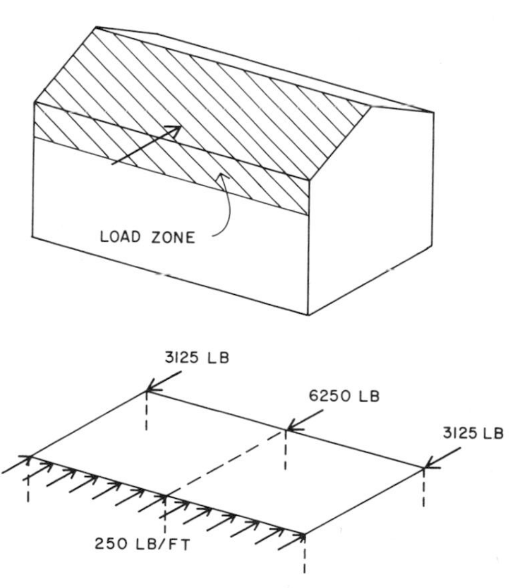

FIGURE 2.1. Wind load to the roof diaphragm.

The load on the center wall is twice this amount.

Considering the roof diaphragm as a horizontal element, its action is as shown in Figure 2.2. The unit shear stress in the plywood is a maximum at the ends and is equal to the total end shear divided by the plywood edge length. The latter is actually twice the true length of the rafter; thus the calculation is

$$v = \frac{3125}{2 \times 14.42} = 108 \text{ lb/ft } [1.58 \text{ kN/m}]$$

Referring to UBC Table 25-J-1 (see Appendix B) this value is less than the lowest rated value for $\frac{3}{8}$-in. plywood, even with unblocked edges for the plywood sheets. The minimum code required nailing of 6-in. spacing at panel edges and a 12-in. spacing at intermediate supports (not at the edge of a plywood sheet) is adequate for the roof sheathing.

The chord forces in the framing at the front and rear edges of the roof (actually at the top of the wall) must resist the moment shown in Figure 2.2. Because the two halves of the gable roof actually act like two separate diaphragms in tandem, there are actually two elements that are each 12 ft deep, rather than one 24 ft deep. However, the numerical result is the same if we divide the moment in two and use the d of 12 ft or use the whole moment with a d of 24 ft. Thus

$$C = T = \frac{M}{d} = \frac{19{,}531}{24}$$

$$= 814 \text{ lb } [3.62 \text{ kN}]$$

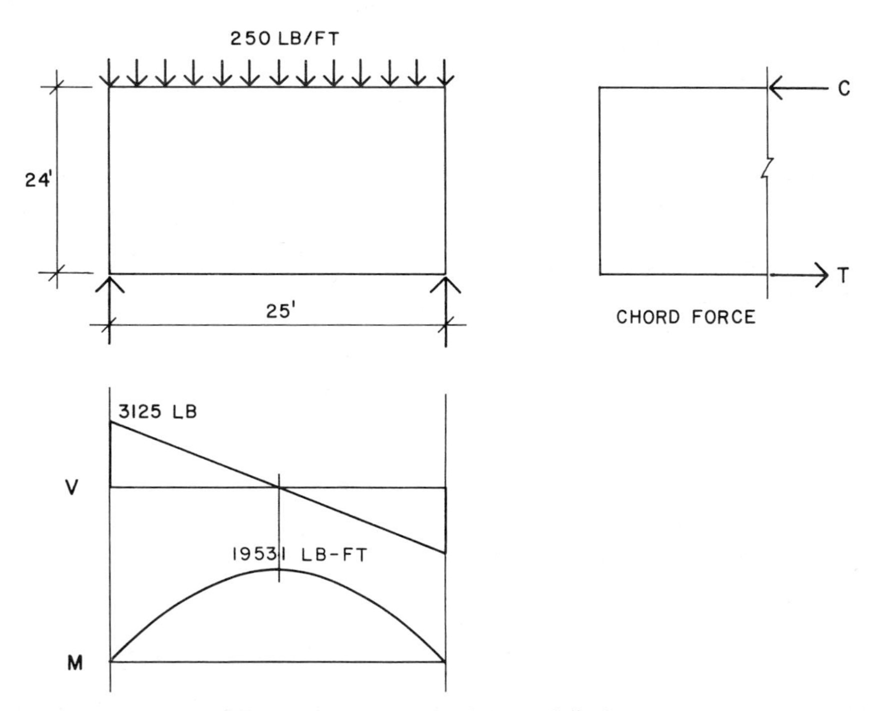

FIGURE 2.2. Analysis of the roof diaphragm.

Referring to Figure 4.2, one may see that the edge force from the roof is transferred through the blocking to the top plates of the stud wall. These continuous elements will actually serve as the edge chord for the roof. From UBC Table 25-A-1 (see Appendix B) the allowable compression for the No. two 2 × 4 members is 1000 psi. Thus the required area for the chord is

$$A = \frac{C}{F_c} = \frac{814}{(1.33)(1000)}$$
$$= 0.612 \text{ in.}^2 \ [395 \text{ mm}^2]$$

This is less than the area of one 2 × 4, and so the plates can easily serve as chords if their continuity is assured. Specifications usually require that the splices of the two plates be staggered a minimum of 4 ft. If this is done, the normal nailing of the plates to each other plus the nailing of the exterior plywood and the interior drywall will provide a reasonable continuity for the plate/chord in this case. Some designers (and some building regulatory agencies) would prefer to ensure a more positive continuity by specifying that the chords be bolted on each side of all splices. If the continuity of only one plate member is required, as in this instance, the plates would be simply bolted to each other with sufficient bolts to develop the chord force. If the continuity of both plates is required, a metal plate would be added to the splice.

From UBC Table 25-F a $\frac{1}{2}$-in. bolt in single shear in the 2× member is good for 470 lb, which may be increased by one third to 626 lb for wind. The number of bolts required is thus

$$N = \frac{C}{p} = \frac{814}{626} = 1.30$$

or two bolts on each side of the splice.

Splicing of the ridge is not necessary because in the tandem action it acts simultaneously as a tension chord for one diaphragm half and a compression chord for the other half.

Transfer of the roof diaphragm forces into the walls is discussed in the wall design and in the development of the construction details.

2.2. THE FLOORS

The second floor acts as a horizontal diaphragm similar to the roof. In this case the load zone, as shown in Figure 2.3, is from midheight of the first story to midheight of the second story, or approximately 9 ft. The load per foot on the edge of the diaphragm is thus

$$w = (20 \text{ psf})(9 \text{ ft})$$
$$= 180 \text{ lb/ft} \ [2.63 \text{ kN/m}]$$

The maximum shear at the ends of the diaphragm is

$$V = (180)\left(\frac{25}{2}\right) = 2250 \text{ lb} \ [10 \text{ kN}]$$

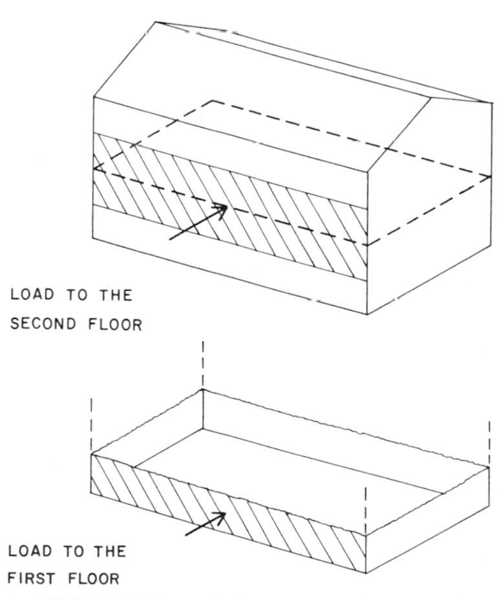

FIGURE 2.3. Wind loads to the floor diaphragms.

The maximum stress in the diaphragm at the building ends will be

$$v = \frac{2250}{24} = 94 \text{ lb/ft } [1.37 \text{ kN/m}]$$

At the center dividing wall the diaphragm is reduced in net width by the stair opening. Assuming a hole of 10-ft length, the net width is thus 14 ft and the stress is

$$v = \frac{2250}{14} = 161 \text{ lb/ft } [2.35 \text{ kN/m}]$$

Referring to UBC Table 25-J-1 (see Appendix B), we see this is still not critical for the $\frac{1}{2}$-in. plywood; thus minimum nailing may also be used for the second-floor deck.

The first-floor deck acts as a horizontal diaphragm, transferring the load from the lower half of the first-story wall to the basement walls. Because this is only half the load on the second-floor deck, it will not be critical.

2.3. THE WALLS

The exterior stud walls have two conditions to consider. The first is a combination of vertical compression due to gravity plus bending due to the direct wind pressure on the wall with the wall spanning vertically, as shown in Figure 2.4. The studs must be checked for this combined load condition. We will assume a design load of wind plus dead load plus one half live load for this condition. The gravity load on the first-story studs on this basis is approximately 1100 lb/stud.

The wind load on the studs is

$$w = (20)\left(\frac{16}{12}\right) = 26.7 \text{ lb/ft } [0.39 \text{ kN/m}]$$

Assuming the studs to span 9 ft in simple span from floor to floor, the maximum moment is

$$M = \frac{wL^2}{8} = \frac{(26.7)(9)^2}{8}$$

$$= 270.3 \text{ ft-lb } [367 \text{ N-m}]$$

The allowable compression stress, as calculated earlier for the interior wall, is 623.5 psi. From UBC Table 25-A-1 (see Appendix B) the allowable bending stress for the No. two 2 × 4 stud is 1650 psi for a repetitive stress member. The interaction of compression plus bending is thus considered as

$$\frac{P/A}{F'_c} + \frac{M/S}{F_b} = \frac{1100/5.25}{1.33(623)} + \frac{270.3(12)/3.063}{1.33(1650)}$$

$$= 0.253 + 0.483$$

$$= 0.736 < 1.0$$

The second condition for the wall involves its function as a shear wall for transfer of the loads from the roof and floor to the basement. The highest stressed wall is the first-story end wall that carries the edge loads from the roof and second floor, as shown in Figure 2.5. The total shear force in the wall is 5375

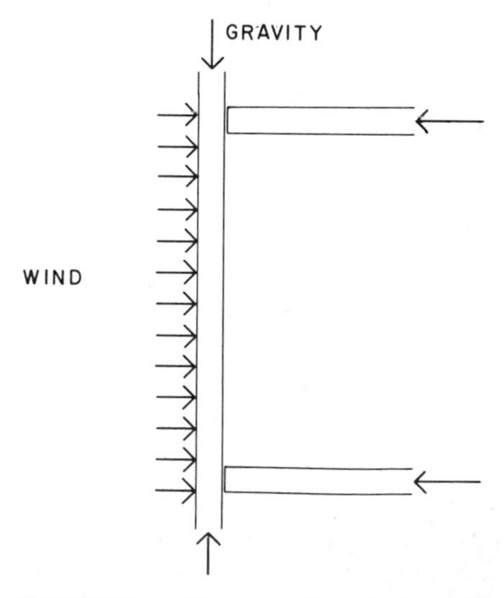

FIGURE 2.4. Forces on the exterior walls.

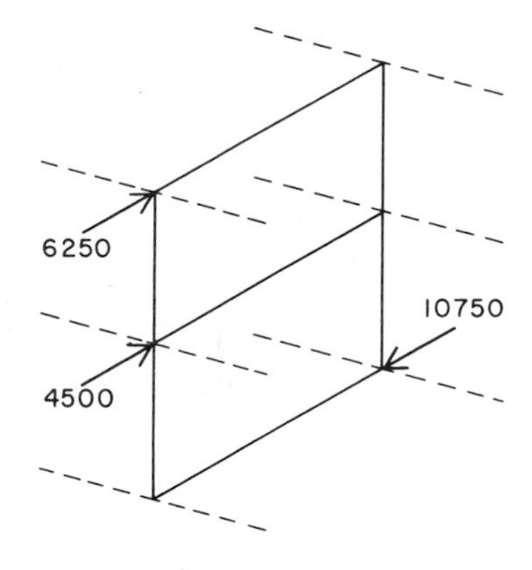

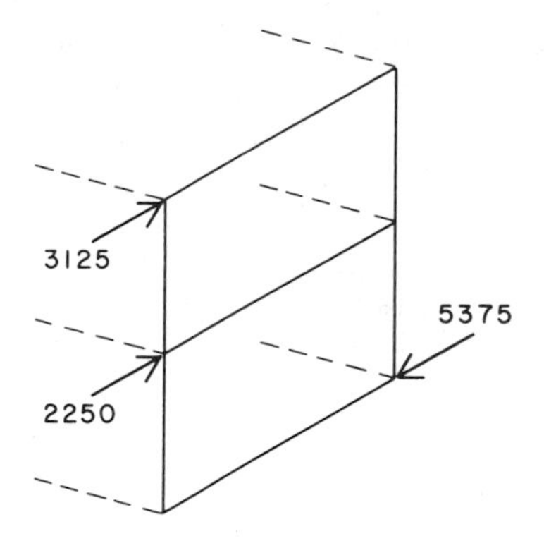

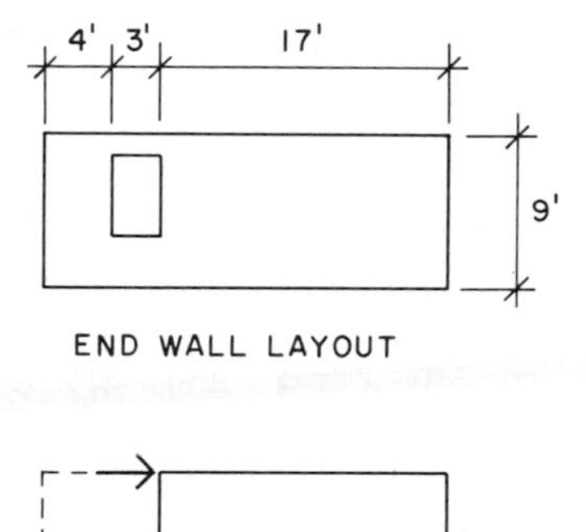

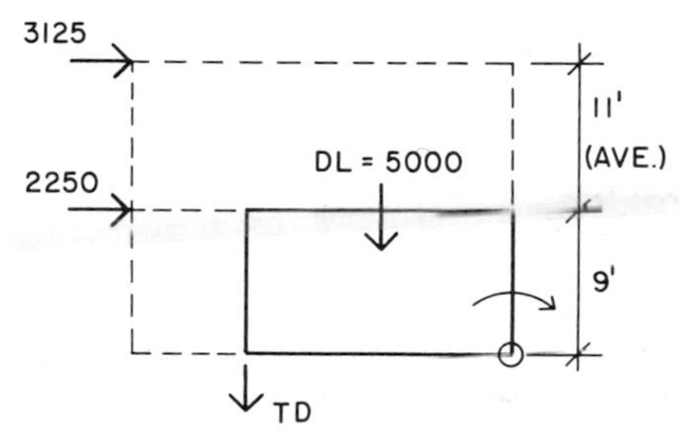

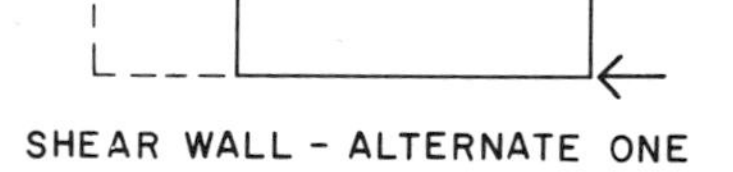

FIGURE 2.5. Analysis of the shear walls.

lb, and if the wall is a continuous surface the average shear is

$$v = \frac{5375}{24} = 224 \text{ lb/ft } [3.27 \text{ kN/m}]$$

If the wall has an opening, as shown in Figure 2.5, there are several approaches to its design. One alternative would be to ignore all but the 17-ft-long solid portion of the wall and consider it as a single panel for resistance of the entire force on the wall. The unit shear would thus be

$$v = \frac{5375}{17} = 316 \text{ lb/ft } [4.61 \text{ kN/m}]$$

By referring to UBC Table 25-K-1 (see Appendix B) the possible options for the wall are the following:

$\frac{1}{2}$-in. Structural I plywood with 6-in. edge nailing.

$\frac{1}{2}$- or $\frac{3}{8}$-in. C-D or Structural II plywood with 4-in. edge nailing.

In addition to the shear stress, the wall must be investigated for sliding and overturn. For the overturn analysis the loads must be considered at their points of application, that is, the roof load at the roof level and the second-floor load at the second-floor level. The overturning moment on the wall due to these forces must be resisted by the dead load on the wall with a safety factor of 1.5. Estimating the dead load at 5000 lb, due to the weight of the wall plus a small portion of the roof and floor, the analysis is as follows:

Overturning M:

$$3125 \times 20 = 62{,}500 \text{ ft-lb}$$
$$2250 \times 9 = \underline{20{,}250}$$
$$\text{Total} = 82{,}750 \text{ ft-lb}$$

Resisting M:

$$5000 \times 8.5 = 42{,}500 \text{ ft-lb}$$

Because the safety factor against overturn is clearly less than 1, an anchorage force called a tiedown is required. The required magnitude for this force is found as follows:

Overturning M × safety factor:
$$82{,}750 \times 1.5 = 124{,}125 \text{ ft-lb}$$
Deducting for dead load moment
$$= \underline{-42{,}500}$$
Net required M for tiedown
$$= 81{,}625 \text{ ft-lb}$$
Required tiedown force:
$$\frac{81{,}625}{17} = 4{,}802 \text{ lb [21.4 kN]}$$

For a conservative design an anchorage device, such as that shown in Figure 2.6, would be provided at each end of the shear wall panel. This device is bolted to a 4× or double 2× member at the shear panel edge and is secured by an anchor bolt in the foundation, in this case the basement wall.

Anchorage is also provided by the sill bolts ordinarily used to secure the stud wall sill member to the concrete. The UBC 2907(f) calls for these bolts to be a minimum of $\frac{1}{2}$-in. bolts at 6-ft centers with one bolt not more than 12 in. from each end of the sill. The use of these bolts for overturn resistance is not generally permitted at the present, however, because the stress condition involves cross-grain bending in the sill.

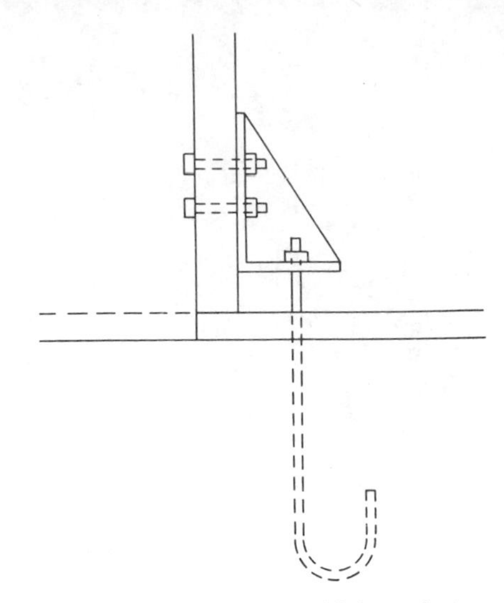

FIGURE 2.6. Typical hold-down device.

At the building corner the plywood on the two intersecting wall surfaces will be nailed to a common framing member. This means that the overturn of one wall requires the lifting of the end of the other wall. If the nailing at the corner is sufficient to develop this interaction of the two surfaces, it is probably redundant to provide an anchorage device for the tiedown force at this point. If we consider the two-story wall approximately 18-ft high, the total tiedown force produces an average shear stress at the corner of

$$v = \frac{4802}{18} = 267 \text{ lb/ft [3.90 kN/m]}$$

If the plywood previously determined is used, this is within the capacity of the edge nailing.

Another possibility for the first-story end wall is to consider the short 4-ft-long section to act in tendem with the longer

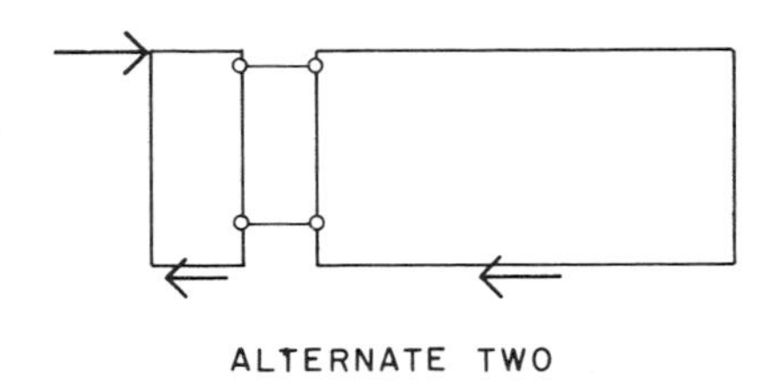

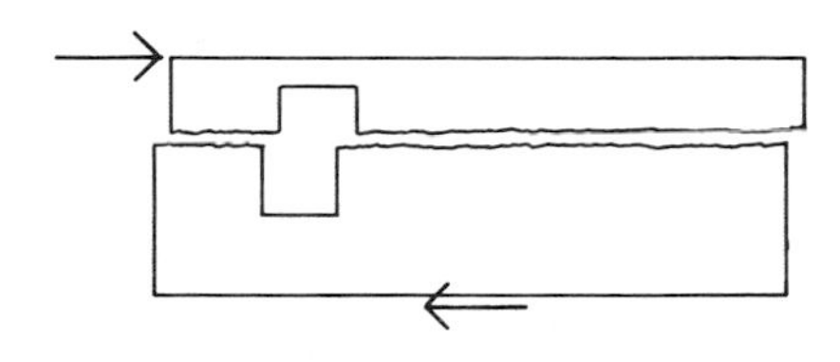

FIGURE 2.7. Alternate design assumptions for the end shear walls.

17-ft panel. This will slightly reduce the load on the 17-ft panel, but will produce considerable overturn for the 4-ft panel. The result would be to add more anchorage to the wall with only a slight reduction in the plywood shear stress (see Fig. 2.7).

A third alternative is to consider the entire 24-ft-long wall as a single panel and to develop the necessary stresses around the hole. The main requirement for this would be to add horizontal blocking at the level of the top and bottom of the window with metal straps to carry the force into the solid wall portions. The additional framing and metal ties would be a trade off against the close nail spacing and the tiedowns required for the first alternative. Of the three alternatives the first is probably the simplest for construction and the most economical.

From the earlier analysis, it was determined that the center dividing wall must resist shear forces twice those in the end walls. Assuming a continuous uninterrupted wall, the shear stresses will be (see Figure 2.5 for loading)

$$v = \frac{6250}{24}$$

$= 260$ lb/ft at the second story [3.79 kN/m]

$$v = \frac{10{,}750}{24}$$

$= 448$ lb/ft at the first story [6.54 kN/m]

Design of this wall must include the consideration that it is required to provide good acoustic separation between the two housing units. One solution for this is to supply two separate, complete stud framing systems with a small separation between them. In effect, this provides two walls, although the construction at the floors and the second floor ceiling will generally give sufficient tying to consider them as a single structural wall. Assuming this construction and the shear stresses just determined, some alternatives for the wall surfacing are:

1. Gypsum wallboard (drywall) on both sides. The UBC Table 47-I (Appendix B) permits a shear of 125 lb/ft on $\frac{1}{2}$-in. drywall with 5d nails at 7-in. spacing. This is just a little short, so it would be necessary to use 4-in. nail spacing, for which the table allows 150 lb/ft. The total resistance for the wall is thus 300 lb/ft, which is sufficient for the second story but not for the first.
2. A $\frac{7}{8}$-in. cement plaster on metal lath. The UBC allows 180 lb/ft for this, making the total for the wall 360 lb/ft. Again, this is adequate for the second story but not for the first.
3. Plywood on one or both sides is an alternative for the second floor and the only choice for the first floor. Note that UBC 4713(a) states that when different materials are applied to the same wall their shear resistance is not cumulative. Thus if ply-

wood is used, its shear resistance cannot be added to that of the finish materials placed over it.

On the basis of these considerations, we would recommend the following construction for the center wall:

At the second floor: $\frac{1}{2}$-in. drywall with 5d nails at 4-in. spacing on both sides of the wall.

At the first floor: $\frac{3}{8}$-in. C-D plywood with 8d nails at 6-in. spacing at edges on both sides of the wall with drywall applied as a finish.

Transfer of the shear force from level to level is relatively simple at the end walls, because the plywood on the exterior is continuous from top to bottom of the wall. Assuming ordinary platform-type construction, this continuity does not exist at the center wall, making some special consideration for the transfer necessary.

Figure 2.8 shows details of the floor and wall framing at the center dividing wall. The routing of the wind shear from the second floor to the basement is:

The second-floor wall surfacing is nailed to the sill.

The sills are nailed to the continuous joists.

The joists are nailed to the spreader block.

The spreader block is nailed to the top plates of the first-story wall.

The first-floor wall surfacing is nailed to the top plates and the sill.

The sill is nailed to the first-floor joists.

The joists are nailed to the spreader.

The spreader is bolted through the sill to the basement wall.

The nailing of the framing members at the second-floor level must transfer the total load of 6250 lb from the second-floor shear wall to the first-floor shear wall. If 16d common nails are used, the required spacing is determined as follows.

The UBC Table 25-G gives an allowable load of 108-lb/nail. This may be increased by one third for the wind loading. Then

required number

$$= \frac{6250}{(1.33)(108)}$$

$$= 44 \text{ or } 22 \text{ on each side}$$

required spacing

$$= \frac{(24 \times 12)}{22} = 13.1 \text{ in.}$$

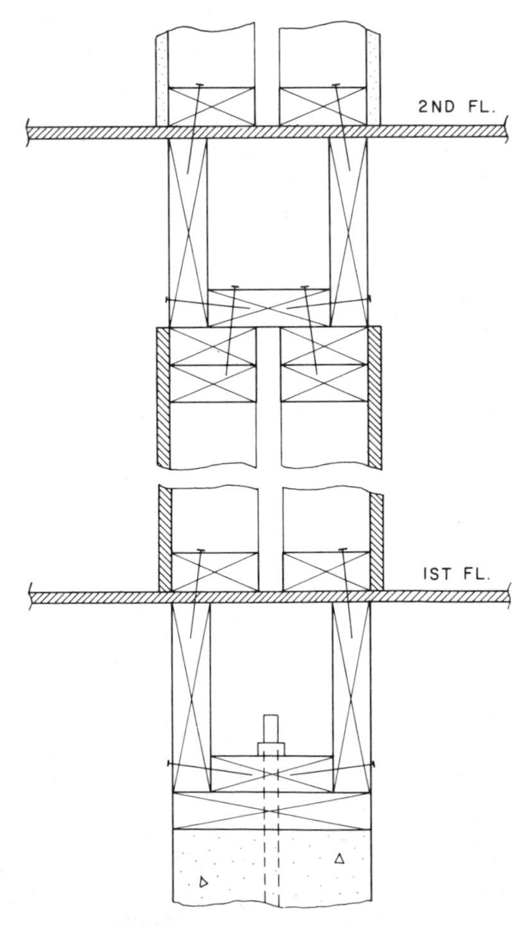

FIGURE 2.8. Details of the center dividing wall.

At the first floor the nails and bolts must transfer the total force of 10,750 lb to the basement wall. The required nailing is thus

$$N = \frac{10{,}750}{(1.33)(108)} = 76$$

$$S = \frac{(24 \times 12)}{38} = 7.58 \text{ in. maximum}$$

If $\frac{3}{4}$-in. bolts are used, UBC Table 25-F gives an allowable load of 710 lb/bolt on the 2× sill in single shear. This should also be checked against the load values in UBC Table 26-G that are for the bolt loads in the concrete wall. In this case, with the f'_c value of 3000 psi for the concrete, the allowable load is 3560 lb, and hence the wood limit is critical. The number of bolts required is thus

$$N = \frac{10{,}750}{(1.33)(710)} = 12$$

Various details of the horizontal diaphragms and shear walls are shown in the construction details that follow. Although there is some standardization by codes and industry recommendations, there is room for considerable variation in these details because of local practices and the judgment of individual designers.

Wind in the east–west direction produces slightly less than half the total load on the building as that in the north–south direction. This will not be critical for stress in the horizontal diaphragms, although attention should be given to the transfer of the edge loads into the chords and shear walls. Stresses in the shear walls will depend on the size and arrangements of openings in the walls. Because the openings will not be the same on the two walls, some consideration may be necessary for the torsional stresses due to the eccentricity of the lateral load from the center of stiffness of the shear walls. An example of this type of analysis will be shown in the design of Building Two for lateral loads.

CHAPTER THREE

Alternate Construction

The basic construction system so far used for this building is the most common for low-rise residential buildings. Numerous variations are possible in the building form and in the finish materials used both inside and outside, but the basic structure is seldom significantly different. In most cases, a change in the structure is justified only when building code regulations for fire resistance require the use of other construction. The material in this chapter presents some of the possible considerations for alternate construction for Building One.

3.1. ALTERNATE ROOF STRUCTURE

The rafter and ceiling joist system previously designed could be replaced by a truss system in which the top chords of the truss provide the roof framing and the bottom chords provide the second-floor ceiling framing. The trusses would span the full width of the building, shifting some additional load to the outside walls but reducing slightly the load on the first-floor beam.

For this span the truss members would consist of single 2× pieces arranged in a single plane with gussets of either plywood or metal. The trusses would usually be factory assembled and may use a patented jointing system. In many areas these trusses could be selected from the stock designs of a manufacturer, thus reducing the building designer's work to determining the span and load. The calculations that follow illustrate the design based on the truss member arrangement shown in Figure 3.1. Analysis is shown for the load condition of dead load plus live load only, assuming this to be critical for design.

The loads on the trusses consist of the dead load of the truss plus the dead and live loads of the roof and ceiling. The roof load is actually applied as a uniform load on the top chord, whereas the ceiling is applied as a uniform load on the bottom chord. These loads are translated into panel, or joint, loadings for the truss analysis, as shown in Figure 3.1. The design of the chord members, however, must consider the truss axial forces plus the bending due to the actual uniform loading.

With the trusses at 24-in. centers, the loadings are as follows:

Roof live load:

30 psf (2 ft)
  = 60 lb/ft on the chord.

60 (6 ft)
  = 360 lb on the top chord joints.

Roof dead load:

5 psf(2)
  = 10 psf(1.2) = 12 psf/horizontal foot.

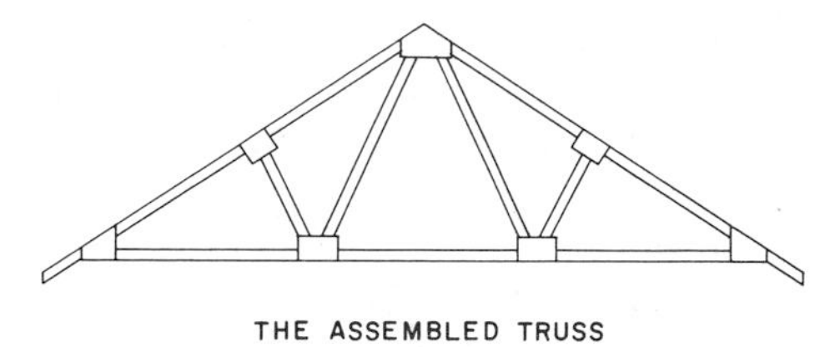

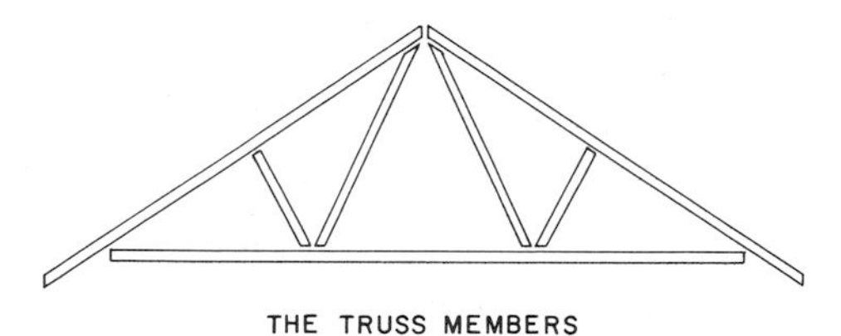

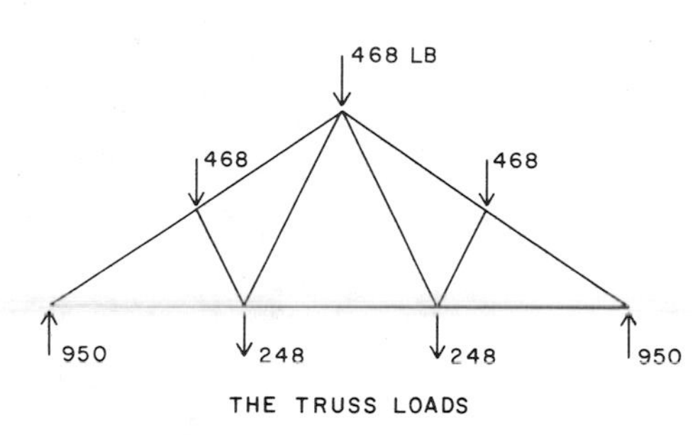

FIGURE 3.1. The alternate roof truss.

12(6)
 = 72 lb on the top chord joints.
Ceiling live load:
10 psf(2)
 = 20 lb/ft on the chord.
20(8)
 = 160 lb on the bottom chord joints.
Ceiling dead load:
2.5 psf(2)(8)
 = 40 lb on the bottom chord joints.

Assuming the trusses to weigh approximately 12 lb/ft of span, a load is added to the joints as follows:

6 plf (6 ft)
 = 36 lb at the top chord joints

6 plf (8 ft)
 = 48 lb at the bottom chord joints

The totals of these loadings are shown as the truss loads in Figure 3.1. The equilibrium diagrams for the individual joints of the truss are shown in Figure 3.2. Since the truss and the loads are symmetrical, only half of the truss is shown. Forces in the members are first found as vertical and horizontal components using the typical solution by joints. These are then translated into the actual vector forces in the members as shown between the joints. The symbol T is used to designate a tension member and C is used for compression.

The top chord must be designed for the maximum compression force of 1712 lb plus the bending due to the uniform load of the roof. From Figure 1.4, with the trusses at 24 in., this load becomes

$$w = 0.832 \times 2 \times 37 = 61.6 \text{ lb/ft}$$

The span is the true member length, which is 1.2 times the 6-ft horizontal dimension, or 7.2 ft. For the continuous two-span chord member the maximum moment is thus

$$M = \frac{wL^2}{8} = \frac{(61.6)(7.2)^2}{8} = 399 \text{ ft-lb}$$

With a 2 × 4 of No. 1 Douglas Fir–Larch, UBC Table 25-A-1 (see Appendix B) specifies that

$$F_b = 2050 \text{ psi} \qquad F_c = 1250 \text{ psi}$$

$$E = 1{,}800{,}000 \text{ psi}$$

Because the roof deck braces the weak axis, $L/d = 7.2(12)/3.5 = 24.7$. Then the allowable compression stress is determined as

$$K = 0.671 \sqrt{\frac{E}{F_c}}$$

$$= 0.671 \sqrt{\frac{1{,}800{,}000}{1250}} = 25.46$$

$$F'_c = F_c \left\{1 - \frac{1}{3}\left(\frac{L/d}{K}\right)^4\right\}$$

$$= 1250\left\{1 - \frac{1}{3}\left(\frac{24.7}{25.46}\right)^4\right\}$$

$$= 881 \text{ psi } [6.07 \text{ MPa}]$$

The combined compression and bending is thus

$$\frac{P/A}{F'_c} + \frac{M/S}{F_b} = \frac{1712/5.25}{881} + \frac{(399 \times 12)/3.06}{2050}$$

$$= 0.370 + 0.763 = 1.133$$

Because this is over 1.0, the 2 × 4 is theoretically not adequate. However, the overstress is only 13% and a more accurate analysis may show it to be less. The true truss span is slightly less than 24 ft, and the gussets tend to reduce the actual unbraced length of the members.

For a conservative design the member size may be increased to a 2 × 6 or the stress grade raised by using a higher quality wood.

The bottom chord must be designed for the maximum tension force of 1424 lb plus the bending due to the ceiling load. The load for bending will be 2(12.5) = 25 lb/ft. Thus

$$M = \frac{wL^2}{8} = \frac{(25)(8)^2}{8} = 200 \text{ ft-lb}$$

Trying a 2 × 4 as before, we may use the previous data, adding that the allowable tension stress (F_t) from Table 25-A-1 is 1050 psi. Then

$$\frac{P/A}{F_t} + \frac{M/S}{F_b} = \frac{1424/5.25}{1050} + \frac{200(12)/3.06}{2050}$$

$$= 0.258 + 0.383 = 0.641$$

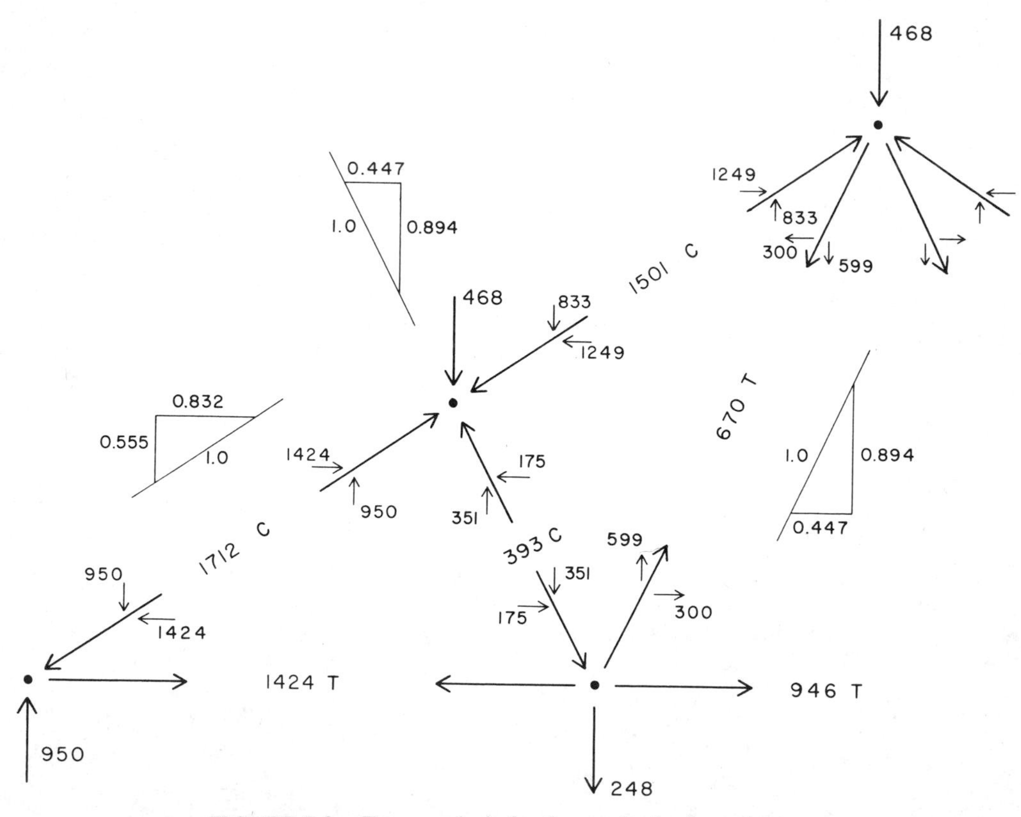

FIGURE 3.2. Force analysis for the gravity loads on the truss.

The stresses are very low on the interior members because no bending is involved. For connection purposes these need to be 2× members and would probably be 2 × 3 or 2 × 4 for development of the joint forces at the gussets.

One disadvantage of the trusses is that the second-floor ceiling drywall would have to be heavier to span the 24-in. distance. Also the attic space becomes less functional for storage space when it is filled with the forest of truss members.

3.2. ALTERNATE FLOOR CONSTRUCTION

For this type of building there is seldom consideration of alternates to the plywood deck and solid wood joist system. This implies that the conditions of use are:

1. Some type of finish (carpet, wood flooring, tile, etc.) will be applied to the top of the plywood because the plywood does not provide a finished quality surface.
2. Some type of ceiling will be used, if the space below is other than basement or storage, because the joists and the underside of the plywood do not present a very attractive exposed structure.
3. Spans are modest, thus making the small size 2× joists a feasible structural choice.
4. The general construction of the building is a light wood frame.
5. Acoustic separation between the space above and the space below the floor structure is not of a critical nature.

If other conditions exist, it may be necessary to consider other alternatives.

3.3. ALTERNATE WALL CONSTRUCTION

As with the floor construction, the 2 × 4 stud wall is the definitive choice unless there are special circumstances. Some conditions that may modify the choice are:

1. The wall unbraced height exceeds the code limits. These limits depend on the grade of the studs; the maximum height permitted is usually 14 ft for the 2 × 4 studs.
2. The bearing load on the wall exceeds the capacity of the studs as axially loaded columns or the compression plus bending for the exterior wall is excessive. Modifications for these situations include the possibilities of increasing the size of the stud, increasing the number of studs by reducing the spacing, or increasing the grade of the wood.
3. The spanning roof or floor structure uses steel or concrete members for which a wood support is not appropriate.
4. It is necessary to have acoustic separation or a high resistance to thermal flow that requires more wall thickness or special construction.

CHAPTER FOUR

Construction Drawings

The structural designer ordinarily documents the design in the form of three elements: the structural calculations, the structural working (or construction) drawings, and the written specifications. The calculations should be written in reasonably readable form, for they must often be submitted for review when obtaining a building permit. The specifications are written in fairly legal form because they are actually parts of a contract. For a building of this type the construction drawings for the structure are often incorporated in the general architectural drawings with a minimum of separate, purely structural details. Many of the items usually covered by lengthy specifications in larger buildings are often covered with abbreviated notes on the construction drawings for this type of building.

The drawings that follow are typical "structural" details and are intended primarily to illustrate the design of the systems just discussed. They are not intended as models for construction drawings, although that style is used in the illustrations. They do not, in most cases, show the complete architectural details in terms of the finish materials because the focus has been on the structure.

4.1. STRUCTURAL PLANS

Basement and Foundation Plan (Fig. 4.1)

This is the conventional form for this type of plan, showing the footings, the floor slab, and the walls and columns. The footings are shown by dotted lines where they are beneath the floor slab and by solid lines where they extend outside the walls and are covered only by the soil.

First-Floor Framing Plan (Fig. 4.1)

This is one convention for this type of drawing. Individual joists or beams are shown by a solid line indicating their approximate location in plan. Some of the members in a repetitive series are often omitted to simplify the drawing, with the extent of the series shown by including at least the end members on the drawing. Beams and headers are sometimes indicated by a heavier line or by a broken line of some form to make them more obvious.

Openings are normally shown with an X. The floor sheathing is shown in terms of the arrangement of the 4-ft by 8-ft plywood sheets. The staggering shown slightly increases the strength and stiffness of the horizontal diaphragm.

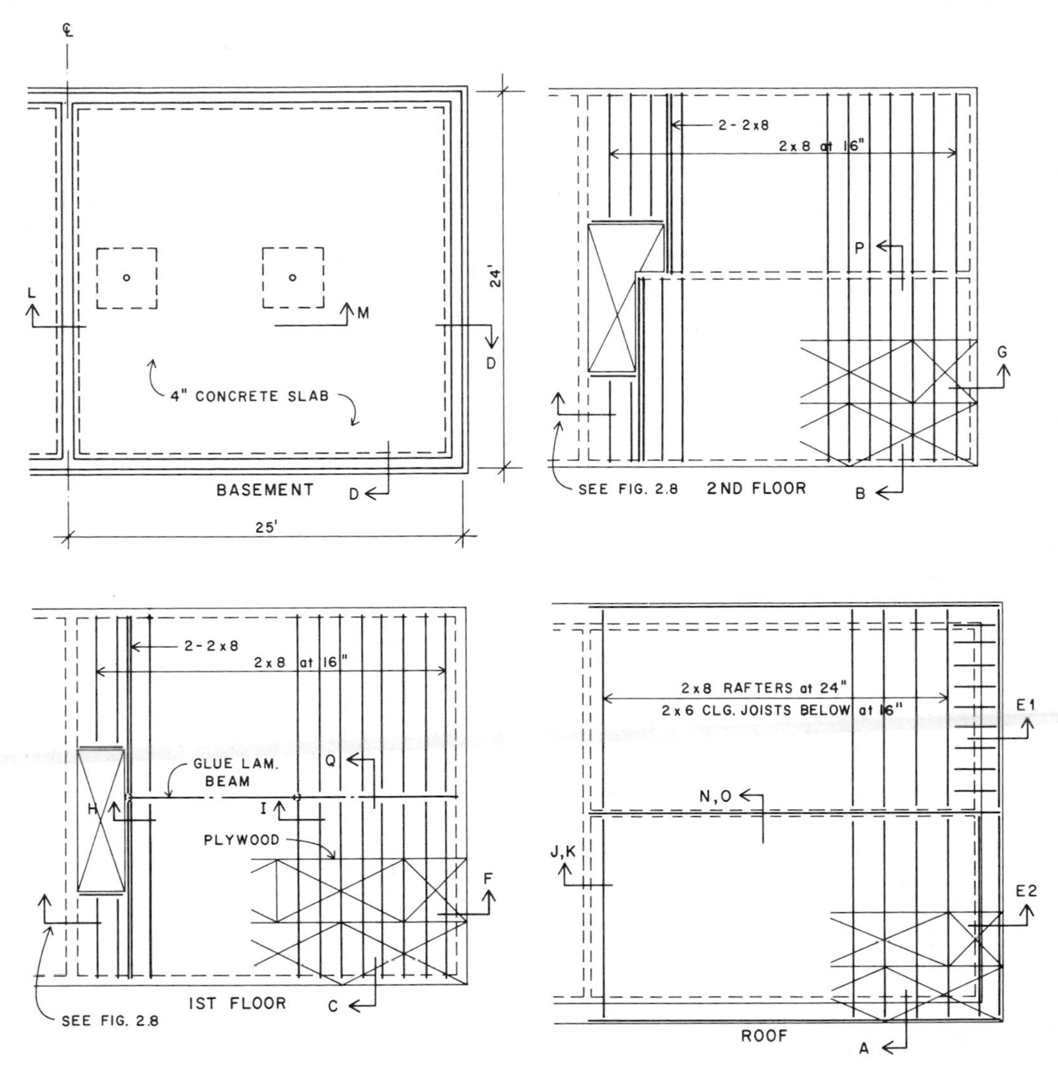

FIGURE 4.1. Structural plans.

Walls below the framing level, especially bearing walls, are usually shown by dotted lines.

Note that the joists are doubled at the edge of the stair opening. They would also be doubled under the partitions that are parallel to the joists.

Second-Floor Framing Plan (Fig. 4.1)

This is essentially similar to the first-floor plan except for the slightly different stair opening and the lack of the beam at the center.

Roof Framing Plan (Fig. 4.1)

This shows the rafter and roof deck layout. The ceiling joists are covered by a note because their arrangement is similar to the rafters except for spacing. Two options are shown for the extended gable end in the two variations of Detail E.

4.2. CONSTRUCTION DETAILS

Detail A (Fig. 4.2)

This shows the typical roof edge condition at the front and rear. The rafter is notched to provide full bearing on top of the plate and is normally toenailed to the plates. The sheet metal tie shown is one type that may be used for a more positive anchorage of the roof to the walls.

The vertical blocking is used to transfer the roof wind load into the wall. The horizontal blocking on top of the plates serves two purposes: It assists in the transfer of wind shear from the vertical blocking and also provides a backup for the edge of the ceiling drywall.

The roof edge, facia, and soffit have many variations, depending on the desired architectural detailing.

Detail B (Fig. 4.2)

This shows the typical condition at the edge of the second floor at the front and rear. Note the transfer of shear from the floor plywood through the continuous header and into the wall plywood. The block on top of the plate is strictly for backup of the drywall in this case.

Detail C (Fig. 4.2)

This is the typical detail at the edge of the first floor. The wind load transfer is primarily from the exterior plywood to the sill and into the concrete through the bolts. In this structure there is no wind load on the first-floor diaphragm, although the wall plywood should be nailed to the continuous edge header as a positive tie to the floor. At the walls parallel to the joists this header should be doubled to provide support for the wall above.

Most codes require that the wood construction be kept some distance above the exterior grade, usually a minimum of 6 in.

Detail D (Fig. 4.2)

Some codes have minimum requirements for this footing as well as for the minimum wall and basement floor slab thicknesses. The need for various waterproofing details will depend on the specific site conditions. Reinforcing is not shown, although we recommend a minimum of one layer of wire fabric in the slab and the minimum wall reinforcing as discussed in the calculations.

Detail E (Fig. 4.3)

At the low edge of the roof the rafters can simply be cantilevered to form an overhang, as shown in Detail A. At the gable end the only cantilevered parts of the construction are the roof plywood, the ridge member, and the facia. If an overhang is desired, two possible ways of achieving it are shown in Figure 4.3.

Detail E1 illustrates the use of outriggers that rest on the top of the wall and are carried back to the first spanning rafter. The rafter that forms the roof edge is carried on the cantilevered ends of these outriggers. The blocking shown is used to carry the wind load from the roof plywood into the shear wall.

Detail E2 shows the condition if the roof edge rafter is made to span from the ridge to the facia. This assumes that the ridge member shown in Detail P and the facia member shown in Detail A are both cantilevered to hold the ends of the rafter. The wall is then literally built up to the underside of the roof deck. This may be done as shown or in other ways, depending on the soffit details, the wall finish, and so on.

The choice of one of these options, or of others, is usually primarily dependent

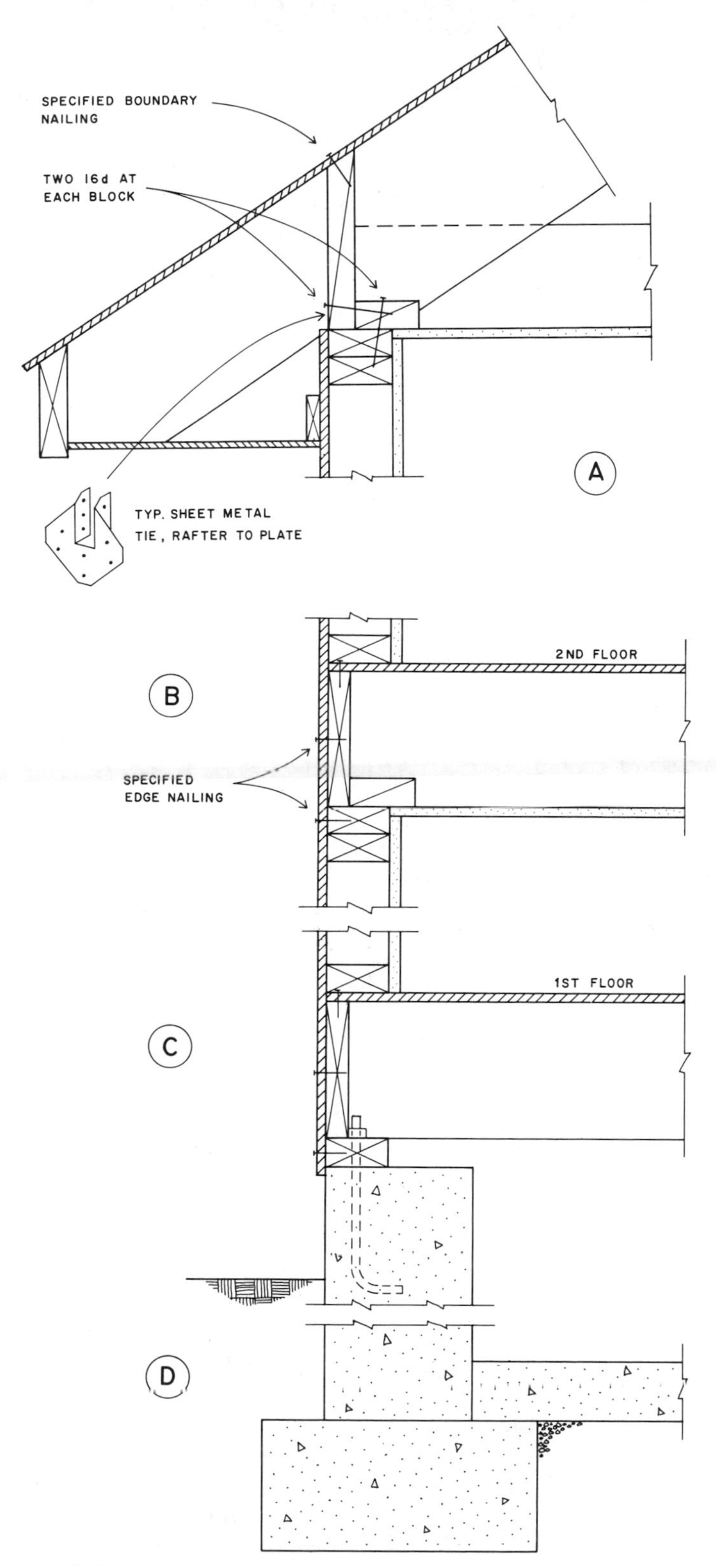

FIGURE 4.2. Construction details—front and rear walls.

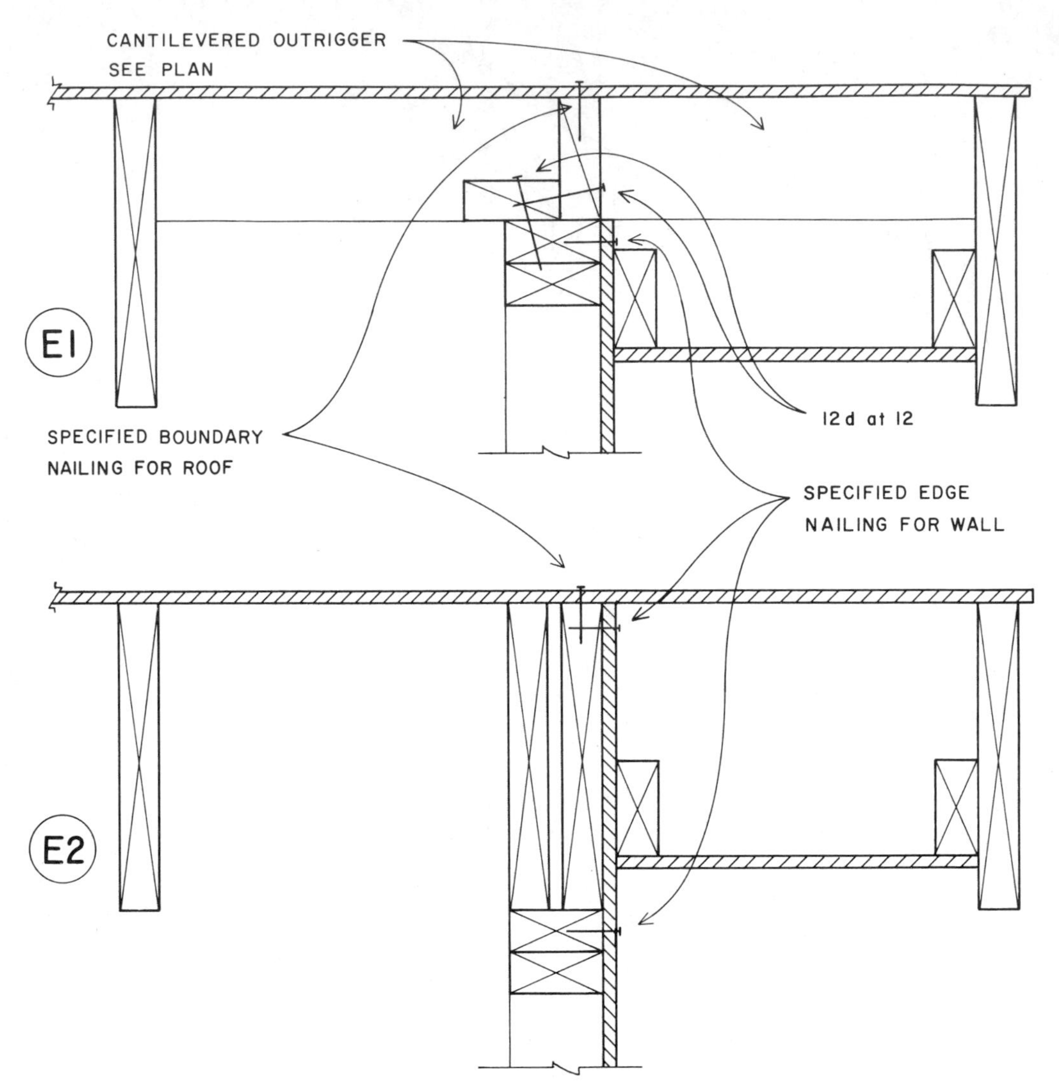

FIGURE 4.3. Construction details—gable end.

on considerations of architectural detailing rather than on structural necessity.

Detail F (Fig. 4.4)

This is the second-floor edge condition at the wall that is parallel to the joists. Because the floor plywood transfers its edge load into the edge joist that rests on the wall, the wall plywood should also be edge nailed to this member. Although we have not shown it, there is also a joint in the wall plywood here somewhere, depending on the sheet size used.

The second joist on top of the wall helps to support the wall above and is offset slightly to provide a backup for the ceiling drywall.

Detail G (Fig. 4.4)

This shows the floor edge detail and the support for the end of the floor beam. The wind load transfer from the wall ply-

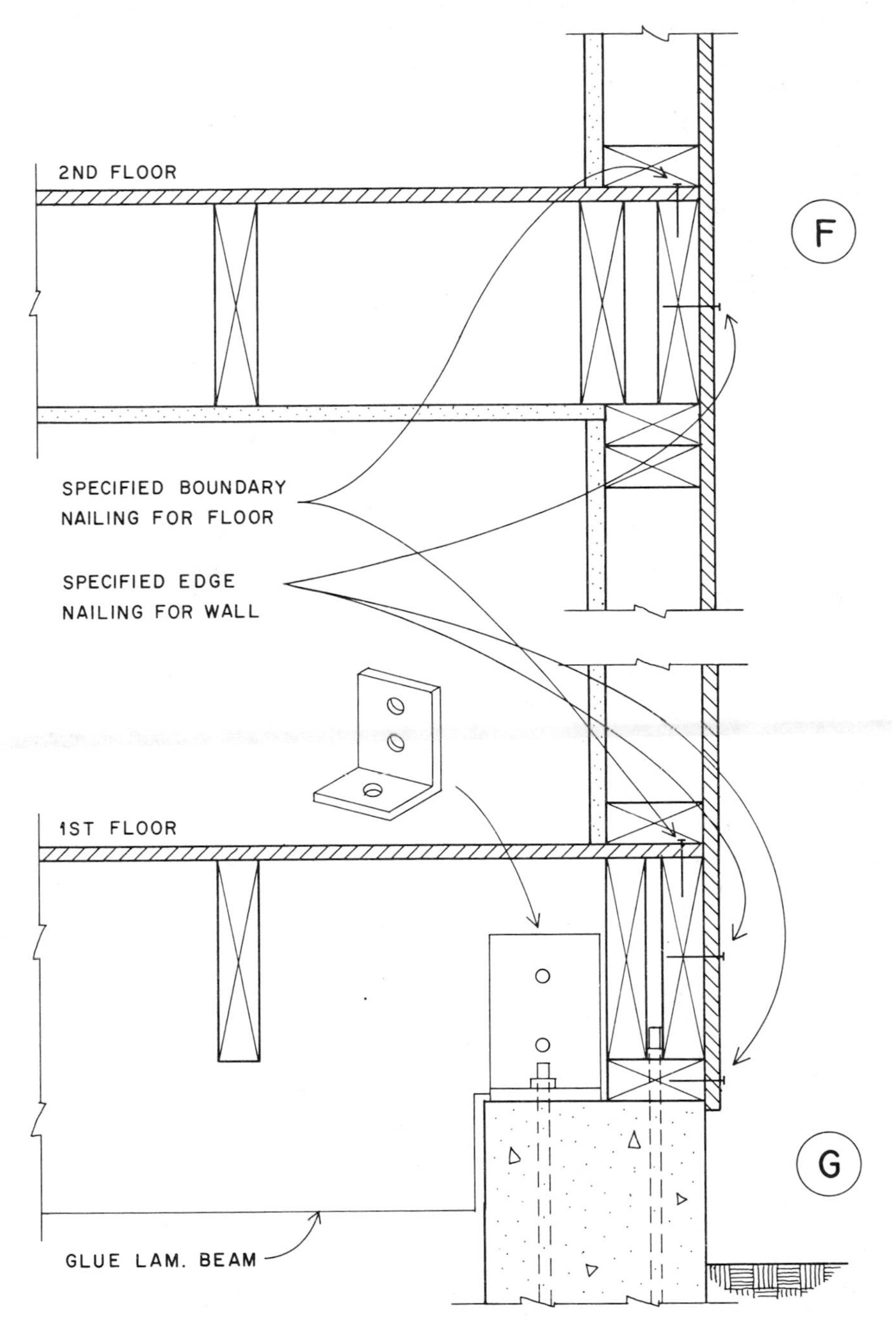

FIGURE 4.4. Construction details—end walls.

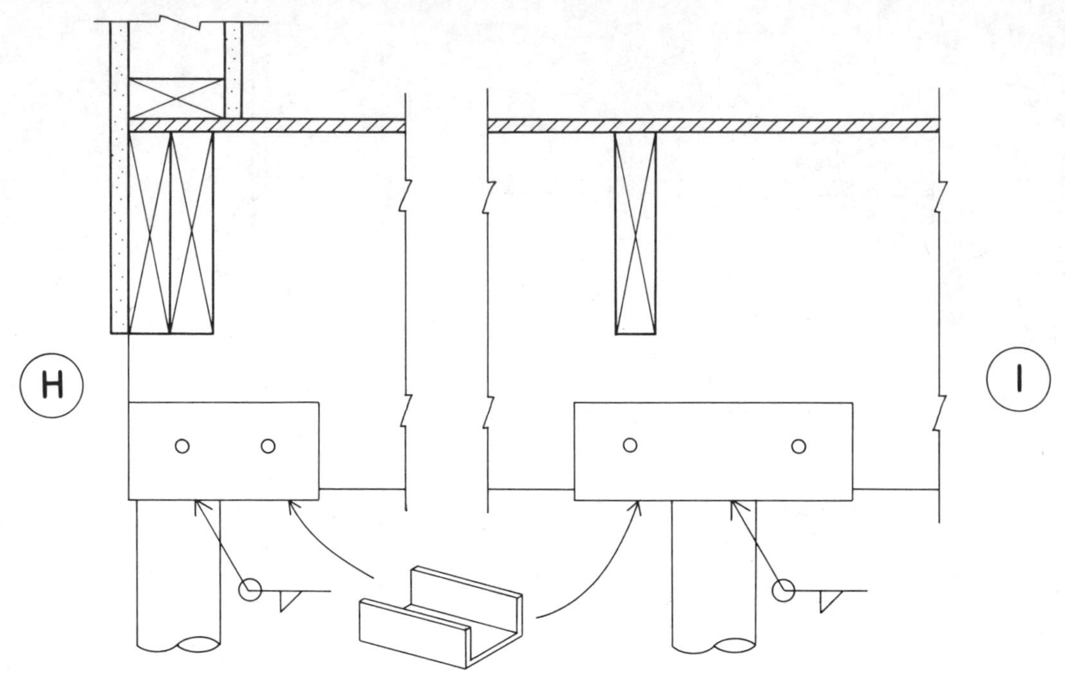

FIGURE 4.5. First-floor beam.

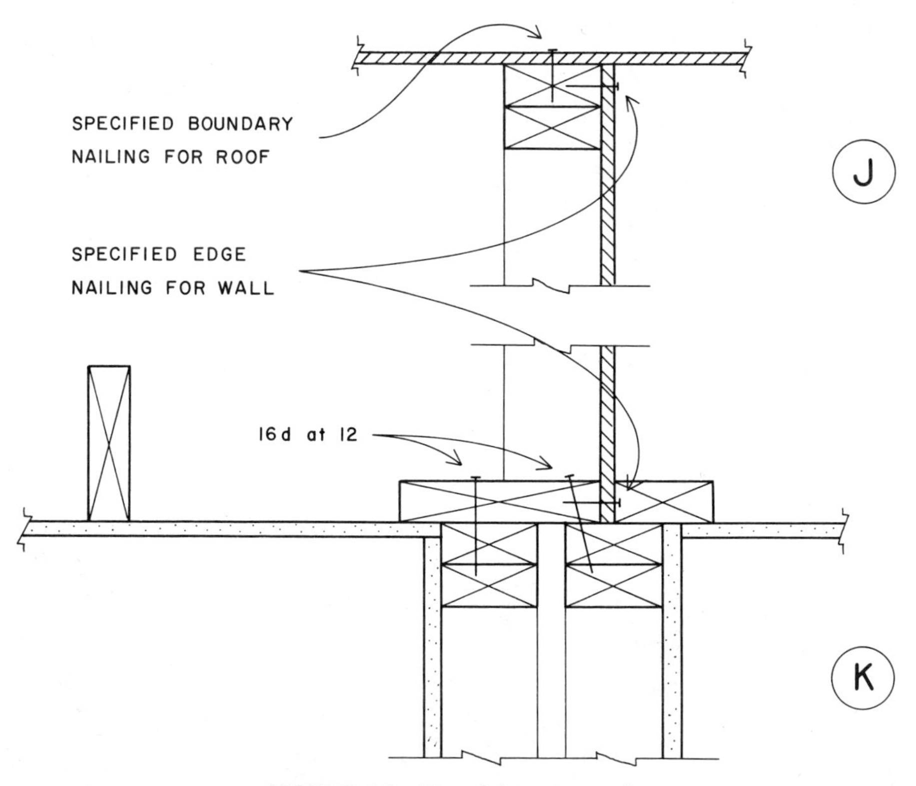

FIGURE 4.6. Top of the center wall.

wood to the basement concrete wall is essentially the same as in Detail C. As in Detail F, the second joist at the edge is used to support the wall above.

The end of the beam must be supported for vertical load as well as held in place during construction. This is usually achieved by using a steel member of some kind to facilitate the connection of the wood beam to the concrete. A simple pair of angles may be used as shown, or a single horizontal plate may be welded to two vertical plates to form a double T. Since the top of the wall is higher than the bottom of the beam, the beam must be notched as shown or the wall must be pocketed. The latter is probably simpler for construction and is dependent only on the ability of the reduced cross section to take the end shear in the beam.

Details H and I (Fig. 4.5)

Details H and I show the connections of the floor beam to the steel pipe columns using a U-shaped bent steel plate. A similar connection could be used with wood columns, with plates welded to the bottom of the U-shaped plate for attachment to the wood column.

Details J and K (Fig. 4.6)

This shows the cripple wall that sits on top of the double-stud dividing wall. It serves to divide the space in the attic for the two housing units and also carries the roof diaphragm wind load down to the dividing wall. The remainder of this wall, showing the details from the second floor down to the basement, was illustrated in Figure 2.8.

Detail L (Fig. 4.7)

This shows two options for the wall footing for the basement wall at the dividing wall. The difference between the two has to do with the sequence of construction of the wall and floor slab. The upper detail would be used if the wall is poured first. This is sometimes done to protect the slab during construction, where the pour is delayed until the first floor is in place. If the slab is poured first, the lower detail could be used, in which the footing and the slab are poured together.

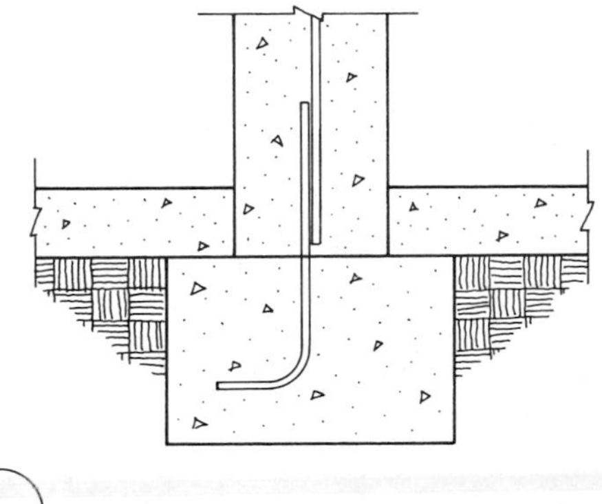

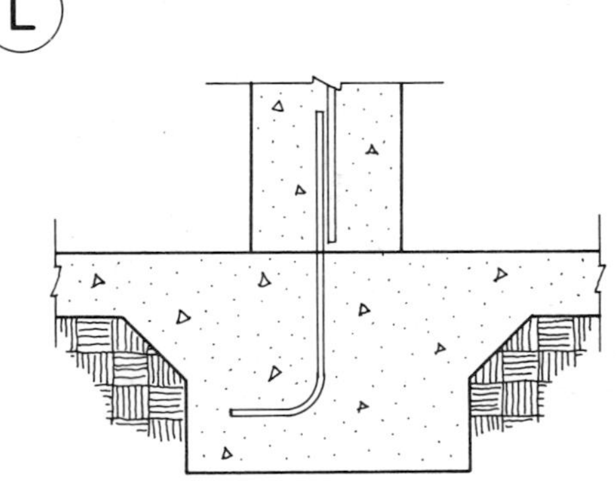

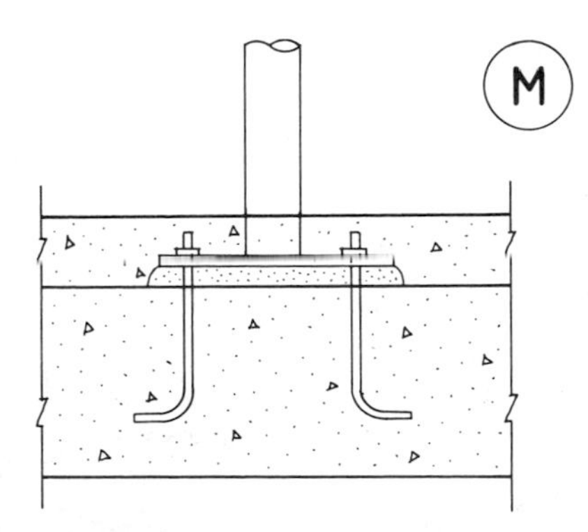

FIGURE 4.7. Foundation details.

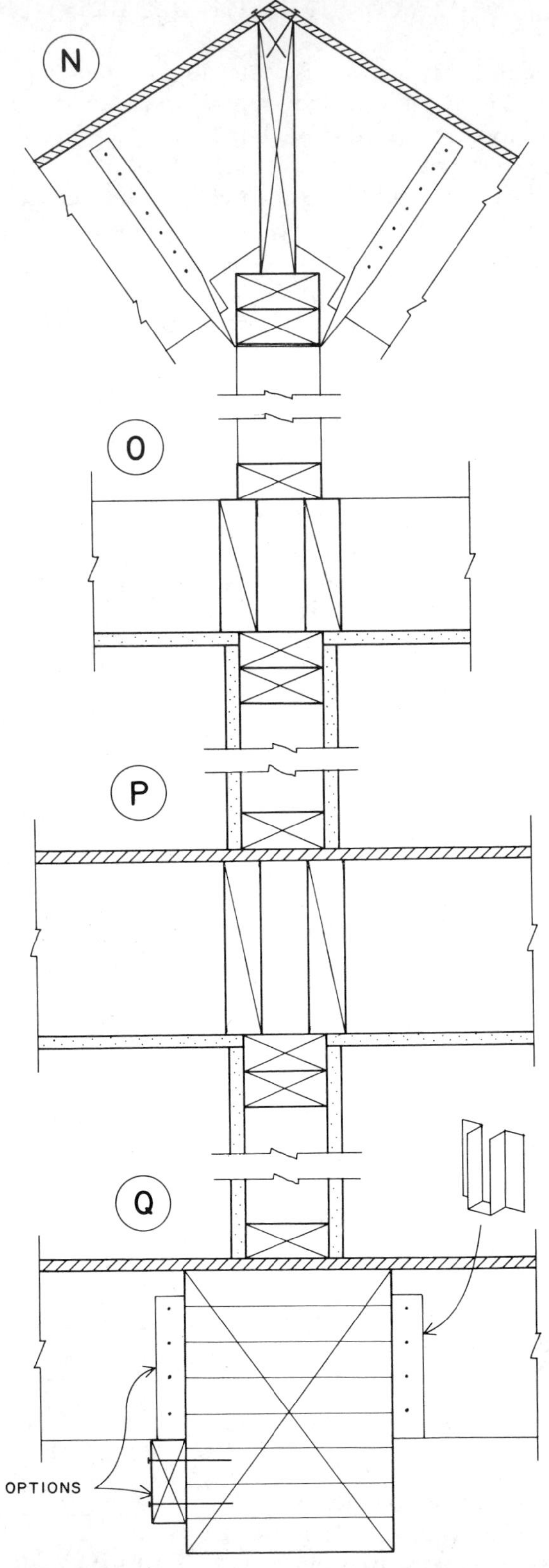

FIGURE 4.8. Construction details—interior bearing wall.

Detail M (Fig. 4.7)

This shows the base detail for the steel column. A steel base plate is welded to the bottom of the column and rests on a grout bed on top of the footing. The anchor bolts may be minimal in size, for they merely serve to hold the column in position during construction. If the slab and footing are poured at the same time, as in the lower illustration for Detail L, a pocket would be provided in the top of the footing so that the base plate and anchor bolts may be kept below the top of the slab.

Detail N (Fig. 4.8)

This shows the typical ridge detail for the roof. The ridge member is used to facilitate the joining of the rafters as well as to provide edge nailing for the plywood. The rafters are normally toenailed to the ridge. The strap shown is used to tie the rafters together and provide for some resistance to uplift. An alternative to the strap would be to use a 2 × 4 just under the plate of the wall and nailed to the rafter on each side.

Detail O (Fig. 4.8)

This shows the seating of the attic cripple wall on top of the center partition wall below. Because the cripple wall serves only for vertical load transfer, it could be an open stud wall with no surfacing. If so, the studs should be braced, possibly with 1 × 4 diagonal braces nailed to the studs, plates, and sill.

Although it is theoretically not required that the ceiling joists serve to tie the building against the thrust of the rafters (see Fig. 1.5), it adds generally to the structural integrity of the building if they do so. This may be done by lapping them or using metal straps as shown in the sketches.

SHEAR WALL AND HORIZONTAL DIAPHRAM NAILING SCHEDULE					
LOCATION	SURFACING	NAILS	NAIL SPACING AND LOCATION		
			AT BOUNDARIES	AT OTHER EDGES	AT INTERMEDIATE SUPPORTS
ROOF	3/8" C-D PLYWOOD	6 d COMMON	6	6	12
FLOORS	1/2" C-D PLY.	8 d COM.	6	6	12
WALL A	1/2" C-D PLY.	8 d COM.	6	6	12
WALL B	3/8" C-D PLY.	6 d COM.	6	6	12
WALL C	1/2" GYP. DRYWALL	5 d COOLER	4	4	4
WALL D	1/2" GYP. DRYWALL	5 d COOLER	7	7	7

FIGURE 4.9. Nailing for the Building One diaphragms.

Detail P (Fig. 4.8)

This shows the floor at the center wall. The blocking serves the dual purpose of vertical support and backup for the ceiling drywall. The floor joists may be lapped or tied as with the ceiling joists. However, the floor plywood serves adequately for tie purposes at this location.

Detail Q (Fig. 4.8)

This shows the floor beam at the center wall. Since the top of the floor joists is at the same level as the top of the beam, the ends of the joists would be supported with metal joist hangers or a ledger bolted to the side of the beam.

4.3. OTHER INFORMATION

Diaphragm Nailing

Figure 4.9 shows a typical schedule for the nailing of the wall and floor paneling to the framing for diaphragm action. Because of the relatively low shear stresses in most cases, the nail spacing shown is the minimum code required nailing for most of the diaphragms. Shear walls are usually labeled as such on the plans with number or letter designations for identity in the schedule.

On larger construction projects considerable information is placed in table form. On projects of this scale, however, most information required on the construction drawings is placed directly on the plans or the construction details.

PART TWO

BUILDING TWO

Building Two is a simple box—a single-story, flat-roofed, single-space building. The possible variations for the structural system and for the materials and details of individual components are quite extensive. If the building is built essentially for investment purpose, dictates of economy, local codes, and available local materials would probably narrow the range of choice. We will show the design for two different solutions. The first is an all wood structure. The second is a structure with masonry walls and a steel framed roof.

CHAPTER FIVE

Design of the Wood Structure

5.1. THE BUILDING

The general configuration of the building is shown in Figures 5.1 and 5.2. For maximum flexibility in the arrangement of interior walls, it is desired that there be no interior structural walls or columns. The roof therefore requires a clear span of 60 ft.

For the 9000-ft^2 building the UBC requires a one-hour fire rating for the walls and roof. This could be eliminated if a fire resistive partition is used to divide the interior, but we assume that this is not desired.

Some of the design criteria are:

Roof live load: 20 psf (reduced as per the UBC) [0.96 kN/m^2].

Lateral loads: 20 psf [0.96 kN/m^2] wind on vertical surfaces.

Soil capacity: 2000 psf maximum for shallow spread footings [96 kN/m^2].

5.2. THE WOOD STRUCTURE

The plans in Figure 5.3 show the layout for the wood structure consisting of plywood roof deck, wood rafters, glue-laminated girders, and wood stud walls. Girder spacing relates to the module of the plan and not necessarily to any structural optimization.

The following materials will be used:

2× and 3× framing: No. 2 Douglas Fir–Larch.

4× framing and larger: No. 1 Douglas Fir–Larch.

Glue-laminated members: Douglas Fir–Larch, 2400f grade.

Structural steel: A36, F_y = 36 ksi [248 MPa].

Concrete: F'_c = 3000 psi [20.7 MPa].

5.3. DESIGN OF THE WOOD ROOF STRUCTURE

Dead load on the structure consists of

Roofing: tar and gravel at 6.5 psf [0.31 kPa].

Insulation, lights, ducts, and so on: assume 5 psf [0.24 kPa].

Ceiling: assume 10 psf total [0.48 kPa] finish plus suspension system.

Total dead load: 21.5 psf [1.03 kPa] plus structure.

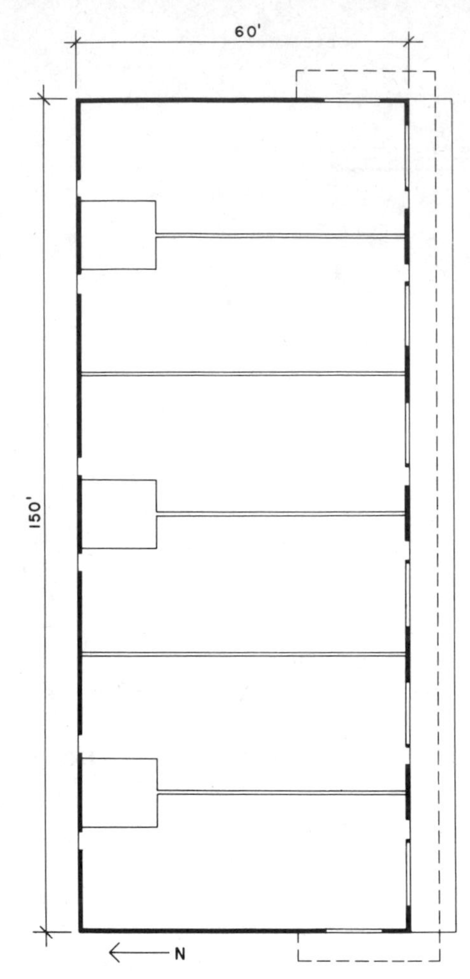

FIGURE 5.1. Floor plan—Building Two.

Roof Deck

The UBC Table 25-S-1 (see Appendix B) permits $\frac{1}{2}$-in. plywood for rafters spaced 24 in. on center. The selection of the plywood grade and the required nailing will be done as part of the design for seismic load.

Rafters

Allowable stresses for the rafters depend on their size. Our procedure is to make an initial size assumption, perform the analysis and size the rafter, and verify the original assumption. Note that UBC 2504c4 permits a 25% increase in allowable stress for roof loads.

From UBC Table 25-A-1 for 2 × 6 and larger (see Appendix B):

F_b = 1450 psi (repetitive use) [10 MPa]

E = 1,700,000 psi [11.7 GPa]

Live load: = 20 psf

Dead load:
Applied = 21.5 psf
$\frac{1}{2}$-in. deck = 1.5
Rafters + blocking = 4.0 (estimate)
Total dead load = 27.0 psf [1.29 kPa]

With rafters at 24 in., the total design load as a uniformly distributed load on one rafter is

$$2(20 + 27) = 94 \text{ lb/ft } [137 \text{ kN/m}]$$

Then

$$\text{maximum } M = \frac{wL^2}{8} = \frac{94(25)^2}{8} = 7344 \text{ ft-lb } [9.96 \text{ kN-m}]$$

and

$$\text{required } S = \frac{M}{F_b} = \frac{7344(12)}{(1.25)(1450)} = 48.6 \text{ in.}^3 \; [797 \times 10^3 \text{ mm}^3]$$

This requires a 3 × 12 rafter. If the stress grade is increased to No. 1, the allowable F_b is 1750 psi [12.1 MPa] and the required S is

$$S = \frac{1450}{1750}(48.6) = 40.3 \text{ in.}^3 \; [661 \times 10^3 \text{ mm}^3]$$

This permits a 2 × 14 rafter. The choice is somewhat arbitrary and would

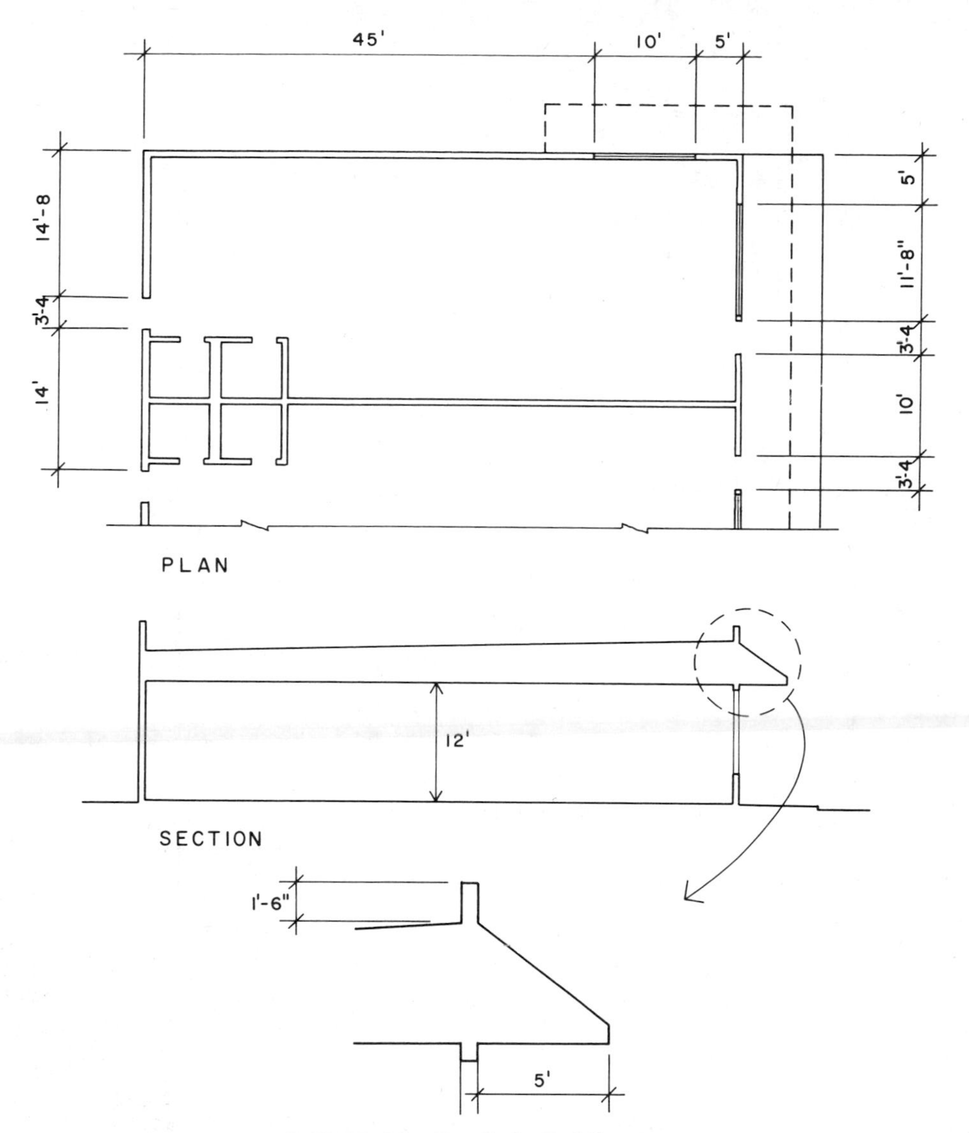

FIGURE 5.2. Details for Building Two.

probably be made on the basis of lumber prices and available sizes and lengths. A possible consideration is that of the rafter width required for the plywood nailing, which is discussed with regard to the design for lateral forces.

Allowable total load deflection is $\frac{1}{180}$ of the span or

$$\frac{24(12)}{180} = 1.60 \text{ in. [41 mm]}$$

With the 3 × 12 the deflection will be

$$\Delta = \frac{5WL^3}{384EI} = \frac{5(94 \times 25)(25 \times 12)^3}{384(1{,}700{,}000)(297)}$$

$$= 1.64 \text{ in. [42 mm]}$$

This is quite close to the limit and the choice of the 3 × 12 may be questioned. However, the true span for the rafter is probably somewhat less than 25 ft, and the choice may be deferred until the ex-

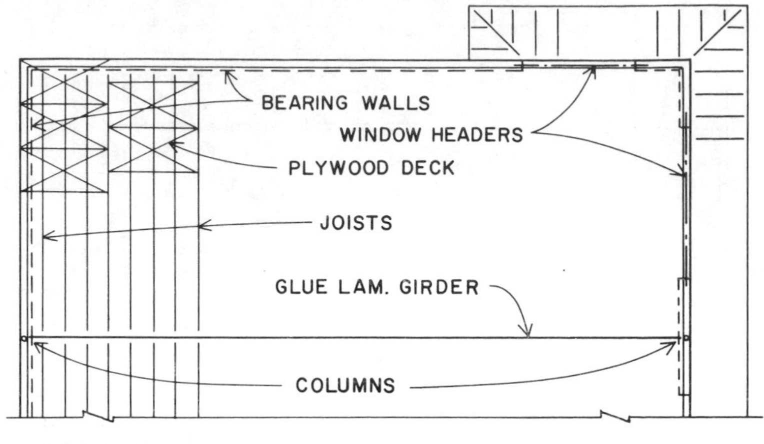

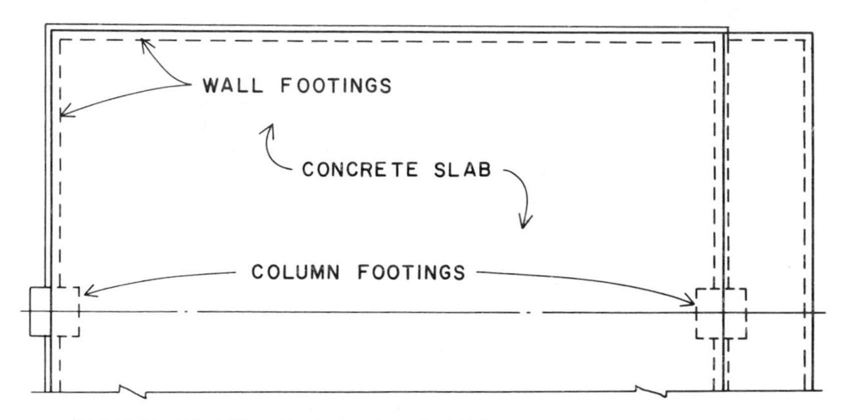

FIGURE 5.3. Structural plans for the wood structure.

act nature of the construction details is determined. The rafters will likely be supported by metal hangers on the face of the girders and a ledger on the face of the outside walls, which will result in a clear span of less than that assumed.

The situation for the 2 × 14 is similar. Its I value is less than that for the 3 × 12, but the E for the higher stress grade is 1,800,000 psi.

The code has requirements for lateral bracing for these rafter sizes. The feasibility of developing this bracing must be studied with all of the considerations of the construction requirements. These include the blocking and nailing for the plywood, load transfers to the supports, and the development of the ceiling structure.

The Glue-Laminated Girder

From UBC Table 25-C-1 for DF-L, 24f grade:

$$F_b = 2400 \text{ psi } [16.5 \text{ MPa}]$$

$$F_v = 165 \text{ psi } [1.14 \text{ MPa}]$$

$E = 1{,}800{,}000$ psi [12.4 GPa]

For the girder tributary area of 1500 ft^2, the UBC Table 23-C permits a reduction of the live load to 12 psf. The uniformly distributed load on the girder is thus

$$(25 \text{ ft})(12 + 27 \text{ psf}) = 975 \text{ lb/ft [14.2 kN/m]} + \text{the girder weight}$$

Assuming a total load of 1075 lb/ft [15.7 kN/m],

$$\text{maximum } M = \frac{wL^2}{8} = \frac{1075(60)^2}{8} = 483{,}750 \text{ ft-lb [656 kN-m]}$$

$$\text{required } S = \frac{M}{F_b} = \frac{483{,}750(12)}{(1.25)(2400)} = 1935 \text{ in.}^3 \; [31.7 \times 10^6 \text{ mm}^3]$$

Glue-laminated timbers are mostly made up of multiples of standard 2× members. Their depths are therefore some multiple of 1.5 in. Widths are slightly less than the standard lumber widths because some grinding or planing of the surface is done after the gluing is finished. Actual sizes should be obtained from local suppliers. The following options were chosen from the catalog of one timber fabricator.

$8\frac{3}{4} \times 39$, $C_F = 0.88$

effective $S = 0.88(2218) = 1952$ in.^{3}

$10\frac{3}{4} \times 36$, $C_F = 0.88$

effective $S = 0.88(2322) = 2043$ in.^{3}

$12\frac{3}{4} \times 33$, $C_F = 0.89$

effective $S = 0.89(2223) = 1978$ in.^{3}

Checking the shear for the narrowest beam with the least cross-sectional area,

$$\text{maximum } V = \frac{wL}{2} = \frac{1075(60)}{2} = 32{,}250 \text{ lb [143 kN]}$$

$$\text{maximum } F_v = \frac{3V}{2A} = \frac{3(32{,}250)}{2(341.3)} = 141.7 \text{ psi [977 kPa]}$$

which is less than the allowable of 165 psi.

Allowable deflection under total load is $L/180$, or $60(12)/180 = 4$ in. [100 mm]. Deflection calculations will show that all the optional sections previously listed will deflect slightly more than this. Because it is common procedure to camber (curve slightly upward) the glue-laminated beam, this is probably not a real concern. However, if the requirement is to be strictly adhered to, the procedure would be to find the I value required as

$$\text{required } I = \frac{5WL^3}{384E\,\Delta} = \frac{5(1075 \times 60)(60 \times 12)^3}{384(1{,}800{,}000)(4.0)} = 43{,}530 \text{ in.}^4 \; [18 \times 10^9 \text{ mm}^4]$$

The lightest section with this I is a $8\frac{3}{4} \times 40.5$, which is only slightly deeper.

With the rafter system, consideration must be made for the edges of the plywood sheets in the direction perpendicular to the rafter span. If these are allowed to deflect freely, damage to the membrane roofing will occur. Options for dealing with this are:

1. Use tongue–and–groove plywood panels, which may be available from suppliers.
2. Use metal clips between the abutting plywood panels if the roofers will accept them.
3. Provide blocking (short pieces of rafter fit between the spanning rafters) for support and nailing at the unsupported edges.

4. Develop a framing system that does not require any of the preceding but supplies support for all panel edges.

Option 1 in Figure 5.4 shows the layout for the rafter system just designed, with blocking indicated for the plywood. Option 2 consists of a three-part system that provides full edge support without the use of blocking. This system uses a very short rafter for which a 2 × 4 is adequate. The purlins would be as follows.

Purlins

Because the purlins carry about four times the load of the single rafters as previously designed, one may observe that the required size is likely to be in the stress grade category of "Beams and Stringers" in UBC Table 25-A-1 (see Appendix B) for which the design values for Douglas Fir–Larch, No. 1 grade are

F_b = 1300 psi [9 MPa] (not repetitive)

E = 1,600,000 psi [11GPa]

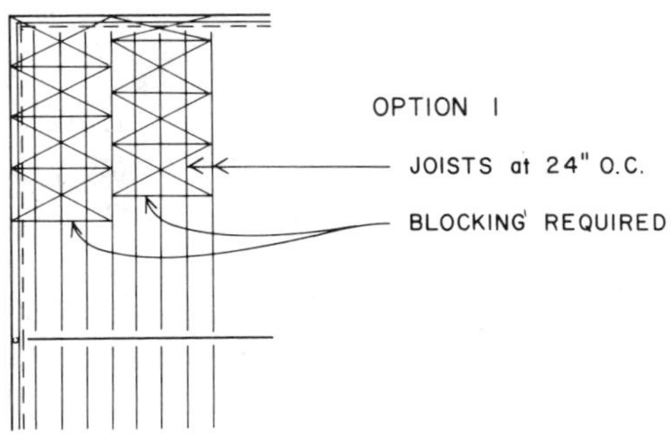

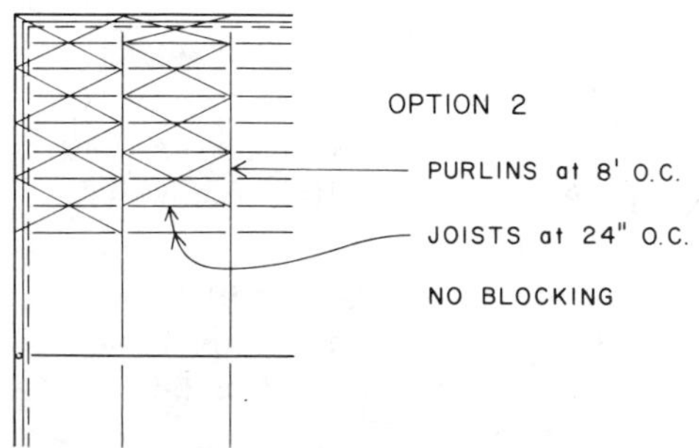

FIGURE 5.4. Optional roof systems.

Using the loading determined previously for the rafter design,

$$w = 8(20 + 27)$$

$$= 376 \text{ lb/ft } [5.49 \text{ kN/m}]$$

$$M = \frac{wL^2}{8} = \frac{376(24)^2}{8}$$

$$= 27{,}072 \text{ ft-lb } [36.7 \text{ kN-m}]$$

$$\text{required } S = \frac{M}{F_b} = \frac{27{,}072(12)}{1.25(1300)}$$

$$= 200 \text{ in.}^3 \; [3278 \times 10^3 \text{ mm}^3]$$

For the total load deflection limit of 1.60 in.,

$$\text{required } I = \frac{5WL^3}{384E\,\Delta}$$

$$= \frac{5(376 \times 24)(24 \times 12)^3}{384(1{,}600{,}000)1.60}$$

$$= 1096 \text{ in.}^4 \; [456 \times 10^6 \text{ mm}^4]$$

These requirements can be met with a 6 × 16 with S = 220 in.^{3} and I = 1707 in.^{4} An alternative glue-laminated member with 24f grade would require

$$S = \frac{M}{F_b} = \frac{27{,}072(12)}{1.25(2400)}$$

$$= 108 \text{ in.}^3 \; [1770 \times 10^3 \text{ mm}^3]$$

$$I = \frac{5WL^3}{384E\,\Delta} = \frac{5(376 \times 24)(24 \times 12)^3}{384(1{,}800{,}000)(1.60)}$$

$$= 975 \text{ in.}^4 \; [406 \times 10^6 \text{ mm}^4]$$

These requirements can be met by a member with the dimensions $5\frac{1}{8}$ × 13.5, with S = 0.99(155.7) = 154 in.^{3} effective and I = 1050.8 in.^{4}

5.4. DESIGN OF THE WOOD STUDS AND COLUMNS

Studs

The end walls of the building carry the ends of the 25-ft span rafters or purlins.

Because of the sloping roof surface, it is easier for construction to make these studs continuous to the top of the parapet and to carry the rafters or purlins on a ledger bolted to the face of the studs. The unbraced height of the studs is thus the distance from the floor to the bottom of the ledger. With the roof slope, this distance varies 15 in. from front to rear of the building. To attain the desired 12-ft clear ceiling height under the deep girder, the bottom of the rafters at the front of the building will be at approximately 15.5 ft above the floor. For this height the UBC requires that the studs be 2 × 6.

The studs should be checked for the combined compression plus bending due to the wind load and gravity load.

Gravity load equals:

Roof:
12.5 ft (20 + 27) = 588 lb/ft of wall
Canopy:
Assume = 100
Total DL + LL = 688 lb/ft [10 kN/m]
Or
688(1.33) = 915 lb/stud at 16-in. centers [4.07 kN]

For the 15.5-ft high stud, $h/t = 15.5(12)/5.5 = 33.8$,

$$F'_c = \frac{0.3\ E}{(h/t)^2} = \frac{(0.3)(1{,}700{,}000)}{(33.8)^2}$$

$$= 446 \text{ psi } [3.08 \text{ MPa}]$$

With the wind load the allowable load/stud is thus

$$\text{load} = 446(8.25)(1.33)$$

$$= 4894 \text{ lb } [21.8 \text{ kN}]$$

With wind pressure of 20 psf on the wall, the moment is

$$M = \frac{wL^2}{8} = \frac{20(1.33)(15.5)^2}{8}$$

$$= 799 \text{ ft-lb } [1.08 \text{ kN-m}]$$

The combined effect is thus

$$\frac{P/A}{F'_c} + \frac{M/S}{F_b} = \frac{915}{4894} + \frac{799(12)/7.56}{1.33(1450)}$$

$$= 0.187 + 0.658$$

$$= 0.845 < 1$$

Because the canopy is cantilevered from the wall, the studs should also be checked for this load unless the cantilever forces are carried directly back into the roof construction with struts and ties. Since we are not detailing the canopy construction, we will assume the 2 × 6s to be adequate for this condition.

We have checked the heaviest loaded stud so that we may safely use the 2 × 6s for the other walls. Actually, 2 × 4s can probably be used for the shorter rear wall.

Columns

The girder ends bring large concentrated loads to the front and rear walls, making it desirable to build a column into the wall at this location. Three options are possible: multiple 2 × 6 studs, a solid 6× member, or a steel post inserted in the hollow wall space. From the girder calculations the end load is approximately 32 kips. At the front wall the column height is approximately 13.5 ft to the bottom of the girder.

For No. two 2 × 6 studs, UBC Table 25-A-1 gives (see Appendix B)

$$F_c = 1050 \text{ psi } [7.24 \text{ MPa}]$$

$$E = 1{,}700{,}000 \text{ psi } [11.7 \text{ GPa}]$$

Then

$$\frac{h}{t} = \frac{13.5(12)}{5.5} = 29.45$$

$$F'_c = \frac{0.3\ E}{(h/t)^2} = \frac{(0.3)(1{,}700{,}000)}{(29.45)^2}$$

$$= 588 \text{ psi } [4.05 \text{ MPa}]$$

required area

$$= \frac{32{,}000}{588}$$

$= 54.4$ in.^{2} (or seven 2 × 6s) [35×10^3 mm^2]

If a solid timber is used, E drops to 1,600,000 psi and a 6 × 12 would be required. If a steel column is used, options are a 4-in. standard round pipe or a 4-in. square tube with $\frac{1}{4}$-in. wall thickness, both of which will fit in the wall space.

Design of the wall details and of the girder connection and the foundations may determine the desirability of one of these options over the other.

Header at Wall Opening

Load on the headers consists of the weights of the wall and canopy and part of the roof load. With the purlin system the load on the front wall headers will be:

Roof DL + LL:
4(47) = 188 lb/ft
Wall and parapet:
5(15) = 75
Canopy (estimate)
= 100
Header (estimate)
= 25

Total load = 388 lb/ft [5.66 kN/m]

At the front wall the header will span approximately 16 ft, the exact dimension depending on the construction details:

$$\text{maximum } M = \frac{wL^2}{8} = \frac{(388)(16)^2}{8}$$

$= 12{,}416$ ft-lb [16.8 kN-m]

If a solid 6×, No. 1, Douglas Fir–Larch, with $F_b = 1300$ psi [9 MPa] is used,

required S

$$= \frac{M}{F_b} = \frac{(12{,}416)(12)}{1300}$$

$= 114.6$ in.^{3} [1878×10^3 mm^3]

A 6 × 12 can be used to provide this section modulus. Since the percentage of live load is small, deflection should not be critical for the window construction detailing. The true loading and span conditions should be verified when the final details of the construction are developed. The actual span will be from center to center of the posts if a direct bearing is used for the end connection of the header. If the header spans from face to face of the posts, the span will be slightly less.

The header at the end wall carries more roof load because of the purlin span. This higher load on the shorter span will probably result in approximately the same size header.

Checking shear for the 16-ft span,

maximum $V = 388(7) = 2716$ lb

(approximately critical shear)

$$F_v = \frac{3V}{2A} = \frac{3(2716)}{2(63.25)}$$

$= 64.4 \text{ psi} < 85 \text{ psi}$

[444 kPa < 586 kPa]

5.5. DESIGN OF THE FOUNDATIONS

Stud Wall Foundation

At the solid end wall this load is approximately as follows:

Roof dead load:
27 psf × 12 ft = 324 lb/ft of wall
Wall:
20 psf × 17 ft = 340
Grade wall and footing (estimated)
= 450

Total dead load
= 1114 lb/ft

Roof live load:
20 psf × 12 ft = 240 lb/ft
Total DL + LL
= 1354 lb/ft [19.8 kN/m]

With the allowable pressure of 2000 psf [96 kPa] this requires less than a foot of width. Depending on the depth required for frost or for adequate soil bearing, there are several options for detailing of the grade wall and footing. We will assume that a depth of 3 ft [1 m] is required and will design for a separate wall and footing, as shown in the construction details.

This footing should be slightly wider than the wall for construction purposes. If a 14-in.-wide footing is used, the dead load pressure will be slightly less than 1000 psf. To equalize settlements the rest of the footings should be designed for this dead load pressure rather than for the maximum total load limit of 2000 psf.

Column Footing

Because the column occurs in the wall, there are several options for this footing. As shown in Figure 5.5, three possibilities are as follows:

1. The grade wall may be designed as a continuous beam, distributing the loads to a constant width footing.

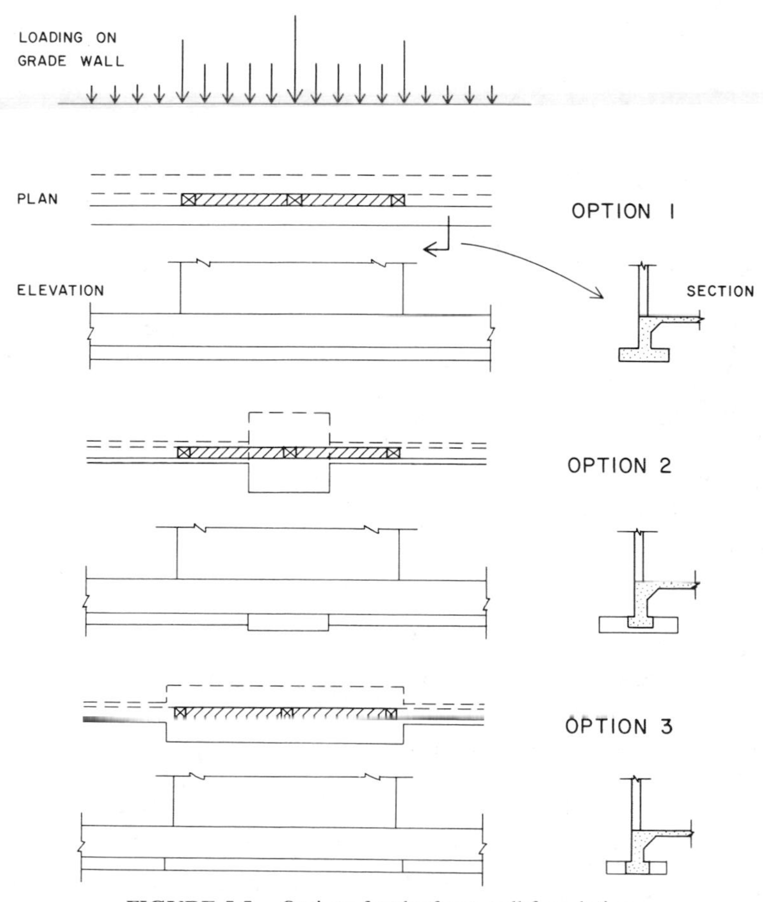

FIGURE 5.5. Options for the front wall foundation.

2. A separate square column footing may be designed to carry the column plus only the wall directly over the column footing. The remainder of the wall length would be directly carried by a narrow wall footing.
3. The footing under the 10-ft solid wall portion may be designed for the column plus the wall plus the header post loads, and a minimal footing provided under the remainder of the grade wall.

All three options can be adequately designed. Option 1 is the simplest in detail and easiest to build, but requires a reasonably deep grade wall for the beam action. We will design the system for Option 3 with a wide footing 12-ft long under the solid wall. The total load on this footing will be:

Header post load:
308 plf (dead load) × 16 ft
= 4,928 lb
Roof dead load on wall:
27 psf × 4 ft × 10 ft
= 1,080
Wall dead load:
20 psf × 17 ft × 10 ft
= 3,400
Grade wall:
300 plf (estimate) × 10 ft
= 3,000
Girder dead load:
775 plf × 30 ft
= 23,250
Total dead load
= 35,658 lb + footing [159 kN]

Using the dead load pressure determined for the end wall,

$$\text{area required} = \frac{35{,}658}{950} = 37.5 \text{ ft}^2 \ [3.48 \text{ m}^2]$$

If 12-ft long,

$$\text{width required} = \frac{37.5}{12} = 3.13 \text{ ft} \ [0.95 \text{ m}]$$

We will use a 3-ft-wide footing for which the actual dead load pressure will be:

Assume a 10-in.-thick footing:
Weight of footing: 3.0 × 12 × 0.83 × 150 pcf = 4482 lb.
Total dead load: 35,658 + 4482 = 40,140 lb.
Dead load pressure:

$$\frac{40{,}140}{3 \times 12} = 1115 \text{ psf} \ [53.4 \text{ kPa}].$$

The continuous grade wall will distribute some of this pressure to the narrow foot-

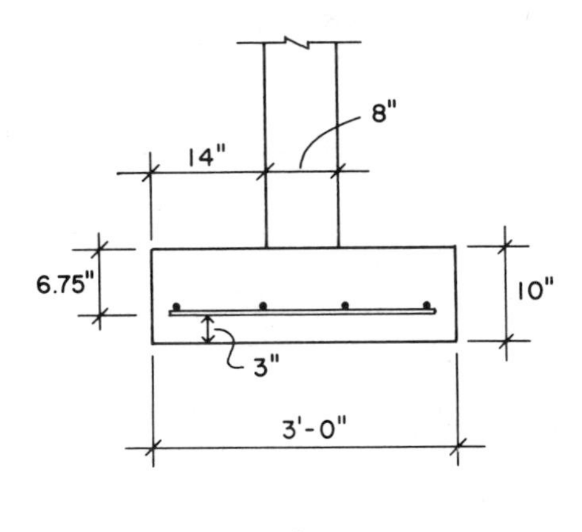

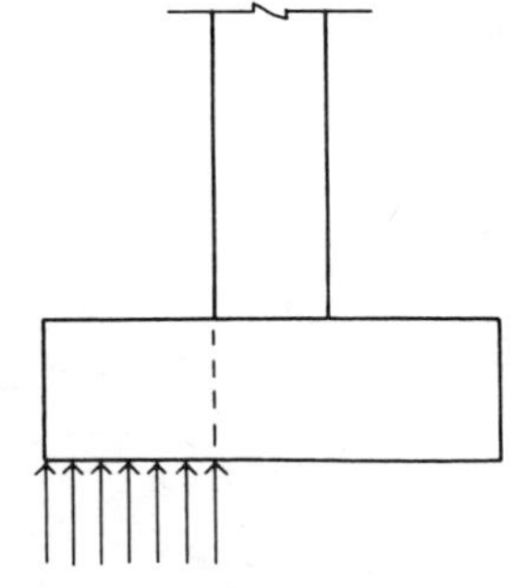

FIGURE 5.6. Details for the front wall foundation.

ing under the windows, which will otherwise be quite lightly loaded.

By adding the roof live load at 12 psf, the total pressure will be

roof live load

$$= 30 \times 25 \times 12 \text{ psf} = 9000 \text{ lb}$$

total pressure

$$= \frac{49{,}140}{36} = 1365 \text{ psf [65.4 kPa]}$$

This is well below the allowable of 2000 psf.

The reinforcing in the short direction should be designed for the total pressure less the weight of the footing, or approximately 1240 psf. Referring to Figure 5.6,

$$M = \frac{1240(14/12)(7)}{12}$$

$$= 844 \text{ ft-lb [1.14 kN-m]}$$

$$\text{required } A_s = \frac{M}{f_s jd} = \frac{844(12)}{16{,}000(0.9)(6.75)}$$

$$= 0.104 \text{ in.}^2/\text{ft [220 mm}^2/\text{m]}$$

Or, for the entire 12-ft-long footing:

$$A_s = 0.104(12) = 1.25 \text{ in.}^2 \text{ [807 mm}^2]$$

This may be provided by seven No. 4 bars in the short direction.

In the long direction a minimum shrinkage reinforcement of 0.02% of the cross section will be used:

$$A_s = 0.002(10 \times 36)$$

$$= 0.72 \text{ in.}^2 \text{ [465 mm}^2]$$

This may be provided by using four No. 4 bars, two of which can be made continuous with the reinforcing in the narrow footing.

CHAPTER SIX

Design for Seismic Force on the Wood Structure

Design for the seismic forces on the building includes the following considerations:

1. Design of the roof diaphragm for forces in both directions.
2. Design of the vertical shear walls.
3. Development of the various construction details for transfer of the forces from the horizontal to the vertical diaphragms and for transfer of the forces from the shear walls to the foundations.

6.1. PLANNING THE LATERAL RESISTIVE SYSTEM

A critical preliminary decision is the identification of the walls to be used as shear walls. Some of the considerations in this decision are:

1. The actual magnitude of force that the walls in each direction must resist. A quick estimate of the load should be made to determine this.
2. Which walls lend themselves to being used. This has to do with their plan location, their length, and the type of bracing used. The code establishes maximum height-to-length ratios for various bracing: up to $3\frac{1}{2}:1$ for plywood, $1\frac{1}{2}:1$ for plaster or drywall.
3. What materials are planned for the wall surfaces that may be used for their shear resistance, and what materials or bracing can be added where the surfacing materials are not adequate.

We assume that the ordinary construction to be used is drywall on the interior and cement plaster (stucco) on the exterior of the walls. This means that only one of these surfacings can be used for the exterior walls because the code does not permit addition of dissimilar materials [UBC Section 4713(a)]. Where the stucco is not sufficient, we add plywood to the wall, ignoring the surfacing materials. Where plywood is not required for the full length of a wall it sometimes simplifies the detailing if it is added to the interior, rather than to the exterior, of the wall.

Figure 6.1 shows the proposed layout of the shear wall system. For load in the short direction the roof will span from end to end of the building, transferring the shear to the two 45-ft-long end walls.

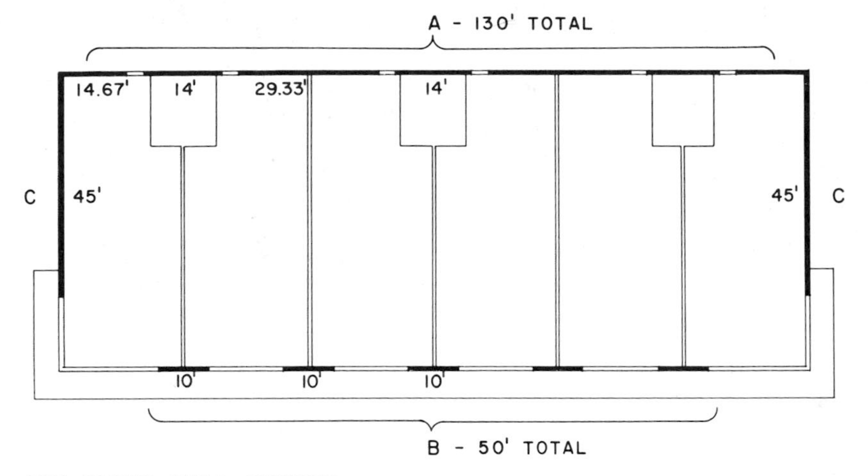

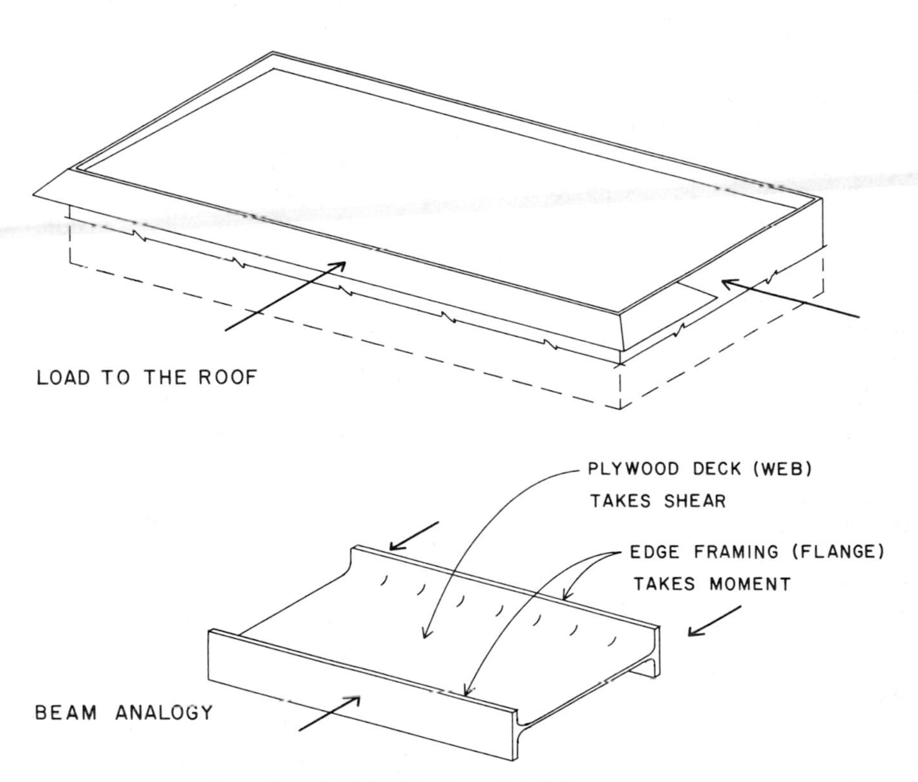

FIGURE 6.1. The lateral resistive system.

For the load in the long direction the five 10-ft-long walls will be used on the front and the entire wall will be used on the rear, that consists of a net wall length of 130 ft. The latter will result in some eccentricity between the load and the centroid of the walls in the long direction, thus requiring an investigation of the torsional effect.

6.2. DESIGN OF THE ROOF DIAPHRAGM

The dead loads to be used for the seismic force to the roof diaphragm are shown in Table 6.1. Wall loads are taken as the weight of the upper half of the walls, ignoring openings that are generally in the bottom portion. In each direction the wall loads considered are only those of the walls perpendicular to the load direction. The canopy load is assumed to be taken by the roof in both directions. A nominal load is assumed for rooftop HVAC units.

TABLE 6.1. Loads to the Roof Diaphragm

Load Source and Calculation	Loads (kips) N–S	Loads (kips) E–W
Roof dead load		
150 × 60 × 27 psf	243	243
East and West exterior walls		
$\frac{1}{2}$ × 60 × 17 × 20 psf × 2	0	20.4
North and South exterior walls		
$\frac{1}{2}$ × 150 × 17 × 20 psf × 2	51	0
Interior dividing walls		
$\frac{1}{2}$ × 60 × 15 × 8 psf × 5	0	18
Toilet walls		
$\frac{1}{2}$ × 190 × 15 × 8 psf	11.4	11.4
Canopy		
190 ft × 100 lb/ft	19	19
Rooftop HVAC units (estimate)	5	5
Total load (*W* for seismic calculation)	329.4	316.8

The total load in both directions is reasonably symmetrically placed. The canopy load on the front wall is offset by the toilet walls and the heavier rear wall in the long direction loading. The seismic design load is determined as (see UBC 2312)

$$V = ZIKCSW$$

where Z = 1.0 for UBC zone 4 (assumed condition)
I = 1.0 for this building occupancy
K = 1.0 for the all wood framed box system
CS = 0.14 as a product, if S is not established by geological studies of the site
W = the building's gravity weight that affects seismic action in the direction considered

Thus

$$V = (1.0)(1.0)(1.0)(0.14)W = 0.14W$$

Using the loads from Table 6.1, we therefore determine the seismic loads to the roof diaphragm as

$$V = 0.14(329.4)$$
$$= 46.1 \text{ k in the N–S direction [205 kN]}$$

$$V = 0.14(316.8)$$
$$= 44.4 \text{ k in the E–W direction [198 kN]}$$

If the load in the north–south direction is considered to be uniformly distributed and symmetrically placed, the maximum shear in the deck will be one half of the total load, and the unit shear in the roof plywood is

$$v = \frac{\text{maximum shear}}{\text{diaphragm width}} = \frac{46{,}100/2}{60}$$
$$= 384 \text{ lb/ft [5.6 kN/m]}$$

For the membrane roofing it is usually required to have a minimum plywood thickness of ½ in. Assuming the use of the 2× rafters, Structural II grade plywood, and 8d nails, UBC Table 25-J-1 (see Appendix B) yields the following requirements for the nail spacing.

Nails at 2.5 in. at the diaphragm boundary and at continuous panel edges parallel to the load.

Nails at 4 in. at other panel edges.

In the center of the plywood panels other code requirements usually ask for nails at a minimum of 12 in. along supports. This is usually called field nailing or nailing at intermediate supports.

We will assume the use of the rafter and purlin system (Option 2 in Fig. 5.4) for which the layout of the plywood panels will be Case 5 as shown in the footnotes for UBC Table 25-J (see Appendix B). For this situation we would most likely specify the required boundary nailing for all panel edges.

If the seismic load is assumed to be uniformly distributed along the length of the single-span diaphragm, the magnitude of the shear will diminish continuously from the maximum value at the end to zero at the center of the span. It is therefore possible to consider a reduction in the amount of nailing for portions of the diaphragm nearer the center of the building. An example of this so-called zoned nailing is shown in Figure 6.2. We have used only the values for a blocked diaphragm in the example because the purlin system provides a blocked diaphragm throughout the roof. If only the rafters are used (Option 1 in Fig. 5.4), it would be possible to consider the use of an unblocked diaphragm for the center portion of the building where the shear is lowest.

For the chord force (see Fig. 5.7) we first find the maximum moment for the spanning diaphragm. Then, assuming the chords to act with a moment arm equal to the width of the building, the analysis is as follows:

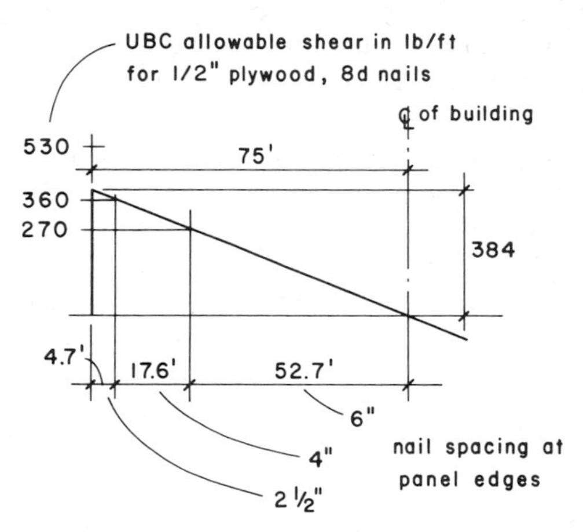

FIGURE 6.2. Zoned nailing for the roof.

$$M = \frac{WL}{8}$$

$$\text{chord force: } T = C = \frac{M}{d}$$

where d is the building width. Combining these,

$$T = C = \frac{WL}{8d} = \frac{46{,}100 \times 150}{8 \times 60}$$

$$= 14{,}400 \text{ lb [65 kN]}$$

If the double top plate of the stud wall is used for the chord (see Fig. 7.2), an analysis will show that this is too much force for the member comprised of two 2 × 6s. Alternatives are to raise the usually low stress grade of the plates (Douglas Fir–Larch, No. 2) or to increase the size to 3 × 6.

Because it is not possible to use continuous members for the top plates (150 ft long), the splicing of the plates must be done in a manner that preserves the tension capacity of the chord. For low chord forces the usual practice of staggering the splices of the two plate members and the minimum required nailing between them (see UBC Table 25-P in Appendix B) may be sufficient. For a force of the magnitude of that required for our example, it will be necessary that the splicing use steel straps or steel bolts.

If the studs run full height to the top of the parapet, the edge of the roof deck would be nailed directly to a ledger fastened to the face of the studs. (See construction of the end walls in Chapter 7.) If the ledger serves as the chord, it must be designed for the chord forces and its splicing must be developed to preserve the tension capacity required.

The load in the east–west direction will produce less stress in the deck and is therefore not critical for the selection of the deck or its nailing. The maximum stress at the deck boundary is

$$v = \frac{44{,}400/2}{150} = 148 \text{ lb/ft [2.16 kN/m]}$$

This is less than the minimum capacity produced by using the minimum nailing (8d at 6 in.), but the load represented must be collected along the wall. If the deck is nailed to supporting header beams over large openings, the forces collected by these members must be transferred into the shear walls. Development of the construction detailing must include these considerations.

6.3. DESIGN OF THE SHEAR WALLS

The shear walls carry the edge shear forces from the roof diaphragm. In addition, they resist the full force of the wall weight in the plane of the shear wall, for this was not included in the load to the roof. The end walls thus resist the load of 23.05 k from the roof plus the following:

Wall weight:

$20 \text{ psf} \times 17 \text{ ft} \times 45 \text{ ft} = 15{,}300 \text{ lb}$

$20 \text{ psf} \times 7 \text{ ft} \times 15 \text{ ft} = 2{,}100$

$5 \text{ psf} \times 10 \text{ ft} \times 15 \text{ ft} = 750$

Total $= 18{,}150$ lb [76.3 kN]

Lateral load:

$0.14\,W = 0.14 \times 18{,}150 = 2540$ lb [11.3 kN]

The loading of the end shear wall is thus as shown in Figure 6.3. The total shear force is 25,590 lb [113.8 kN] and the unit shear in the wall is

$$v = \frac{25{,}590}{45} = 569 \text{ lb/ft [8.30 kN/m]}$$

From UBC Table 25-K (see Appendix B) this requires $\frac{1}{2}$ in. Structural II plywood with 10d nails at 3-in. centers at all panel edges. A footnote to the table requires the use of 3× studs and the staggering of the nails for this nailing.

The overturn analysis is:

Overturn M:

$23.05 \times 15 = 346$ k-ft

$+ 2.54 \times 9 = 23$

Total $= 369$ k-ft [500 kN-m]

Dead load M:

$30 \times 22.5 = 675$ k-ft [915 kN-m]

Safety factor:

$$\frac{\text{dead load } M}{\text{overturn } M} = \frac{675}{369} = 1.83$$

This would generally be considered to be adequate, for the usual requirement is for a minimum safety factor of 1.5. If the safety factor is less than 1.5, a tiedown anchor, as shown in Figure 6.3, must be used.

At the base of the wall the total horizontal sliding force must be transferred to the foundation through the wall sill

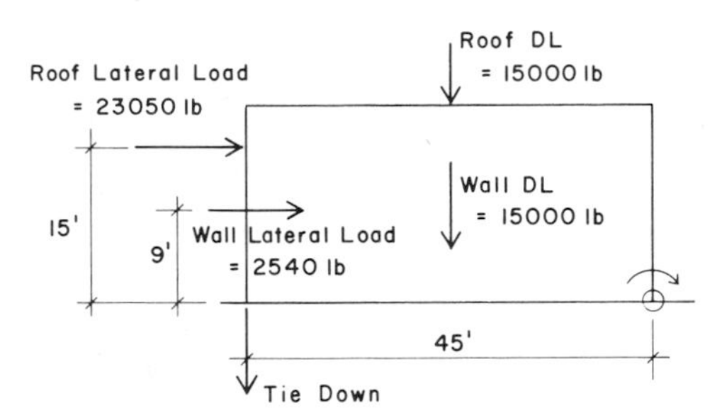

FIGURE 6.3. Stability for wall C.

anchor bolts. Using table 25-F from the UBC and assuming a $2\frac{1}{2}$ in. thick sill, we determine the following:

For $\frac{1}{2}$-in. bolts:

$$\text{allowable load per bolt} = 1.33(630) = 838 \text{ lb } [3.7 \text{ kN}]$$

$$\text{number required} = \frac{25{,}590}{838} = 30.5, \text{ or } 31$$

This requires bolts at approximately 18-in. centers, assuming the first bolt to be 12 in. from the end of the wall.

For $\frac{3}{4}$-in. bolts:

$$\text{allowable load per bolt} = 1.33(1115) = 1483 \text{ lb } [6.6 \text{ kN}]$$

$$\text{number required} = \frac{25{,}590}{1483} = 17.2$$

at approximately 2 ft 4 in. on center.

In most cases the contractors would prefer the smaller number of bolts.

In the east–west direction there is a lack of symmetry in the disposition of the shear walls (the north and south building exterior walls). There are two approaches to the analysis for the seismic force on these walls, as follows:

1. *Analysis by Peripheral Distribution.* In this analysis it is assumed that the roof acts as a simple beam and one half of the total lateral load is thus delivered to each wall. Thus the shear stresses in the walls would be

$$v = \frac{22{,}200}{130} = 171 \text{ lb/ft}$$

[2.49 kN/m] for the north wall

$$v = \frac{22{,}200}{50} = 444 \text{ lb/ft}$$

[6.48 kN/m] for the south wall

The length of wall used for each calculation is simply the total length of shear wall in each of the building walls.

2. *Analysis for Torsion.* In this analysis the north and south walls are considered in terms of their respective proportional stiffnesses, with stiffness of the plywood walls assumed to be proportionate to the wall plan lengths. Stress in the walls is considered as the sum of the direct stress plus the stress due to torsion as follows:

$$\text{direct stress} = \frac{\text{total lateral force}}{\text{sum of all wall lengths}} = \frac{44{,}400}{180} = 247 \text{ lb/ft } [3.6 \text{ kN/m}]$$

Referring to Figure 6.4, the torsional analysis follows.

For the center of stiffness:

$$y = \frac{(50)(60)}{180} = 16.67 \text{ ft from the rear wall}$$

The torsional moment of inertia is as shown in Table 6.2 and the torsional stress on the front wall is

$$v = \frac{Tc}{J} = \frac{(44{,}400 \times 13.33)(43.33)}{636{,}250} = 40 \text{ lb/ft } [0.59 \text{ kN/m}]$$

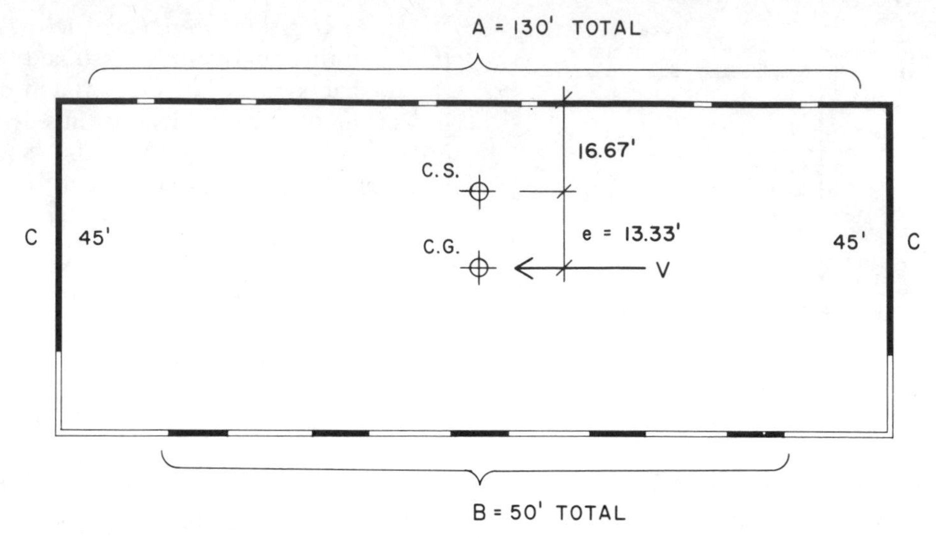

FIGURE 6.4. Torsional analysis: east–west load.

The total stress is thus

$$v = 247 + 40 = 287 \text{ lb/ft } [4.19 \text{ kN/m}]$$

In theory, the torsional shear is negative for the rear wall; that is, the true shear will be the direct shear minus the torsional shear. However, the usual practice is not to deduct but only to add the torsional effect. Thus for the rear wall we use only the direct shared stress of 247 lb/ft.

For the most conservative design we may use the peripheral method for the lateral force in the front walls and the torsional method for the rear walls. If this is done we consider

$$v = 444 \text{ lb/ft for the front wall}$$

$$v = 247 \text{ lb/ft for the rear wall}$$

As with the end shear walls, the weight of the walls in the plane of the shear walls should be added to the loads for the design of the individual piers. We have not done so, as these are quite minor loads in this structure. The procedure is the same as that illustrated for the end wall.

For the rear wall UBC Table 25-K permits the use of $\frac{3}{8}$-in. plywood with 8d nails at the maximum spacing of 6 in. Note that the footnote to Table 25-K permits an increase of 20% in the table values for $\frac{3}{8}$-in. plywood when studs are 16 in. on center.

For the overturn analysis of the rear wall we consider two approaches as shown in Figure 6.5. The first is to regard it as a series of independent piers linked together. For overturn analysis these piers would be considered to have a height from the sill to the roof deck level. The other option is to regard the wall as a continuous diaphragm with piers having a height equal to the door opening height. The latter option results in considerably less overturn, but requires

TABLE 6.2. Torsional Moment of Inertia of the Shear Walls

Wall	Length (ft)	Distance from Center of Stiffness (ft)	$J = L(d)^2$
A	130	16.67	36,126
B	50	43.33	93,874
C	2(45)	75	506,250
Total J for the shear walls			636,250

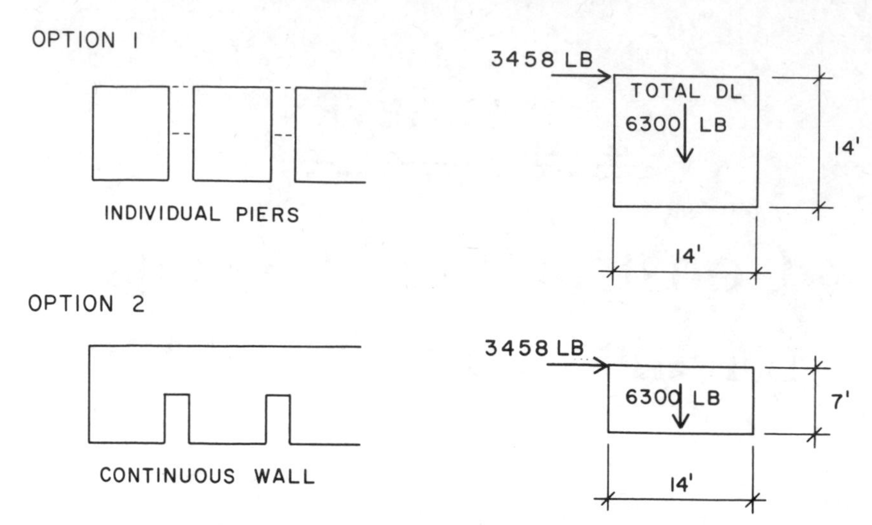

FIGURE 6.5. Options for the rear shear walls.

some additional framing and tying to reinforce the wall at the openings.

In referring to Figure 6.5, the overturn analysis for the short end pier is:

Lateral load:
247 × 14 = 3,458 lb

Overturn M:
3458 × 14 × 1.5 SF = 72,618 ft-lb

DL M:
6300 × 7 = 44,100

Net M for hold-down: 28,518 ft-lb

Required T:

$$\frac{28,518}{14} = 2,037 \text{ lb}$$

This option would therefore require a hold-down device with an ultimate resistance of 2037 lb or more. For the other option, the analysis is

overturn M:
3458 × 7 × 1.5 = 36,309 ft-lb

Because the DL moment is the same, observe that no hold-down is required; that is, the actual safety factor is greater than 1.5 (actually 1.82 if analyzed as for the end wall).

For the front wall the UBC permits a $\frac{3}{8}$-in. plywood with 8d nails at 2.5 in. or $\frac{1}{2}$-in. with 10d nails at 4 in. Overturn for this wall will not be critical because the end reaction of the girder delivers a considerable dead load at the center of the wall. Since it is not possible to tilt the wall without lifting the girder, the situation is quite safe.

The overturn forces on these walls must be transmitted to, and resisted by, the foundations. This requires that there be sufficient dead weight in the grade wall and footings and some bending and shear resistance by the grade wall. Assuming the depth of grade wall as shown in the construction drawings, this resistance can be developed with minimal top and bottom continuous reinforcing. If the grade wall is quite shallow, this problem should be carefully investigated.

Obviously, some reconsideration of the building details could reduce the requirements for lateral load resistance. Use of a lighter roofing and a lighter ceiling material would considerably reduce the roof dead load and consequently the lateral force. Use of one or more permanent cross walls in the interior would reduce the stresses in the roof deck and the end shear walls and eliminate some of the large girders.

CHAPTER SEVEN

Construction Drawings for the Wood Structure

The drawings that follow show the basic construction details for the wood structure. The drawings are essentially intended to show the structural details and are not all fully complete as architectural details. In many instances there are equivalent alternatives for achieving the basic structural tasks, which could be more intelligently evaluated if all information about finish materials and architectural details were known. Some details may also be effected by considerations of the design of the lighting, electrical power, HVAC, and plumbing systems or by problems of security, acoustics, fire ratings, and so on.

7.1. STRUCTURAL PLANS

Note that the roof framing system used is Option 2 as shown in Figure 5.4. If Option 1 with only the rafters and girders is used, some of the wall details would change.

Foundation and roof framing layouts are shown in the partial plans in Figure 7.1. Detail sections shown on the plans are illustrated in Figures 7.2 through 7.6. The following is a discussion of some of the considerations made in developing these details.

7.2. CONSTRUCTION DETAILS

Detail A (Fig. 7.2)

Detail A shows the canopy, parapet, shear wall, and roof at the front of the building. Depending on the height of the parapet and the location of the top of the canopy, it may be advisable to run the wall studs continuously to the top of the parapet. This would permit the top of the canopy to be higher than the roof deck. In any event, if the top of the canopy is not exactly at the level of the roof deck, as shown in Figure 7.2, additional framing would be required for the anchoring of the tie straps.

In the detail shown both the roof deck and wall sheathing are nailed directly to the top plate of the wall. This achieves a direct transfer of load from the horizontal to the vertical diaphragm. If the wall studs were continuous to the top of the parapet, a ledger would be provided at the face of the studs to support the rafters and provide for the edge nailing of the plywood. Transfer of the roof seismic load to the wall would then require the addition of blocking in the wall. This condition is illustrated in Detail C, Figure 7.3, in which the load transfer is from the roof plywood, through boundary

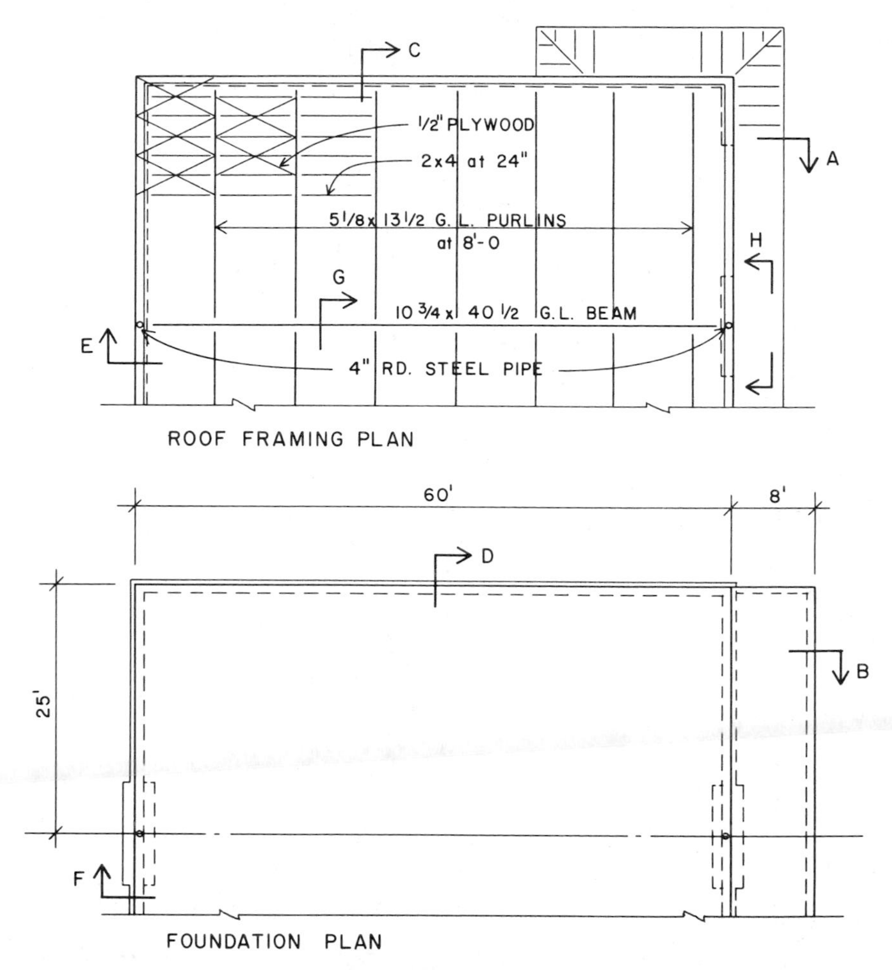

FIGURE 7.1. Structural plans.

nailing to the ledger, then from the ledger to the blocking, and finally from the blocking to the wall plywood.

At the point at which the bottom of the canopy kicks into the wall the strut shown may be required to brace the studs. This may not be required at the solid portion of the wall, but is most likely required at openings, unless the header is designed for the combined vertical and horizontal loads.

If the parapet is simply built on top of the roof deck, as shown, the diagonal struts may be used to brace the canopy and form the cant at the roof edge.

Detail B (Fig. 7.2)

Depending on the level of the exterior grade, the drainage situation and the wall finish materials, the sill plate may be simply put directly on the floor slab, as shown, or may be placed on top of a short curb to raise it above the floor level.

The details of the wall footing, grade wall, and floor slab are subject to considerable variation. Some considerations are:

1. *Depth of the Footing Below Grade.* For frost protection or simply

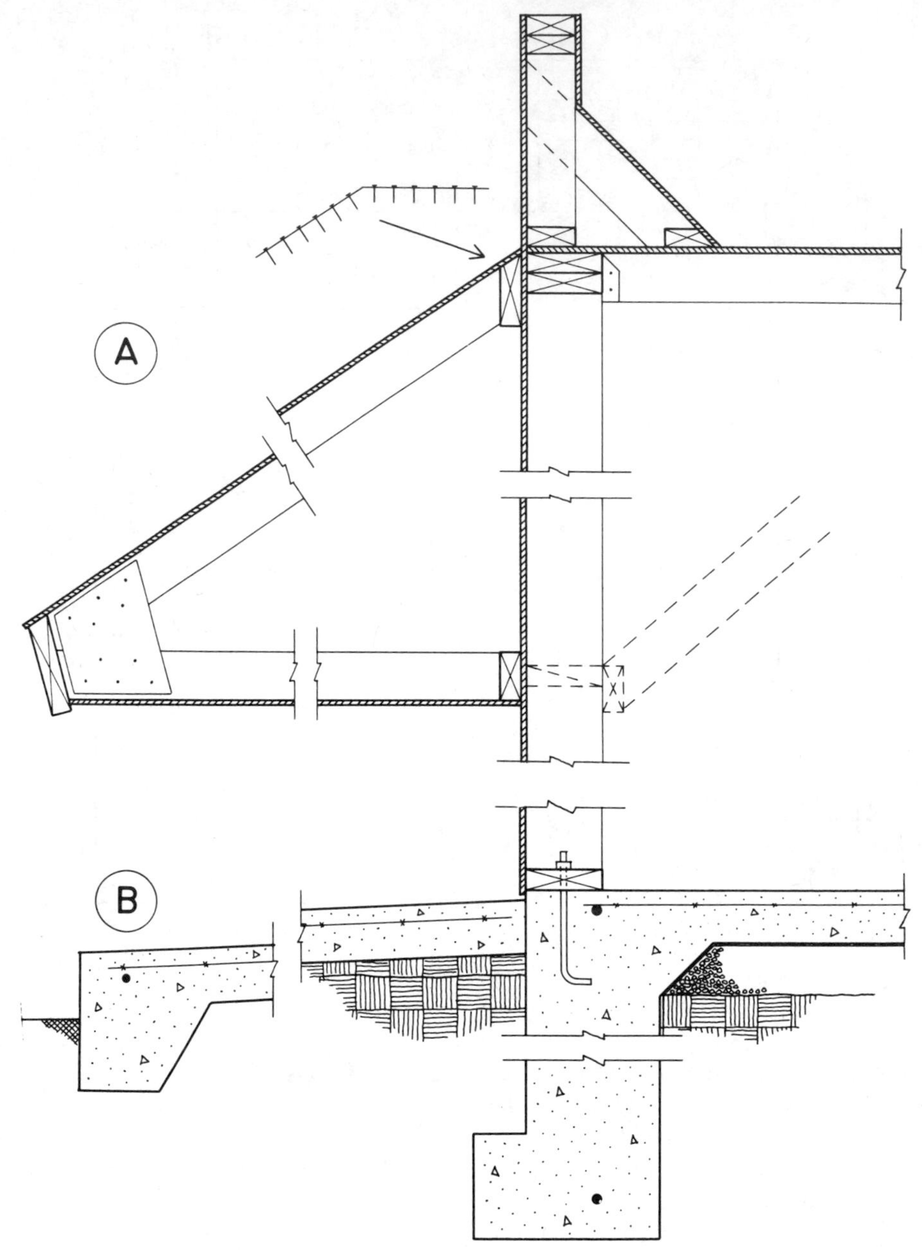

FIGURE 7.2. The front wall.

to reach adequate bearing soil it may be necessary to have a deep grade wall. At some point this requires that it be treated as a true vertical wall with vertical reinforcement, a separate footing with dowels, and so on. In this case there would likely be three concrete pours with a cold joint between the footing, wall, and slab.

2. *Need for Thermal Insulation for the Floor.* In cold climates it is desirable to provide a thermal break between the slab and the cold ground at the build-

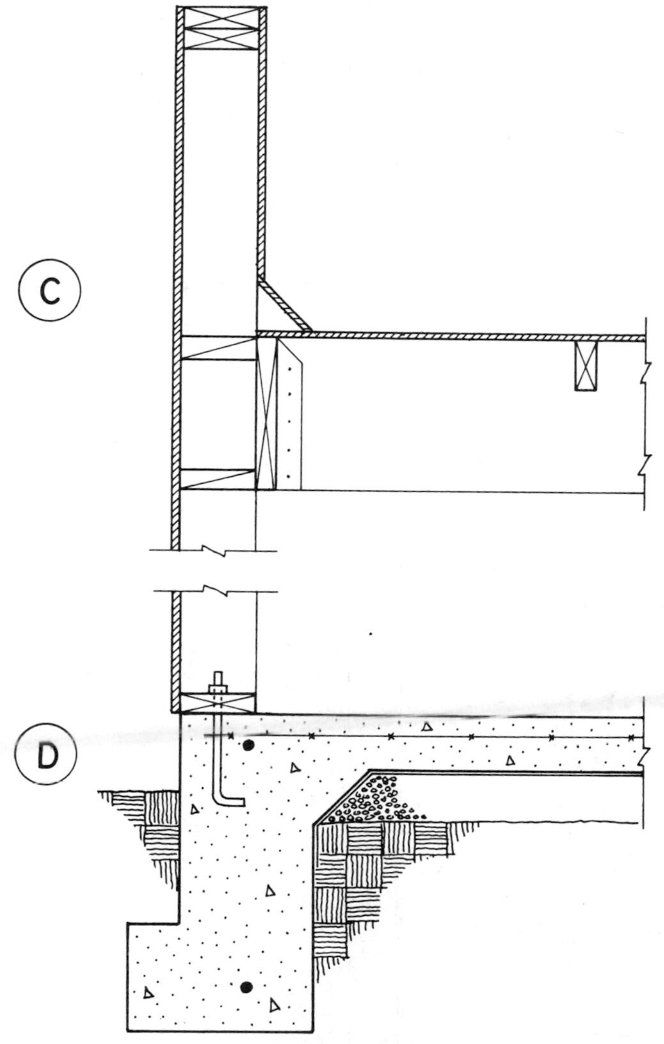

FIGURE 7.3. The end wall.

ing edge. This would be done by placing insulation on the inside of the grade wall and/or by placing it under the slab along the wall.

3. *Cohesive Nature of the Soil.* If the soil at grade level is reasonably cohesive (just about anything but clean sand), and a shallow grade wall is possible, the footing, wall, and slab can be poured in a single pour. In this case the wall and footing are formed by simply trenching and providing a form for the outside wall surface and slab edge.

4. *Beam Action of the Grade Wall.* Because of expansive soil, highly varying soil bearing conditions, or the use of the grade wall for distribution of concentrated forces from posts, tie-downs, and so on, it may be necessary to provide top and bottom reinforcing. In

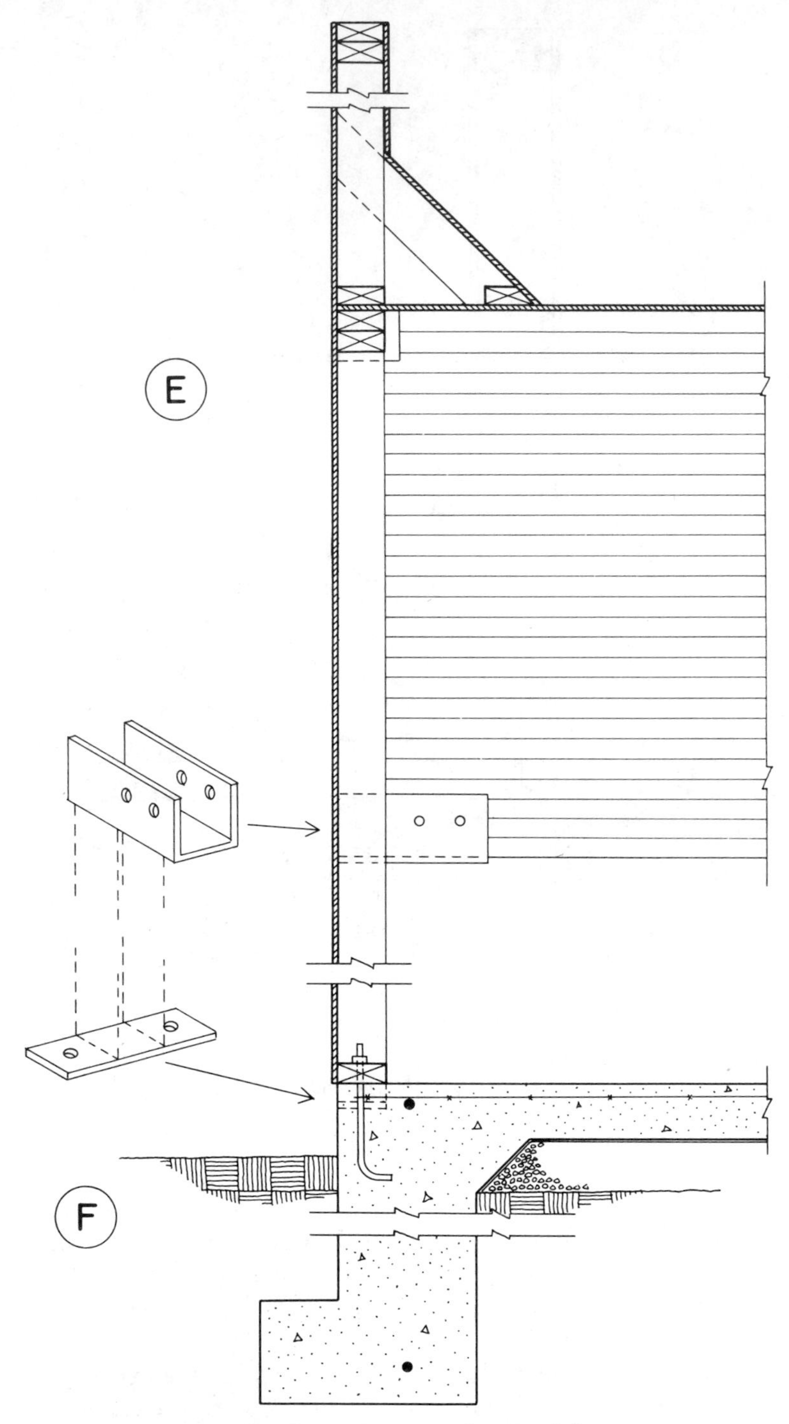

FIGURE 7.4. The rear wall at the girder.

any event, a minimum of one bar should be used at each point for shrinkage and thermal stresses.

Detail C (Fig. 7.3)

Detail C shows the end wall condition where the roof purlins are supported by the wall. In this detail the wall studs are continuous to the top of the parapet and a ledger is provided to which the joist hangers for the purlins are attached. Because of the roof slope, the elevation of the purlins varies. This is simply accommodated by sloping the ledger, whereas if the wall were as in Detail A, all the studs would be different in length and the top plate of the wall would have to be sloped. Then the parapet studs would also be all different in length to achieve the level parapet top.

As discussed for Detail A, the seismic load must be transferred from the roof to the wall by the circuitous route through the ledger and blocking. The vertical roof loading must also be transferred from the ledger to the studs. For a more positive tie it is recommended that the ledger be attached to the studs with lag screws. Two lag screws in each stud and one in each block would probably provide all the load transfers necessary.

Detail D (Fig. 7.3)

This is essentially the same condition as Detail B except for the absence of the exterior paving slab and the posts. Depending on the level of the exterior grade and the exterior wall finish, it may be necessary to raise the sill on a curb, as previously discussed.

Detail E (Fig. 7.4)

This shows the rear wall with the glue-laminated girder framed into the wall. Note that the upper corner of the girder is notched to permit the wall plate to be continuous. The U-shaped bent plate would be welded to the steel column that is encased in the wall. If a wood column is used, the U-shaped plate would be welded to other plates which would be bolted to the wood column.

Detail F (Fig. 7.4)

Most of the comments made for Details B and D are applicable here. The column footing would likely be poured separately with the grade wall and slab poured later. Depending on the construction sequence, as well as the height of the grade wall, it may be possible to set the base of the steel column as shown directly on the footing or on top of the floor slab. The detail shown permits the wall sill to run all the way to the column, providing a maximum nailing edge for the plywood.

Detail G (Fig. 7.5)

In this detail the purlins are framed in metal joist hangers on the sides of the girder. It also shows a detail for the top

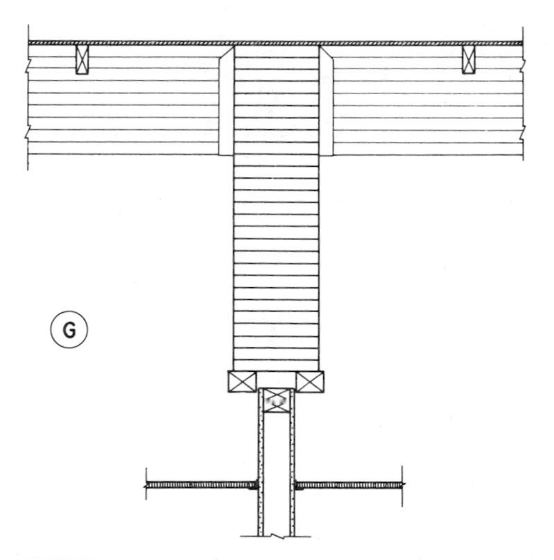

FIGURE 7.5. Details at the girder and interior wall.

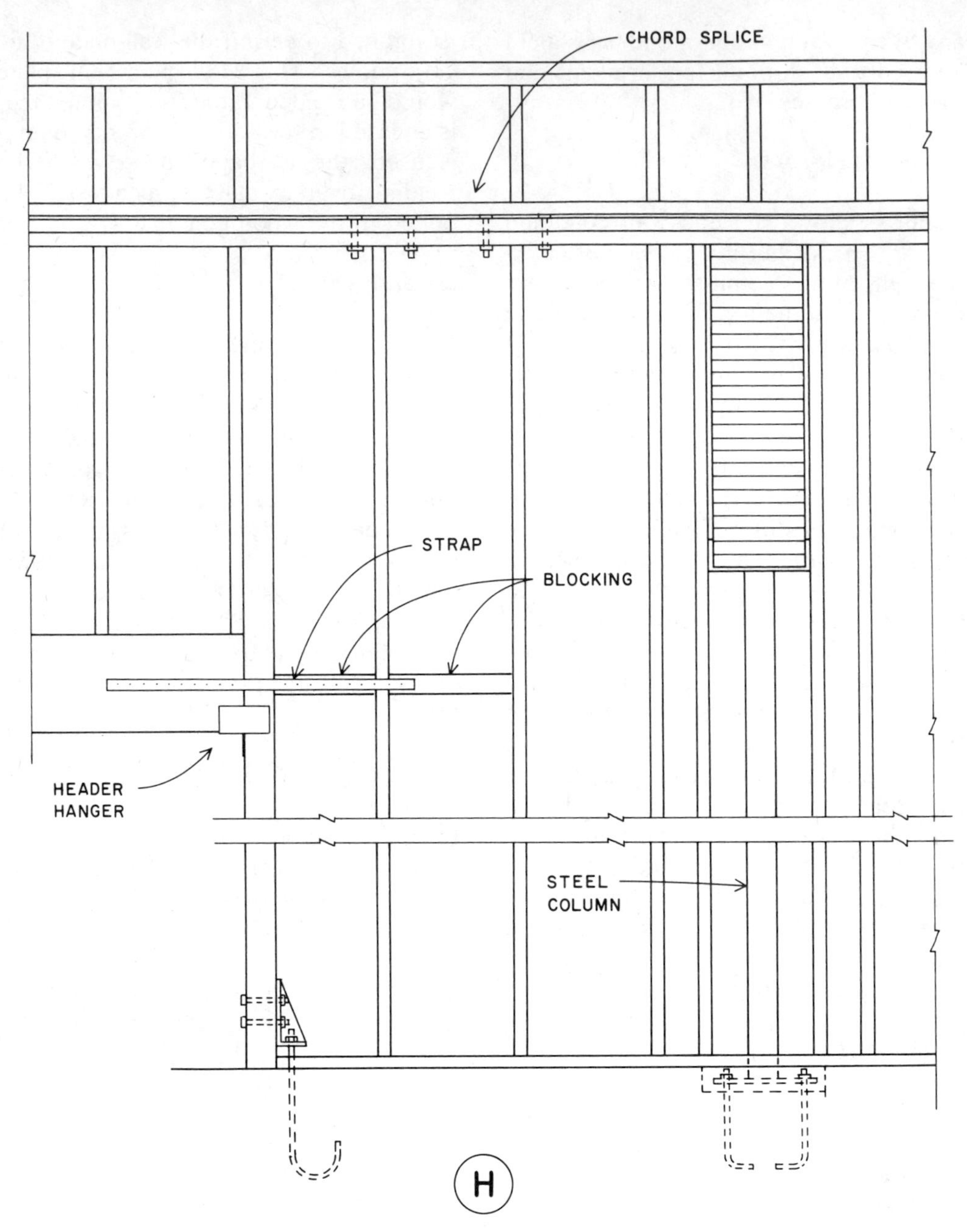

FIGURE 7.6. Framing elevation for the rear wall.

of an interior partition that allows for some live load deflection of the girder.

Detail H (Fig. 7.6)

Detail H is a partial elevation of the framing of the front wall with the plywood removed. Shown are the following:

1. Framing at the end of the girder with the steel post and the bent plate.
2. Support of the window header by a connection attached to the face of the wall end post, which permits the post to be continuous to the top of the parapet.
3. Strapping of the end of the header to blocking in the wall, which reinforces the opening corner and allows the shear wall action as shown in Option 2 in Figure 6.4.
4. Splicing of the top plate with bolts as described in the calculations. For ease of construction it would probably be desirable to oversize the top plate member and recess the bolt heads to clear the roof plywood.

CHAPTER EIGHT

Alternate Solutions for the Wood Structure

There are a number of possible systems for the structure of the roof and walls of Building Two. In this chapter we consider some additional possibilities for wood systems. Solutions utilizing steel and masonry are illustrated in succeeding chapters.

8.1. WOOD TRUSS SYSTEM

Figure 8.1 shows a solution for the wood roof structure that consists of prefabricated trusses and a plywood deck. The trusses are fabricated with wood top and bottom chords and steel web members. A number of manufacturers produce these predesigned trusses and the details vary. They are generally competitive with steel open web joists in the short to medium span range. The actual truss design is usually done by the staff or hired consultants retained by the manufacturer or distributor.

The depth of the trusses can usually be varied to facilitate the sloping roof surface while maintaining a level bottom chord for the direct attachment of the ceiling. Direct attachment of the ceiling is usually possible because the open webs of the trusses permit the passage of ducts and wiring. Although the trusses may be deeper than an all beam system, the resulting total distance from ceiling to roof is usually less with the trusses.

Another potential advantage of the truss system is a freeing of the structural module from the architectural planning. Location of doors, windows, and interior partitions need have no relation to the truss spacing.

Distribution of the roof gravity loads to the walls changes with this system and the walls and headers would be slightly different. The foundations for the end and rear walls would consist of continuous, constant width footings. The front wall loads will be collected into the solid wall sections, so the foundation may be the same here as in the steel column design.

The plywood roof diaphragm and the shear wall designs would be essentially the same. Overturn of the front wall piers is less critical because the header columns will be more heavily loaded in this scheme.

Some investigation of alternatives for the truss spacing would be done to determine the most economical system. There is a trade-off between the size of the trusses on one hand and the roof ply-

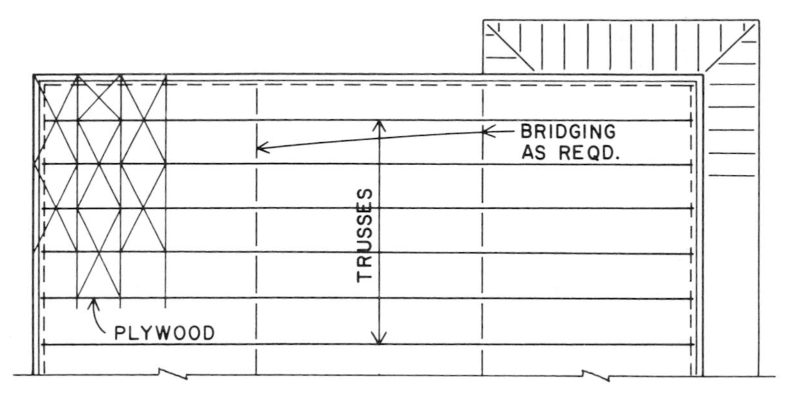

FIGURE 8.1. Structural plan—wood truss system.

wood thickness and ceiling framing on the other hand. Using standard 8-ft plywood sheets, the usual modules would be 24, 32, or 48 in. In some areas longer sheets of plywood are available and would make possible the consideration of 10- or 12-ft module increments. The use of plywood with tongue-and-groove edges may eliminate the necessity for blocking perpendicular to the trusses.

Figure 8.2 shows a modification of Detail A from the previous system. The precise detailing of the truss supports and the details for transfer of the roof diaphragm load to the walls would depend somewhat on the specific manufacturer's products.

8.2. SYSTEMS WITH INTERIOR SUPPORTS

An assumption made originally in developing the roof system for Building Two was that there was a real need for a clear span structure with supports provided only at the outside walls. Although the 60-ft building width is not in the category of a really long span structure, it would nevertheless greatly reduce the scale of the spanning elements required if some interior supports are used. This is, of course, a major architectural planning issue and may indeed relate to the functions required inside the building.

Figure 8.3 shows a structural plan for a framing system that employs interior bearing walls on 25-ft centers supporting a joist system and a plywood deck. If the interior walls are required for architectural planning reasons, their construction would be little different whether they must function as bearing walls or not. If they are not utilized for bearing (and most likely also for lateral shear wall functions), they could be relocated in the future without disturbing the roof structure—an advantage present in the previous solutions.

With a light roof live load the 25-ft span can be achieved with 2 × 12 joists, thus making the system quite easily constructed and very economical. (See UBC Table 25-U-R-1 in Appendix B.)

Figure 8.4 shows a structural plan for Building Two that uses two rows of interior columns. If the columns are placed on 25-ft centers as shown in the long direction of the plan, they could be incorporated into walls as shown in Figure 8.3 and thus not intrude on the interior space. In contrast to the bearing wall scheme, this system permits relocation of the walls, with the worst architec-

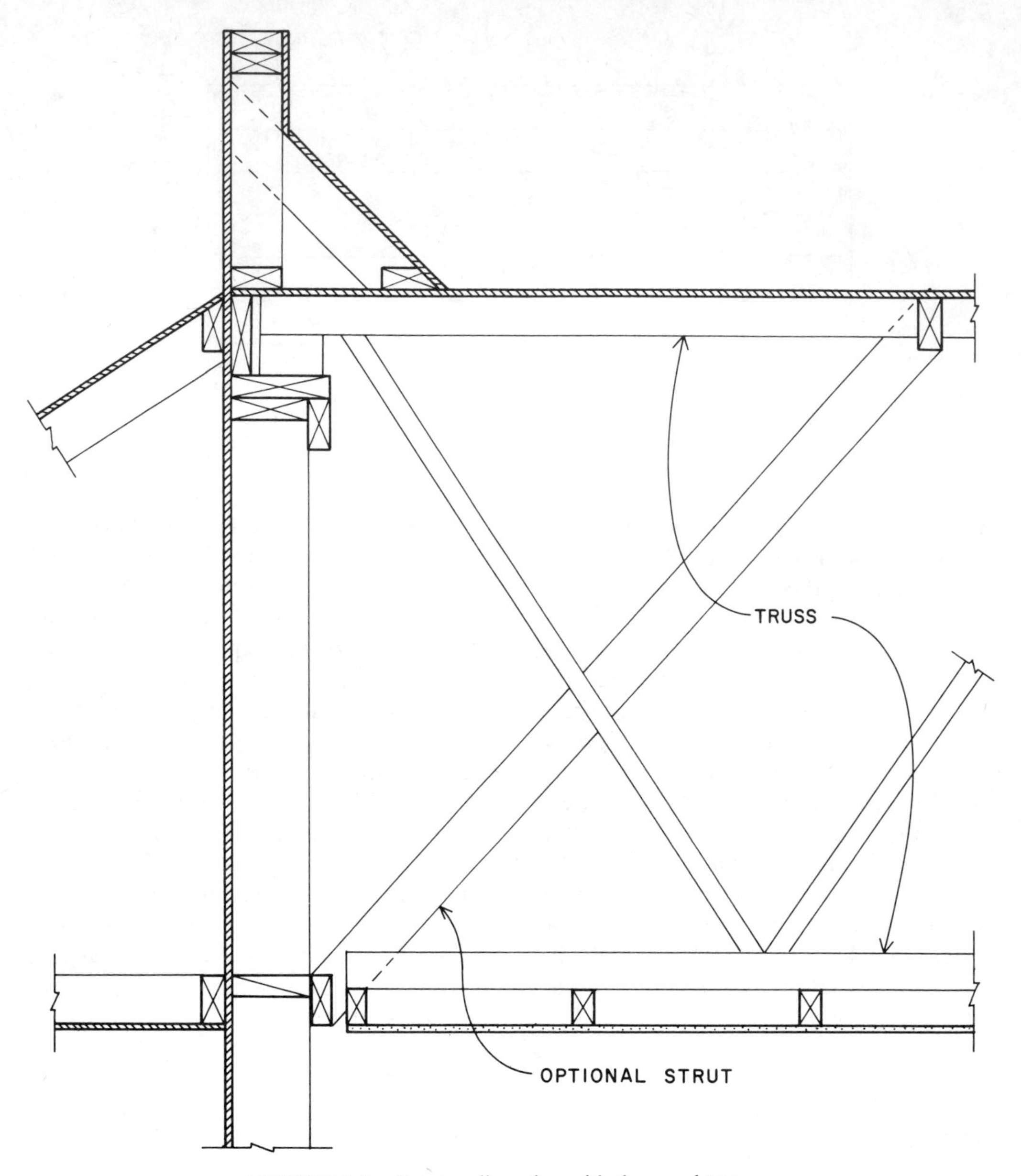

FIGURE 8.2. Front wall section with the wood truss.

tural problem being the possibility of having some freestanding interior columns.

For the scheme in Figure 8.4 the modest spans of both the beams and joists would result in quite light members. This cost gain is offset slightly by the cost of the interior columns with their footings and their connections.

A variation on the system shown in Figure 8.4 would be the use of a single row of columns supporting 30-ft span

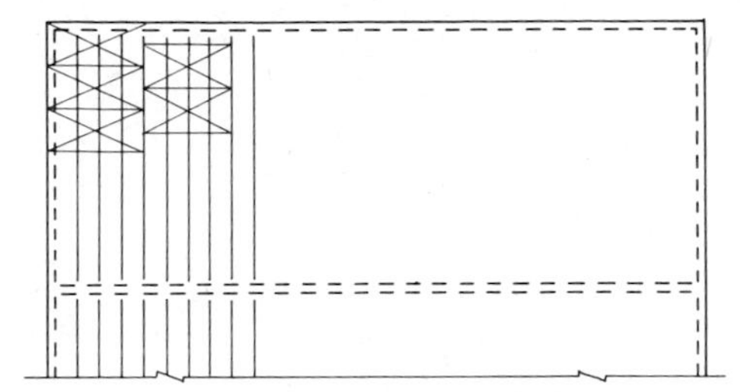

FIGURE 8.3. Structural plan—interior bearing wall and rafter roof system.

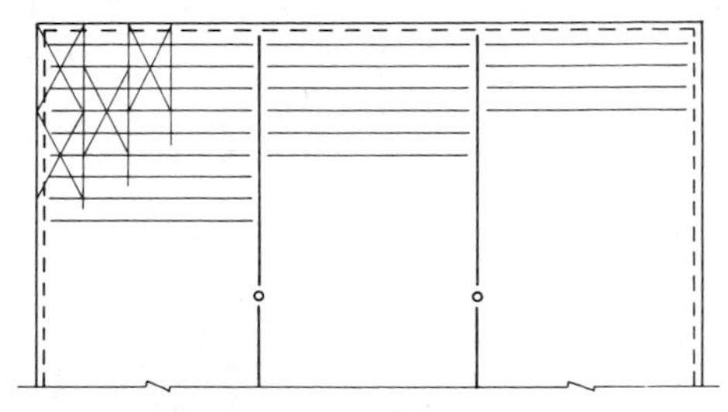

FIGURE 8.4. Structural plan—system with interior columns.

joists. This would require heavier joists, but offers the advantage of only half as many interior columns.

Although construction cost comparisons could be made for all of these schemes, the choice is obviously not just a structural decision. Architectural planning and the development of systems for ceiling construction, lighting, air handling, and roof drainage may well be sources of major influence on the decision.

CHAPTER NINE

Design of the Steel and Masonry Structure

Figure 9.1 shows the layout for a structure for Building Two consisting of a steel framed roof and exterior walls of masonry. To relate to the usual 16-in. horizontal concrete block dimension, the plan dimensions of the walls have been modified slightly, as shown on the partial plans. The roof system is similar in layout to the first wood scheme, but with rolled steel sections replacing the wood joists and the glue-laminated girders and a light gauge steel deck replacing the plywood. Because of the seismic condition, reinforced concrete block is used for the exterior walls.

The following materials are used:

Structural steel: A36, F_y = 36 ksi [248 MPa].

Concrete: sand and gravel aggregate, f'_c = 3 ksi [20.7 MPa].

Masonry: reinforced hollow unit masonry; concrete units of medium weight, Grade N, ASTM C90, f'_m = 1350 psi [9.3 MPa]; mortar type S.

Reinforcing for concrete and masonry: Grade 40, f_y = 40 ksi [276 MPa].

9.1. DESIGN OF THE STEEL ROOF STRUCTURE

Dead loads:

Roofing: tar and gravel, 6.5 psf.

Insulation: rigid type on top of deck, assume 2 psf.

Suspended ceiling: gypsum plaster, assume 10 psf total.

Lights, wiring, ducts, registers: assume average of 3 psf.

Total dead load: 21.5 psf [1.03 kPa] + the structure.

Roof Deck

A number of roof deck systems may be considered, including those with lightweight concrete or gypsum concrete fill. We use an ordinary single sheet, ribbed deck. With the joists at 6-ft centers, a 20-gauge deck will be adequate. This deck will weigh approximately 2 psf, depending on the specified finish.

The deck must also be considered for seismic load. The choice of the deck itself and the details for its installation

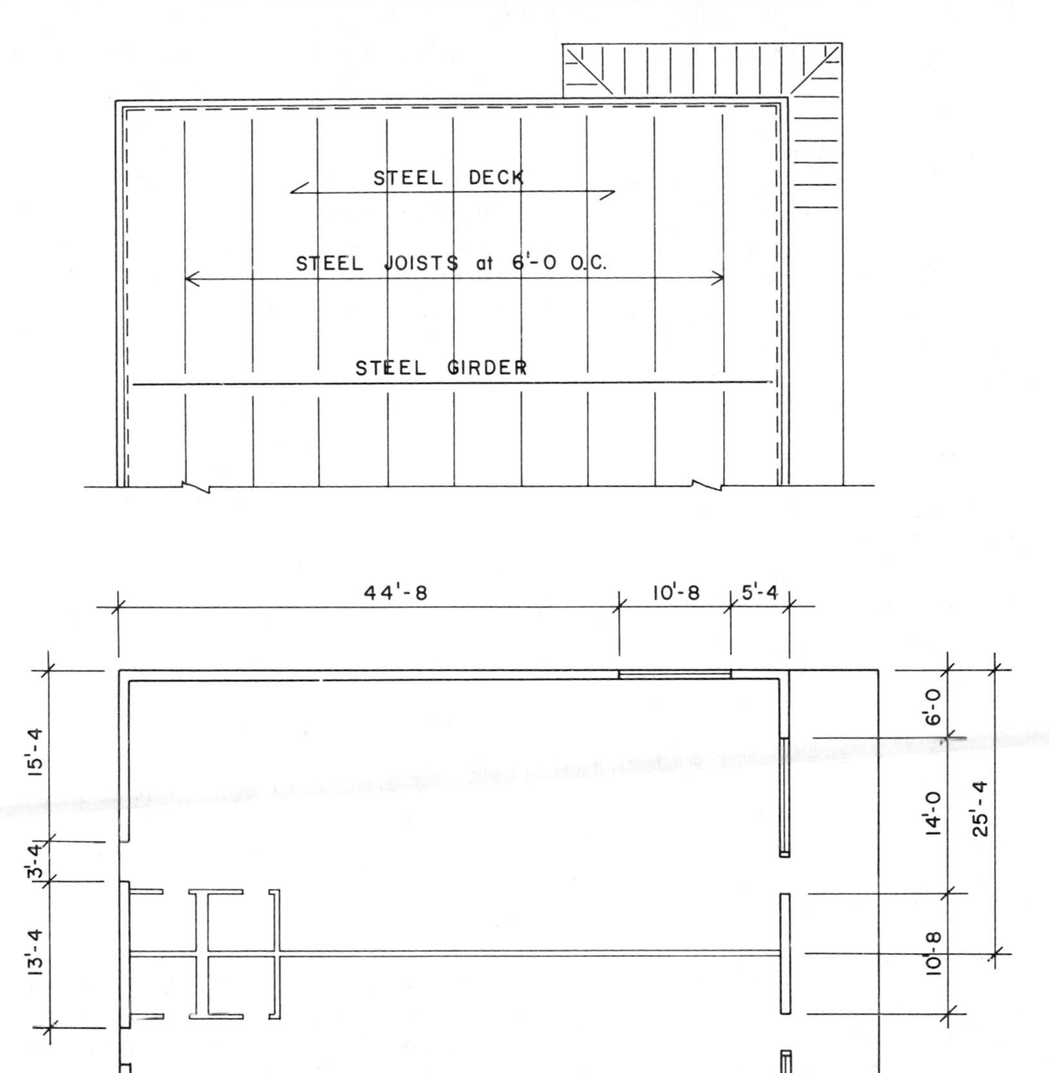

FIGURE 9.1. Plans for the steel roof and masonry walls.

must be considered in terms of the required diaphragm shear and the overall diaphragm deflection.

Joists

Several options are possible for the joists. These may be light-rolled I-shaped sections, open web steel joists, or cold-formed sections of light gauge sheet steel. With joists at 6-ft centers, the loading will be

DL + LL: 23.5 + 20 = 43.5 psf [2.08 kPa].

Load/joist: 43.5 × 6 = 261 lb/ft [3.81 kN/m] + the joist weight/foot.

For design use: total load = 275 lb/ft [4 kN/m].

Using a total load deflection limit of $L/180$, the allowable deflection will be

$$\frac{25 \times 12}{180} = 1.67 \text{ in. [42 mm]}$$

Referring to Appendix A, Table A.4, we first determine the total load for the joist; thus

$$W = 25 \text{ ft} \times 275 \text{ lb/ft}$$

$$= 6875 \text{ lb} \quad \text{or } 6.875 \text{ k [30.6 kN]}$$

For this load possible selections are W 12 × 14 or M 12 × 11.8.

Appendix A, Figure A.2 shows that the 12-in.-deep members are not critical for the deflection limit of $L/180$.

Girder

With the joists at 6-ft centers the actual load on the girder consists of nine joists plus the weight of the girder. Because of the large area supported by the girder, the live load may be reduced to 12 psf. The design load for the girder will thus be

DL + LL: 25.5 + 12 = 37.5 psf [1.8 kPa].

Joists: 9 × 6 × 25 × 37.5 = 50,625 lb [225 kN].

Assumed girder weight: 75 × 60 = 4500 lb.

Total DL + LL: 55,125 lb [245 kN].

The lateral unsupported length is 6 ft (the joist spacing), which should not be critical for this large member. Selection can be made from the load–span tables (Appendix A, Table A.4) or the maximum moment can be determined and a selection made from either the S-listing tables (Appendix A, Table A.1) or the graphs in the AISC Manual (Ref. 10) that incorporate the consideration for lateral unsupported length. To demonstrate the use of the S-listing tables we first determine the maximum moment as

$$M = \frac{WL}{8} = \frac{(55.1)(60)}{8}$$

$$= 413 \text{ k-ft [560 kN-m]}$$

From the S-listing tables (Appendix A, Table A.1) the lightest section for this moment is a W 27 × 84 and the next lightest is a W 24 × 94. Both of these have L_c values over 6 ft, indicating no problem with lateral support. If headroom is considered critical, the 24-in.-deep member may be more desirable, although its deflection should be checked as follows:

Actual deflection for the 24-in.-deep beam on the 60-ft span (from Appendix A, Fig. A.2): 3.7 in. [94 mm].

Allowable deflection under total DL + LL:

$$\Delta = \frac{L}{180} = \frac{60 \times 12}{180}$$

$$= 4.0 \text{ in. [102 mm]}$$

Although the actual deflection is within the specified limit, the dimension of almost 4-in. movement is considerable. Because the live load deflection is the only movement to be experienced after the construction is complete, it may be possible to camber the beam (cold bend it into a permanent upper curvature) by an amount equal to the dead load deflection. Without live load the structure will then be flat, and the construction of interior walls may be able to tolerate the limited deflections due to live load. In any event, the selection of the deeper section will work to the relief of deflection problems.

Column for the Girder

Options for the girder support are to use a steel column at the face of the masonry wall or a reinforced pilaster column built as an extension of the masonry wall. Because the total end reaction of the steel girder is close to that for the wood girder, the steel column would be similar to that previously designed for the wood roof structure. To allow the wall to be continuous for seismic shear resistance, it would probably be placed just inside the wall surface with some ties to the masonry for lateral support.

The design of the pilaster is discussed in the wall design and both options are described in the details in Chapter 11.

9.2. DESIGN OF THE MASONRY WALLS

We assume that the walls will consist of reinforced, hollow concrete blocks with finishes of stucco (cement plaster) on the exterior and gypsum drywall on wood furring strips on the interior. We assume this construction to weigh approximately 70 psf of wall surface.

The exterior walls must be designed for the following combinations of vertical gravity and lateral wind or seismic forces (see Fig. 9.2.):

1. Gravity dead plus live loads.
2. Gravity vertical dead load plus bending due to lateral load.
3. Horizontal shear and overturn due to shear wall actions.

We first consider the long expanses of wall at the building ends and rear. For the end wall the laterally unsupported height varies because of roof slope. We assume it to be a maximum of 15 ft at the end of the solid wall portion nearest the front of the building. With an 8-in. block thickness, the maximum h/t of the wall is thus $(15 \times 12)/7.625 = 23.6$, which is just short of the usual limit of 25.

Assuming that code-required inspection is not provided during construction, the maximum stress for vertical compression is

$$F_a = 0.10 f'_m \left[1 - \left(\frac{h}{42t}\right)^3\right]$$

$$= 0.10(1350) \left[1 - \left(\frac{180}{42 \times 7.625}\right)^3\right]$$

$$= 111 \text{ psi } [0.77 \text{ MPa}]$$

and the maximum allowable bending stress is

$$F_b = 0.166 \quad f'_m = 224 \text{ psi } [1.54 \text{ MPa}]$$

For a total wall height of 18 ft the wall dead load is $18 \times 70 = 1260$ lb/ft [18.4 kN/m]. Assuming the clear purlin span to be 24 ft, the loads from the purlins are

$$\text{dead load} = 12 \times 25 \text{ psf}$$
$$= 300 \text{ lb/ft } [4.38 \text{ kN/m}]$$
$$\text{live load} = 12 \times 20 \text{ psf}$$
$$= 240 \text{ lb/ft } [3.50 \text{ kN/m}]$$

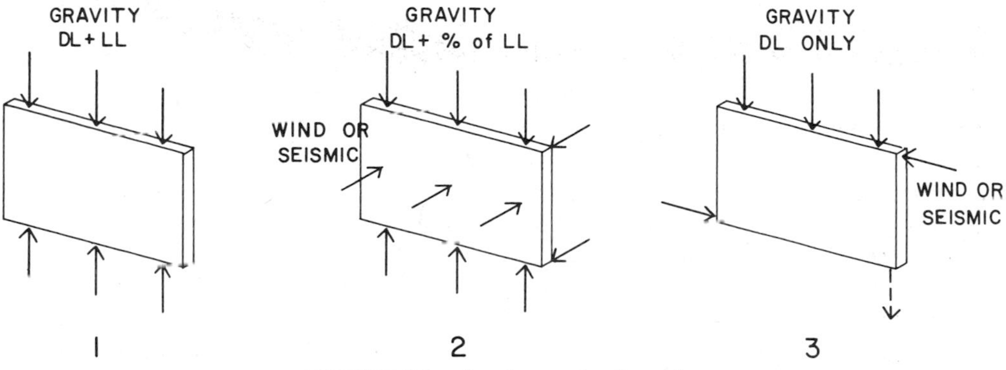

FIGURE 9.2. Load cases for the walls.

The total gravity vertical load on the wall is thus 1800 lb/ft and the average net compression stress assuming the wall to be 65% solid is

$$f_a = \frac{P}{A} = \frac{1800}{0.65(12 \times 7.625)}$$

$$= 30.3 \text{ psi } [0.21 \text{ MPa}]$$

Assuming the purlins to be supported by a ledger that is bolted to the wall surface, the roof loading will cause a bending moment equal to the load times one half the wall thickness; thus

$$M = 540 \times \frac{7.625}{2}$$

$$= 2059 \text{ in.-lb per ft of wall length}$$

Using Figure A.1, Appendix A, we find an approximate bending stress as follows.

Assume an average reinforcing with No. 5 bars at 40-in. centers. Thus

$$p = \frac{(0.31)(12/40)}{12 \times 7.625} = 0.001$$

$$np = 44 \times 0.001 = 0.045$$

$$K = \frac{M}{bd^2} = \frac{2059}{12(3.813)^2} = 11.8$$

From the graph, $f_m = \pm 90$ psi $= f_b$ [0.62 MPa]. Then

$$\frac{f_a}{F_a} + \frac{f_b}{F_b} = \frac{30.2}{111} + \frac{90}{224}$$

$$= 0.27 + 0.40 = 0.67$$

Because this is less than 1.0 the wall is adequate for the vertical gravity load alone. For the case of gravity plus lateral bending we must determine the maximum bending moment due to wind pressure or seismic load. For seismic action the code requires a force in the direction perpendicular to the wall surface equal to 30% of the wall weight, or 0.30 × 70 psf = 21 psf. Because this slightly exceeds the wind pressure of 20 psf, we use it for the bending. The wall spans the vertical distance of 15 ft from the floor to the roof. The dwelling of the reinforcing at the base plus the cantilever effect of the wall above the roof will reduce the positive moment at the wall midheight. We thus use an approximate moment for design of

$$M = \frac{qL^2}{10} = \frac{(21)(15)^2}{10}$$

$$= 473 \text{ ft-lb } [0.64 \text{ kN-m}]$$

To this we add the moment due to the eccentricity of the roof dead load; thus

$$M = 300 \times \frac{7.625}{2 \times 12}$$

$$= 95 \text{ ft-lb } [0.13 \text{ kN-m}]$$

and we now design for a total moment of 568 ft-lb [0.77 kN-m]. Assuming an approximate value of $j = 0.85$, we find that the required reinforcing is

$$A_s = \frac{M}{f_s jd} = \frac{0.568 \times 12}{(1.33 \times 20)(0.85)(3.813)}$$

$$= 0.079 \text{ in.}^2\text{/ft}$$

We try No. 5 at 32 in.

$$A_s = (0.31)(12/32) = 0.116 \text{ in.}^2\text{/ft}$$

Then

$$p = \frac{A_s}{bd} = \frac{0.116}{12 \times 3.813} = 0.0025$$

$$np = 44 \times 0.0025 = 0.112$$

$$K = \frac{M}{bd^2} = \frac{568 \times 12}{12(3.813)^2} = 39$$

From Figure A.1, Appendix A, $f_m = \pm 240$ psi. For axial compression due to dead load only,

$$f_a = \frac{1560}{0.65(12 \times 7.625)}$$

$$= 26.2 \text{ psi } [0.18 \text{ MPa}]$$

and

$$\frac{f_a}{F_a} + \frac{f_b}{F_b} = \frac{26.2}{111} + \frac{240}{224}$$

$$= 0.24 + 1.07 = 1.31$$

This indicates a combination close to the limit of 1.33. However, the analysis is conservative because the axial stress used is actually that at the bottom of the wall and the tension resistance of the masonry is ignored.

The rear walls have less load from the roof and a slightly shorter unsupported height. For these walls it is possible that the minimum reinforcing required by the code is adequate. The code requirements are:

1. Minimum of 0.002 times the gross wall area in both directions (sum of the vertical and horizontal bars).
2. Minimum of 0.0007 times the gross area in either direction.
3. Maximum spacing of 48 in.
4. Minimum bar size of No. 3.
5. Minimum of one No. 4 or two No. 3 bars on all sides of openings.

With the No. 5 bars at 32 in., the gross percentage of vertical reinforcing is

$$p_g = \frac{0.116}{12 \times 7.625} = 0.00127$$

To satisfy the requirement for total reinforcing, it is thus necessary to have a minimum gross percentage for the horizontal reinforcing of

$$p_g = 0.002 - 0.00127 = 0.00073$$

which requires an area of

$$A_s = p_g \times A_g = (0.00073)(12 \times 7.625)$$

$$= 0.067 \text{ in.}^2\text{/ft}$$

This can be provided by

$$\text{No. 4 at 32 in., } A_s = 0.20 \times \frac{12}{32}$$

$$= 0.075 \text{ in.}^2\text{/ft.}$$

$$\text{No. 5 at 48 in., } A_s = 0.31 \times \frac{12}{48}$$

$$= 0.0775 \text{ in.}^2\text{/ft.}$$

Choice of this reinforcing must also satisfy the requirements for shear wall functions, which is discussed in Chapter 10.

At the large wall openings the headers will transfer both vertical and horizontal loads to the ends of the supporting walls. The ends of these walls will be designed as reinforced masonry columns for this condition. Figure 9.3 shows the details and the loading condition for the header columns. In addition to this loading the columns are part of the wall and must carry some of the axial load and bending as previously determined for the typical wall.

Figure 9.4 shows a plan layout for the entire solid front wall section between the window openings. A pilaster column is provided for the support of the girder. Because of the stiffness of the column, it will tend to take a large share of the lateral load. We will thus assume the end column to take only a 2-ft strip of the wall lateral load. As shown in Figure 9.4, the end column is a doubly reinforced beam for the direct lateral load.

The gravity dead load on the header is:

Roof	= 100 lb/ft
Wall	= (70 psf)(6 ft)
	= 420 lb/ft
Canopy	= 100 lb/ft (assumed)
Total load	= 620 lb/ft [9.05 kN/m]

For the lateral load we will use the wind pressure of 20 psf because the weight of the window wall will produce a low seismic force. Assuming the window

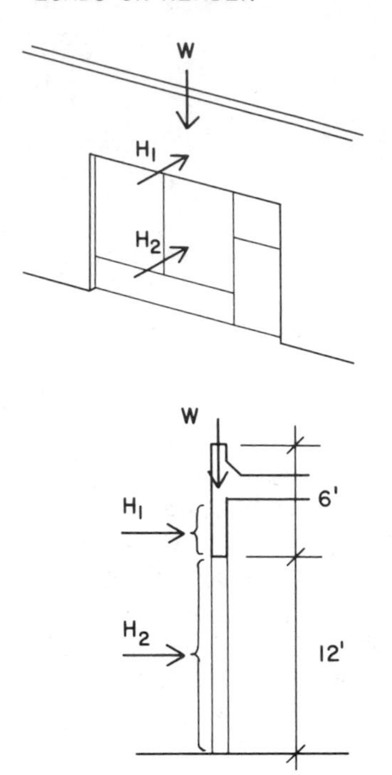

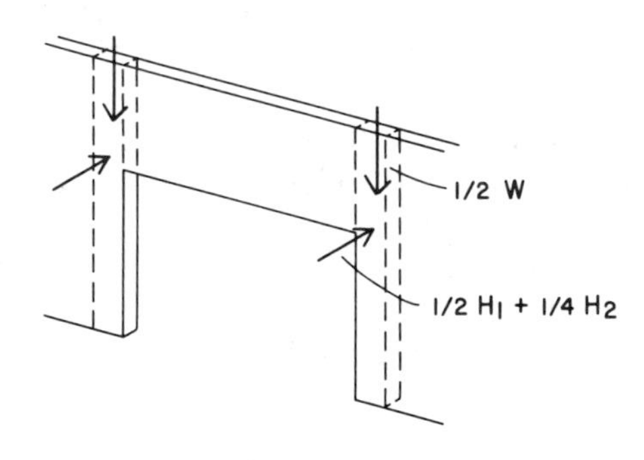

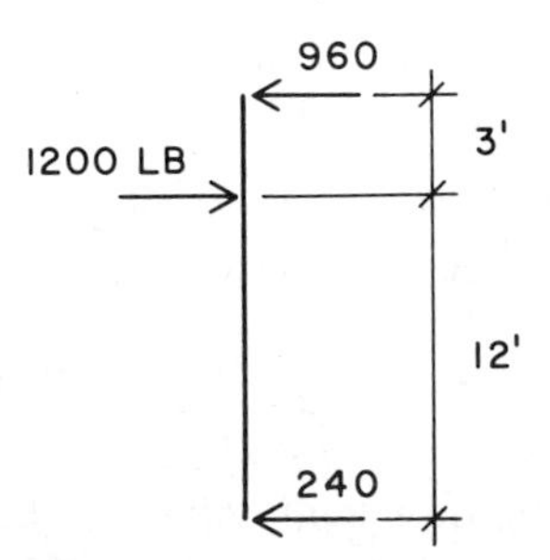

FIGURE 9.3. Loads on the headers and columns.

mullions span vertically, the wind loads are as shown in Figure 9.3.

$$H_1 = (20 \text{ psf})(2 \text{ ft} \times 15 \text{ ft})$$
$$= 600 \text{ lb } [2.67 \text{ kN}]$$
$$H_2 = (20 \text{ psf})(12 \text{ ft} \times 15 \text{ ft})$$
$$= 3600 \text{ lb } [16.0 \text{ kN}]$$

Thus the column loads from the header are:

$$\text{vertical load} = (620 \text{ plf})(15/2)$$
$$= 4650 \text{ lb } [20.7 \text{ kN}]$$
$$\text{horizontal load} = (\tfrac{1}{2}H_1 + \tfrac{1}{4}H_2)$$
$$= 300 + 900$$
$$= 1200 \text{ lb } [5.4 \text{ kN}]$$
$$\text{moment} = 960(3)$$
$$= 2880 \text{ ft-lb } [3.9 \text{ kN-m}]$$

(see Fig. 9.3)

For the direct wind load on the wall we assume a 15-ft vertical span and a 2-ft-wide strip of wall loading. Thus

$$M = \frac{wL^2}{8} = \frac{(20 \text{ psf})(2)(15)^2}{8}$$
$$= 1125 \text{ ft-lb } [1.53 \text{ kN-m}]$$

These two moments do not peak at the same point; thus without doing a more exact analysis we assume a maximum combined moment of 3800 ft-lb. Then, for the moment alone, assuming a j of 0.85,

$$\text{required } A_s = \frac{M}{f_s jd} = \frac{3.8(12)}{26.7(0.85)(5.9)}$$
$$= 0.34 \text{ in.}^2 \text{ } [219 \text{ mm}^2]$$

$$\text{approximate } f_m = \frac{M}{bd^2}\frac{2}{kj}$$
$$= \frac{3800(12)(2)}{(16)(5.9)^2(0.4)(0.85)}$$
$$= 482 \text{ psi } [3.32 \text{ MPa}]$$

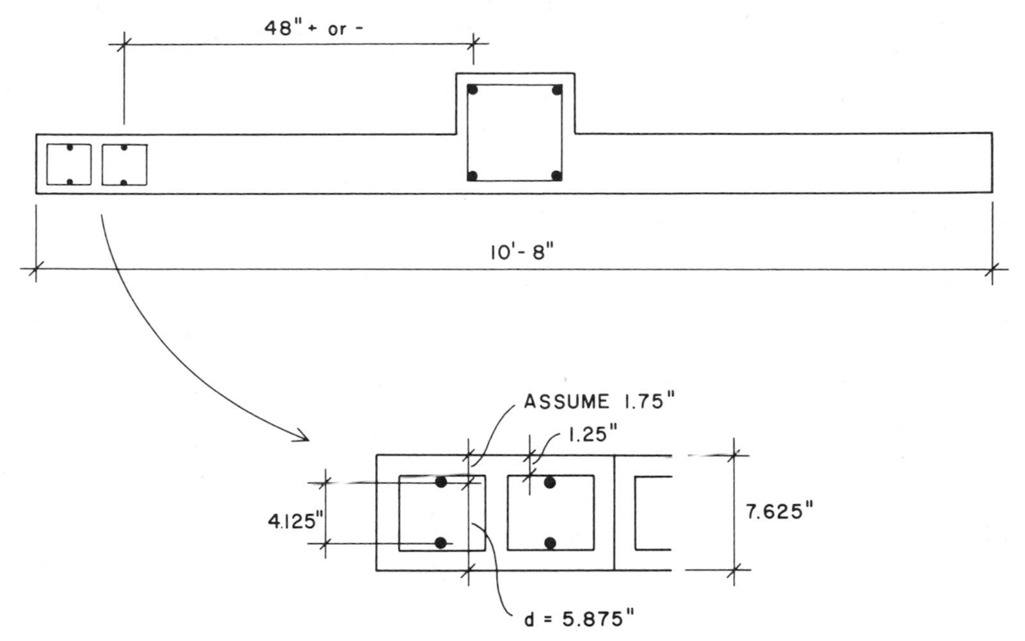

FIGURE 9.4. Details on the front wall.

Although f_m appears high, we have ignored the effect of the compressive reinforcing in the doubly reinforced member. The following is an approximate analysis based on the two moment theory with two No. 5 bars on each side of the column.

For the front wall it is reasonable to consider the use of a fully grouted wall because the pilaster and the end columns already constitute a considerable solid mass. For the fully grouted wall we may use f'_m = 1500 psi, and the allowable bending stress thus increases to

$$F_b = 1.33 \times 0.166 \times 1500$$
$$= 331 \text{ psi [2.28 MPa]}$$

Assuming the axial load to be almost negligible compared to the moment, we analyze for the full moment effect only. With a maximum stress of 331 psi we first determine the moment capacity with tension reinforcing only as

$$M_1 = \frac{f_m(bd^2)(k)(j)}{2}\left(\frac{1}{12}\right)$$
$$= \frac{331(16)(5.9)^2(0.4)(0.85)}{2(12)}$$
$$= 2612 \text{ ft-lb [3.54 kN-m]}$$

This leaves a moment for the compressive reinforcing of

$$M_2 = 3800 - 2600$$
$$= 1200 \text{ ft-lb [1.63 kN-m]}$$

If the compressive reinforcing is two No. 5 bars, then

$$f'_s = \frac{M_2}{A'_s(d - d')} = \frac{1200(12)}{(0.62)(4.125)}$$
$$= 5630 \text{ psi [38.8 MPa]}$$

This is a reasonable stress even with the assumed low k value of 0.4. As shown in Figure 9.5, if k is 0.4 and f_m is 331, the compatible strain value for f'_s will be

$$f'_s = 2n(f_c) = 2(40)(3.31)\left(\frac{0.61}{2.36}\right)$$

$$= 6844 \text{ psi } [47.2 \text{ MPa}]$$

As shown by the preceding calculation, the stress in the tension reinforcing will not be critical. This approximate analysis indicates that the column is reasonably adequate for the moment. The axial load capacity should also be checked, using the procedure shown later for the pilaster design.

Window Header

As shown in Figure 9.6, the header consists of a 6-ft-deep section of wall. This section will have continuous reinforcing at the top of the wall and at the bottom of the header. In addition there will be a continuous reinforced bond beam in the wall at the location of the steel ledger that supports the edge of the roof deck.

Using the loading previously determined, and an approximate design moment of $wL^2/10$, the steel area required for gravity alone will be

$$A_s = \frac{M}{f_s(jd)}$$

where $M = wL^2/10 = 620(15)^2/10$

$= 13{,}950$ ft-lb [18.9 kN-m]

$d =$ approximately 68 in. [1.727 m]

Then

$$A_s = \frac{13.95(12)}{(20)(0.85)(68)}$$

$$= 0.145 \text{ in.}^2 \ [94 \text{ mm}^2]$$

This indicates that the minimum reinforcing at the top of the wall may be two No. 3 bars or one No. 4 bar. This should be compared with the code requirement for minimum wall reinforcing. The UBC 2407(h)4 calls for a minimum of 0.0007

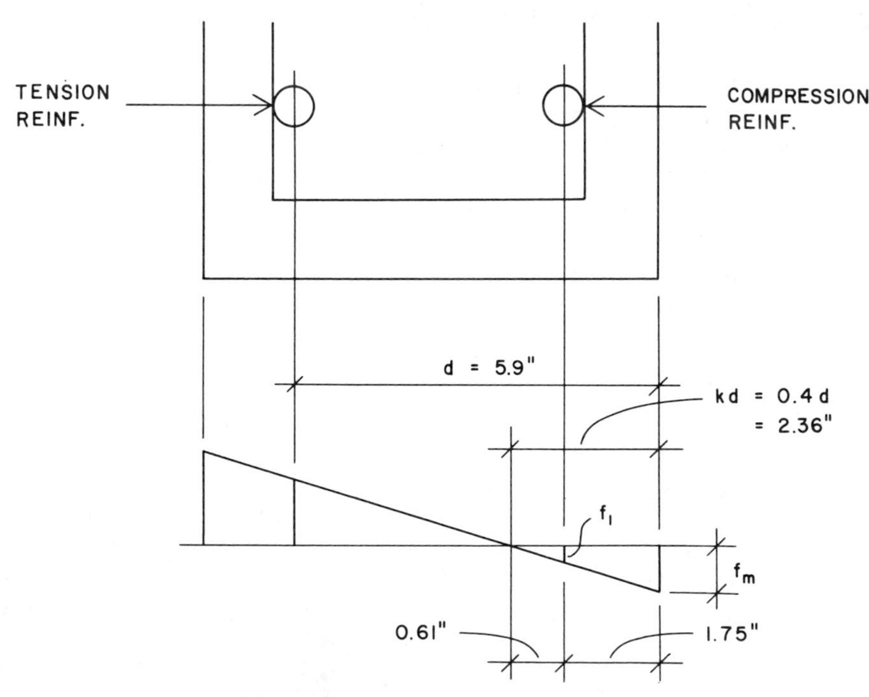

FIGURE 9.5. Stress in the header column.

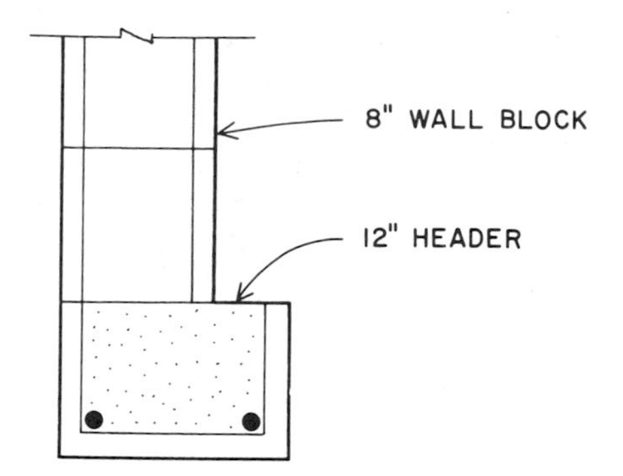

FIGURE 9.6. Alternate header detail.

times the gross cross-sectional area of the wall in either direction and a sum of 0.002 times the gross cross-sectional area of the wall in both directions. Thus

$$\text{minimum } A_s = 0.0007(7.625)(12)$$
$$= 0.064 \text{ in.}^2\text{/ft of width or height}$$

with two No. 3 bars $A_s = 0.22$ in.^{2}

$$\text{required spacing} = \frac{0.22}{0.064}$$
$$= 3.44 \text{ ft or } 41.3 \text{ in.}$$

The minimum horizontal reinforcing would then be two No. 3 bars at 40 in., or every fifth block course.

At the bottom of the header there is also a horizontal force consisting of the previously calculated wind load plus some force from the cantilevered canopy. Estimating this total horizontal force to be 250 lb/ft, we add a horizontal moment as

$$M = \frac{wL^2}{10} = \frac{0.25(15)^2}{10}$$
$$= 5.625 \text{ k-ft [7.63 kN-m]}$$

for which we require

$$A_s = \frac{M}{f_s(jd)} = \frac{5.625(12)}{(26.7)(0.85)(5.9)}$$
$$= 0.504 \text{ in.}^2 \text{ [325 mm}^2\text{]}$$

This must be added to the previous area required for the vertical gravity loads:

$$\text{total } A_s = 0.504 + \frac{(\frac{1}{2})(0.145)}{1.33}$$
$$= 0.504 + 0.055$$
$$= 0.559 \text{ in.}^2 \text{ [361 mm}^2\text{]}$$

The requirement for vertical load is divided by two because it is shared by both bottom bars. It is divided by 1.33, since the previous calculation did not include the increase of allowable stresses for wind loading. If this total area is satisfied, the bottom bars in the header would have to be two No. 7s. An alternative would be to increase the width of the header at the bottom by using a 12-in.-wide block for the bottom course, as shown in Figure 9.6. This widened course would be made continuous in the wall.

The Pilaster-Column

To permit the wall construction to be continuous, the girder stops short of the inside of the wall and rests on the widened portion of the wall called a *pilaster*. As shown in Figure 9.7, the pilaster and wall together form a 16-in. square column. The principal gravity loading on the column is due to the end reaction of the girder. Since this load is eccentrically

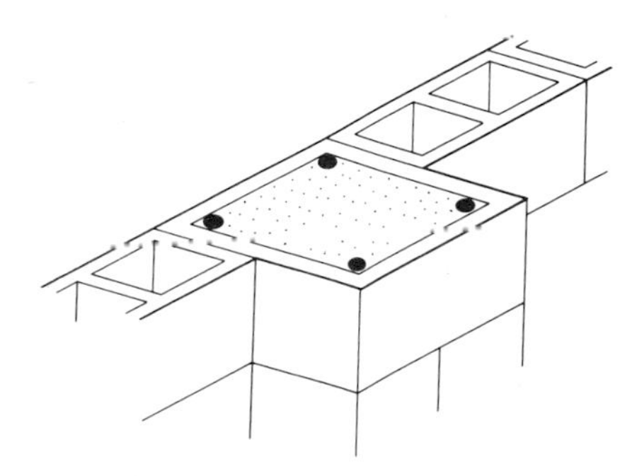

FIGURE 9.7. The pilaster column.

placed, it produces both axial force and bending on the column. The parapet, canopy, and column weight add to the axial compression.

Because of its increased stiffness, the column tends to take a considerable portion of the wind pressure on the solid portion of the wall. We will assume it to take a 6-ft-wide strip of this load. As shown in Figure 9.8, the direct wind pressure on the wall (pushing inward on the outer surface) causes a bending moment of opposite sign from that due to the eccentric girder load. The critical wind load is therefore due to the outward wind pressure (suction force) on the wall. For a conservative design we will take this to be equal to the inward pressure of 20 psf. The combined moments are thus

$$\text{wind moment} = \frac{wL^2}{8} = \frac{(20)(6)(13.33)^2}{8}$$

$$= 2665 \text{ ft-lb } [3.61 \text{ kN-m}]$$

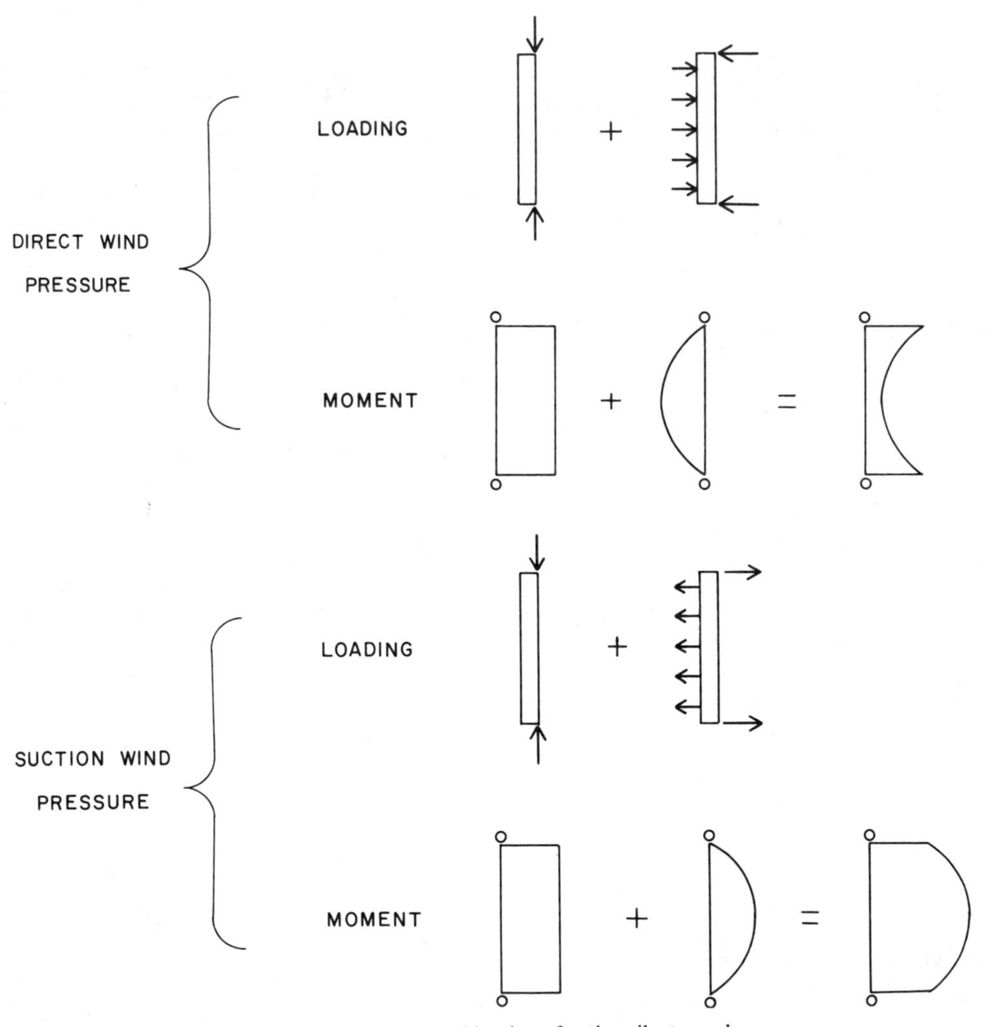

FIGURE 9.8. Load combinations for the pilaster column.

Assuming an eccentricity of 4 in. for the girder (see Fig. 11.2),

$$\text{girder moment} = \frac{23.5(4)}{12} = 7.833 \text{ k-ft or } 7833 \text{ ft-lb } [10.62 \text{ kN-m}]$$

For the combined wind plus gravity loading we have used only half the live load. With the allowable stress increase, it should be apparent that this loading condition is not critical, so we will design for the gravity loads only. For this we will redetermine the girder-induced moment with full live load:

$$\text{girder } M = \frac{27.6(4)}{12} = 9.2 \text{ k-ft } [12.48 \text{ kN-m}]$$

The gravity loads of the canopy, parapet, roof edge, and column must be added. We will therefore assume a total vertical design load of approximately 35 kips. With this total load the equivalent eccentricity for design will be

$$e = \frac{M}{N} = \frac{9.2(12)}{35} = 3.15 \text{ in. } [80 \text{ mm}]$$

The UBC 2418(k)1 requires a minimum percentage of reinforcing of 0.005 of the gross column area. Thus

$$\text{minimum } A_s = 0.005(16)^2 = 1.28 \text{ in.}^2 \ [826 \text{ mm}^2]$$

with four No. 7 bars $A_s = 2.40$ in^2 [1548 mm^2]. Then from UBC 2406(c)2B the allowable axial load is

$$P = \frac{1}{2}(0.20 f'_m A_e + 0.65 A_s F_{sc}) \times \left[1 - \left(\frac{h'}{42t}\right)^3\right]$$

for which

A_e = the total area of the fully grouted column

$= (15.625)^2 = 244$ in.^{2}

$F_{sc} = 0.40 F_y = 16$ ksi

h' = effective (unbraced) height of the column

$= 13.3$ ft or 160 in.

$$P = \frac{1}{2}\{(0.20 \times 1.5 \times 244) + (0.65 \times 2.40 \times 16)\} \times \left[1 - \left(\frac{160}{42(15.6)}\right)^3\right] = 48.4 \text{ kips } [215 \text{ kN}]$$

Ignoring the compression steel, the approximate moment capacity is

$$M = A_s f_s(jd) = \frac{1.20(20)(0.85)(13.5)}{12} = 22.95 \text{ k-ft } [31.1 \text{ kN-m}]$$

and, for the combined effect

$$\frac{\text{actual } P}{\text{allowable } P} + \frac{\text{actual } M}{\text{allowable } M} = \frac{27.6}{48.4} + \frac{9.2}{22.95} = 0.55 + 0.40 = 0.95$$

Note that the $\frac{1}{2}$ factor has been used with the axial load formula, assuming no special inspection during construction. Although a more exact analysis should be performed, this indicates generally that the column is reasonably adequate for the axial load and moment previously determined.

9.3. DESIGN OF THE FOUNDATIONS

The foundations for this structure will be essentially similar to those for the wood structure (See Section 5.5.) Continuous wall footings will be provided under all the exterior walls except at the columns. The same options described for the previous structure are possible for the column footing.

For the end walls the load is:

Roof:
45.5(12.5) = 569 plf

Wall:
80(18) = 1440 plf

Grade wall and footing:
= 300 plf (estimate)

Total load:
= 2309 plf [33.7 kN/m]

Width required:

$$\frac{2309}{2000} = 1.15 \text{ ft or } 14 \text{ in. [350 mm]}$$

At the front wall the column load and header loads are carried by the solid wall portion. If the same scheme used in the previous structure is desired, we would provide a 12-ft-long footing for this total load.

Girder end reaction:
27.6 kips.

Roof edge load:
3(25)(37.5 psf) = 2.8 kips.

Header dead load:
80(6)(14) = 6.7 kips.

Wall dead load:
80(18)(10.67) = 15.4 kips.

Pilaster:
1.8 kips.

Grade wall and footing:
700 plf(12) = 8.4 kips (estimate).

Total footing load:
62.7 kips [279 kN].

Width required:

$$\frac{62.7}{(2)(12)} = 2.61 \text{ ft [796 mm].}$$

This is actually less than the width used for the other, lighter structure, because the design in that case was done for equalized dead load. With the higher proportion of dead to live load in this structure, this equalization is more questionable. If done, however, it would probably result in approximately the same footing as for the wood structure.

CHAPTER TEN

Design for Seismic Force on the Steel and Masonry Structure

The lateral load resistive system for this structure is basically the same as that for the wood structure, in that it consists of the horizontal roof diaphragm and the exterior vertical shear walls. One significant difference is the increased load due to the heavier exterior walls. For determination of the seismic force the code requires a K factor of 1.33 for this building. With other factors the same as for the wood structure, the design load is thus $V = 0.1862\ W$. Reference may be made to the general discussion and to the illustrations for the previous design.

10.1. DESIGN OF THE ROOF DIAPHRAGM

The calculation of the loads applied to the roof diaphragm is shown in Table 10.1. In the north–south direction the load is symmetrically placed, the shear walls are symmetrical in plan, and the long diaphragm is reasonably flexible, all of which results in very little potential torsion. Although the code requires that a minimum torsion be considered by placing the load off center by 5% of the building long dimension, the effect will be very little on the shear walls.

At the ends of the building the shear stress in the edge of the diaphragm will be

$$\text{north–south total } V = 0.1862(535) = 99 \text{ kips [440 kN]}$$

$$\text{maximum } v = \frac{49{,}500}{60} = 825 \text{ plf [12 kN/m]}$$

This is a very high shear for the metal deck. It would require a heavy gauge deck and considerable welding at the diaphragm edge. Although it would probably be wise to reconsider the general design and possibly use at least one permanent interior partition, we assume the deck to span the building length for the shear wall design.

In the other direction the shear in the roof deck will be considerably less.

$$\text{east–west total } V = 0.1862(410) = 76.5 \text{ kips [340 kN]}$$

TABLE 10.1. Loads to the Roof Diaphragm (kips)

Load Source and Calculation	North–South Load	East–West Load
Roof dead load		
150 × 60 × 29 psf	261	261
East and West exterior walls		
50 × 11 × 70 psf × 2	0	77
10 × 6 × 70 psf × 2	0	9
10 × 6 × 10 psf × 2	0	1
North wall		
150 × 12 × 70 psf	126	0
South wall		
65.3 × 10 × 70 psf	46	0
84 × 6 × 70 psf	35	0
84 × 6 × 10 psf	5	0
Interior north–south partitions		
60 × 7 × 10 psf × 5	21	21
Toilet walls		
Estimated 250 × 7 × 10 psf	17	17
Canopy		
South: 150 × 100 plf	15	15
East and west: 40 × 100 plf	4	4
Rooftop HVAC units (estimate)	5	5
Total load	535 [2380 kN]	410 [1824 kN]

$$\text{maximum } v = \frac{38{,}250}{150}$$

$$= 255 \text{ plf } [3.72 \text{ kN/m}]$$

This is very low for the deck, so if any interior shear walls are added, the deck gauge could probably be reduced to that required for the gravity loads only.

10.2. DESIGN OF THE MASONRY SHEAR WALLS

In the north–south direction, with no added shear walls, the end shear forces will be taken almost entirely by the long solid walls because of their relative stiffness. The shear force will be the sum of the end shear from the roof and the force due to the weight of the end wall. For the latter we compute the following:

Wall weight
= 18 ft × 50 ft × 70 psf
= 63,000 lb
6 ft × 10.67 ft × 70 psf
= 4,481 lb
12 ft × 10.67 ft × 5 psf
= 640

Total = 68,121 lb [303 kN]

Lateral force $= 0.1862W = 0.1862 \times 68$

$= 12.7$ k [56.5 kN]

The total force on the wall is thus 12.7 + 49.5 = 62.2 k [277 kN] and the unit shear is

$$v = \frac{62{,}200}{44.67} = 1392 \text{ lb/ft [20.3 kN/m]}$$

The code requires that this force be increased by 50% for shear investigation. (UBC Section 2407 (g) 4.F (ii)). Assuming a 60% solid wall with 8-in. blocks, the unit stress on the net area of the wall is thus

$$v = \frac{1392 \times 1.5}{12 \times 7.625 \times 0.60}$$

$$= 38 \text{ psi [262 kP/a]}$$

From UBC Table 24-H (see Appendix B), with the reinforcing taking all shear and no special inspection, the allowable shear stress is dependent on the value of M/Vd for the wall. This is determined as

$$\frac{M}{Vd} = \frac{(49.5 \times 15) + (12.7 \times 9)}{62.2 \times 44.67} = 0.308$$

Interpolating between the table values for M/Vd of 0 and 1.0,

$$\text{allowable } v = 35 + 0.69(25) = 52 \text{ psi}$$

This may be increased by the usual one third for seismic load to 1.33(52) = 69 psi [476 kPa]. This indicates that the masonry stress is adequate, but we must check the wall reinforcing for its capacity as shear reinforcement. With the minimum horizontal reinforcing determined previously—No. 5 at 48 in. (see Section 9.2)—the load on the bars is

$$V = 1392 \left(\frac{48}{12}\right) (1.5)$$

$$= 8352 \text{ lb [37 kN] per bar}$$

and the required area for the bar is

$$A_s = \frac{V}{f_s} = \frac{8352}{26{,}667}$$

$$= 0.31 \text{ in.}^2 \text{ [200 mm}^2\text{]}$$

This indicates that the minimum reinforcing is just adequate. Some additional stress will be placed on these walls by the effects of torsion, so that some increase in the horizontal reinforcing is probably advisable.

Overturn is not a problem for these walls because of their considerable dead weight and the natural tiedown provided by the dowelling of the vertical wall reinforcing into the foundations. These dowels also provide the necessary resistance to horizontal sliding.

In the east–west direction the shear walls are not symmetrical in plan, which requires that a calculation be made to determine the location of the center of rigidity so that the torsional moment may be determined. The total loading is reasonably centered in this direction, so we will assume the center of gravity to be in the center of the plan.

The following analysis is based on the examples in the *Masonry Design Manual* (Ref. 14.) The individual piers are assumed to be fixed at top and bottom and their stiffnesses are found from Table A.6, Appendix A. The stiffness of the piers and the total wall stiffnesses are determined in Figure 10.1. For the location of the center of stiffness we use the values determined for the north and south walls:

$$\bar{y} = \frac{(R \text{ for the S wall})(60 \text{ ft})}{(\text{sum of the } R \text{ values for the N and S walls})} = \frac{2.96(60)}{17.57}$$

$$= 10.11 \text{ ft [3.08 in.]}$$

The torsional resistance of the entire shear wall system is found as the sum of the products of the individual wall rigidities times the square of their distances from the center of stiffness. This summation is shown in Table 10.2. The tor-

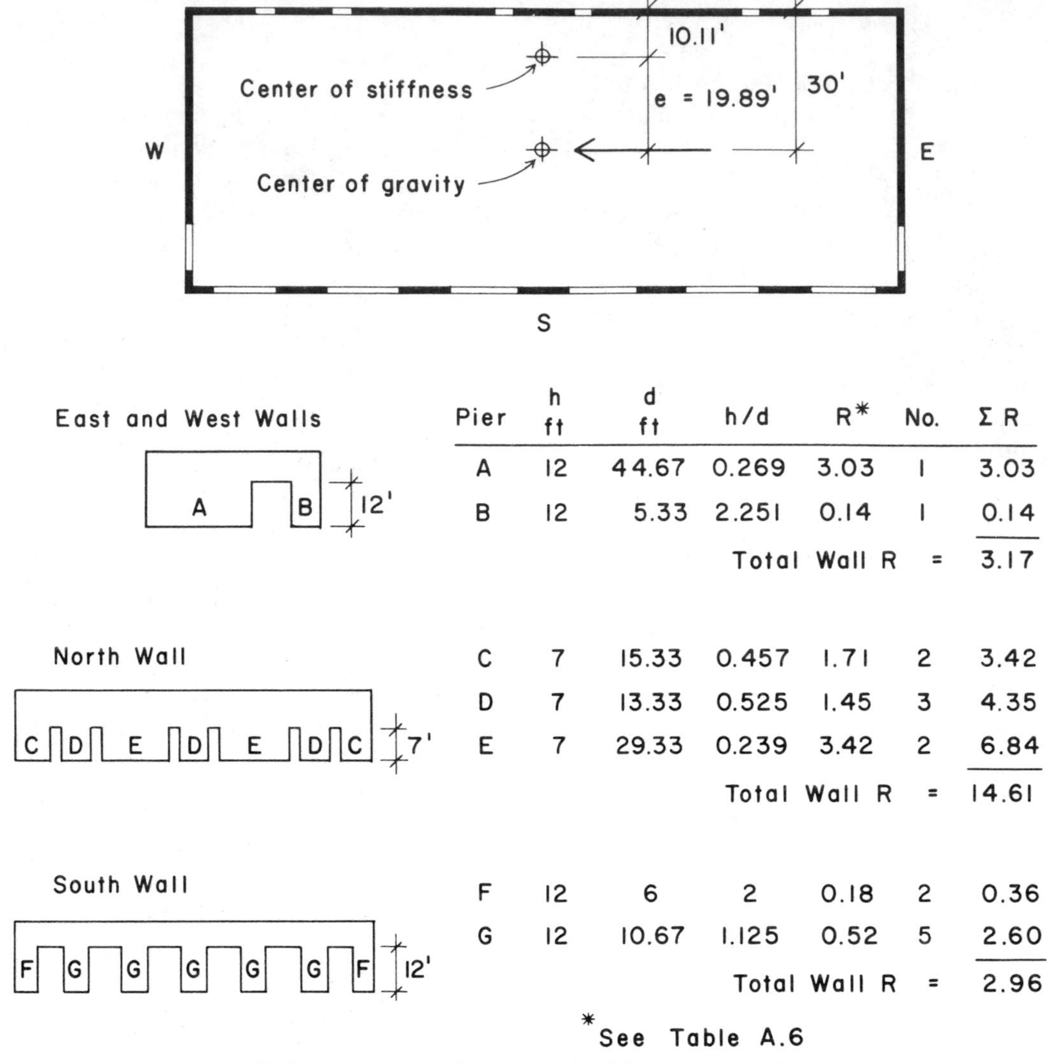

Pier	h ft	d ft	h/d	R*	No.	Σ R
A	12	44.67	0.269	3.03	1	3.03
B	12	5.33	2.251	0.14	1	0.14
			Total Wall R		=	3.17
C	7	15.33	0.457	1.71	2	3.42
D	7	13.33	0.525	1.45	3	4.35
E	7	29.33	0.239	3.42	2	6.84
			Total Wall R		=	14.61
F	12	6	2	0.18	2	0.36
G	12	10.67	1.125	0.52	5	2.60
			Total Wall R		=	2.96

*See Table A.6

FIGURE 10.1. Stiffness analysis of the masonry walls.

sional shear load for each wall is then found as

$$V_w = \frac{Tc}{J}$$

$$= \frac{(V)(e)(c)(\text{the } R \text{ for the wall})}{(\text{the sum of the } Rd^2 \text{ for all walls})}$$

In the north–south direction UBC 2312(e)5 requires that the load be applied with a minimum eccentricity of 5% of the building length, or 7.5 ft. Although this produces less torsional moment than the east–west load, it is additive to the direct north–south shear and therefore critical

TABLE 10.2. Torsional Resistance of the Masonry Shear Walls

Wall	Total Wall R	Distance from Center of Stiffness (ft)	$R(d)^2$
South	2.96	49.89	7,367
North	14.61	10.11	1,495
East	3.17	75	17,831
West	3.17	75	17,831
Total torsional moment of inertia (J)			44,524

for the end walls. The torsional load for the end walls is thus

$$V_w = \frac{(99)(7.5)(75)(3.17)}{44{,}524}$$

$$= 3.96 \text{ kips } [17.6 \text{ kN}]$$

As mentioned previously, this should be added to the direct shear of 49,500 lb for the design of these walls.

For the north wall:

$$V_w = \frac{(76.5)(19.89)(10.11)(14.61)}{44{,}524}$$

$$= 5.05 \text{ kips } [22.5 \text{ kN}]$$

This is actually opposite in direction to the direct shear, but the code does not allow the reduction and thus the direct shear only is used.

For the south wall:

$$V_w = \frac{(76.5)(19.89)(49.89)(2.96)}{44{,}524}$$

$$= 5.05 \text{ kips } [22.5 \text{ kN}]$$

The total direct east–west shear will be distributed between the north and south walls in proportion to the wall stiffnesses:

for the north wall:

$$V_w = \frac{76.5(14.61)}{17.57} = 63.6 \text{ kips } [283 \text{ kN}]$$

for the south wall:

$$V_w = \frac{76.5(2.96)}{17.57} = 12.9 \text{ kips } [57.4 \text{ kN}]$$

The total shear loads on the walls are therefore

north: $V = 63.6$ kips [283 kN]

south: $V = 5.05 + 12.9$

$= 17.95$ kips [79.8 kN]

The loads on the individual piers are then distributed in proportion to the pier stiffnesses (R) as determined in Figure 10.1. The calculation for this distribution

TABLE 10.3. Shear Stresses in the Masonry Walls

Wall	Shear Force on Wall (kips)	Wall R	Pier	Pier R	Shear Force on Pier (kips)	Pier Length (ft)	Shear Stress in Pier (lb/ft)
North	63.6	14.61	C	1.71	7.44	15.33	485
			D	1.45	6.31	13.33	473
			E	3.42	14.89	29.33	508
South	17.95	2.96	F	0.18	1.09	6	182
			G	0.52	3.15	10.67	296

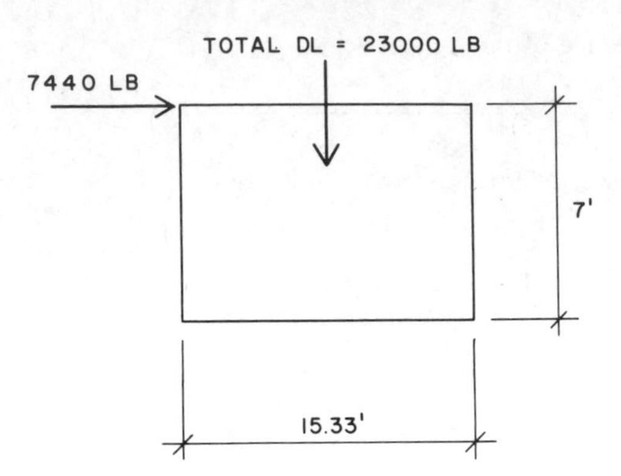

FIGURE 10.2. Stability of wall C.

and the determination of the unit shear stresses per foot of wall are shown in Table 10.3. A comparison with the previous calculations for the end walls will show that these stresses are not critical for the 8-in. block walls.

In most cases the stabilizing dead loads plus the doweling of the end reinforcing into the foundations will be sufficient to resist overturn effects. The heavy loading on the header columns and the pilasters will provide considerable resistance for most walls. The only wall not so loaded is wall C, for which the loading condition is shown in Figure 10.2. The overturn analysis for this wall is as follows:

overturn M

$$= (7440)(7.0)(1.5)$$

$$= 78{,}120 \text{ ft-lb } [106 \text{ kN-m}]$$

stabilizing M

$$= 23{,}000 \left(\frac{15.33}{2}\right)$$

$$= 176{,}295 \text{ ft-lb } [239 \text{ kN-m}]$$

This indicates that the wall is stable without any requirement for anchorage even though the wall weight in the plane of the shear wall was not included in computing the overturning moment.

CHAPTER ELEVEN

Construction Drawings for the Steel and Masonry Structure

The drawings that follow illustrate the construction of the steel and masonry structure for Building Two. There are numerous alternatives for many of the details shown, some of which are discussed in the design work in the preceding chapters. Some alternate possibilities for construction systems for this building are also shown in Chapter 12. The reader is reminded that many of the details shown are essentially limited to the illustration of the structure, and thus there are additional elements of the building construction required for finishes, waterproofing, and so on. Some details may also be affected by considerations for lighting, electrical power, HVAC, and plumbing systems, or by problems of building security, acoustics, fire resistance, and so on.

11.1. STRUCTURAL PLANS

Figure 11.1 shows partial plans for the roof structure and the foundation systems. Two options for the support of the roof purlins are shown. The first of these is the clear span girder with the pilaster columns and the widened wall footing, which is the system that was discussed in the design development in Chapter 9. Also shown is an option for a bearing wall that would replace the girder. If the bearing wall is used, the pilaster and widened wall footing would be omitted, as shown in the drawings.

11.2. CONSTRUCTION DETAILS

Detail A (Fig. 11.2)

This shows the typical front wall condition at the solid wall. The girder, pilaster, pilaster pier, and widened footing are seen in the background. A steel channel is bolted to the masonry wall to receive the end of the steel deck, which in this view is seen at right angle to the corrugations. The deck would be welded to the channel and the channel bolted to the wall to transfer the shear load from the roof diaphragm to the wall.

For the reinforced masonry wall the code requires a minimum vertical spacing of solid horizontal reinforced bond courses. In addition to the minimum

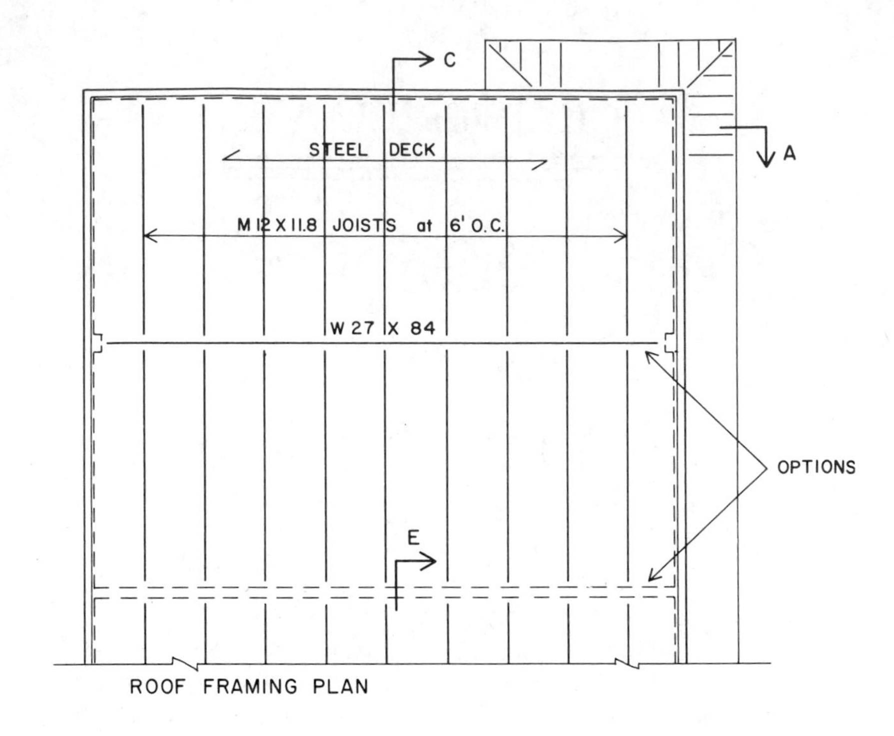

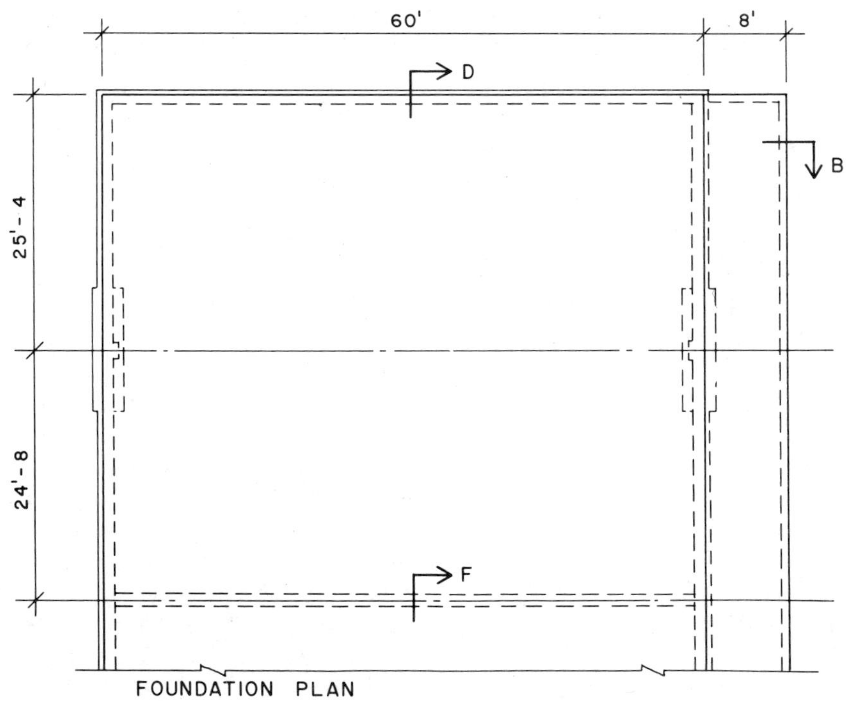

FIGURE 11.1. Structural plans.

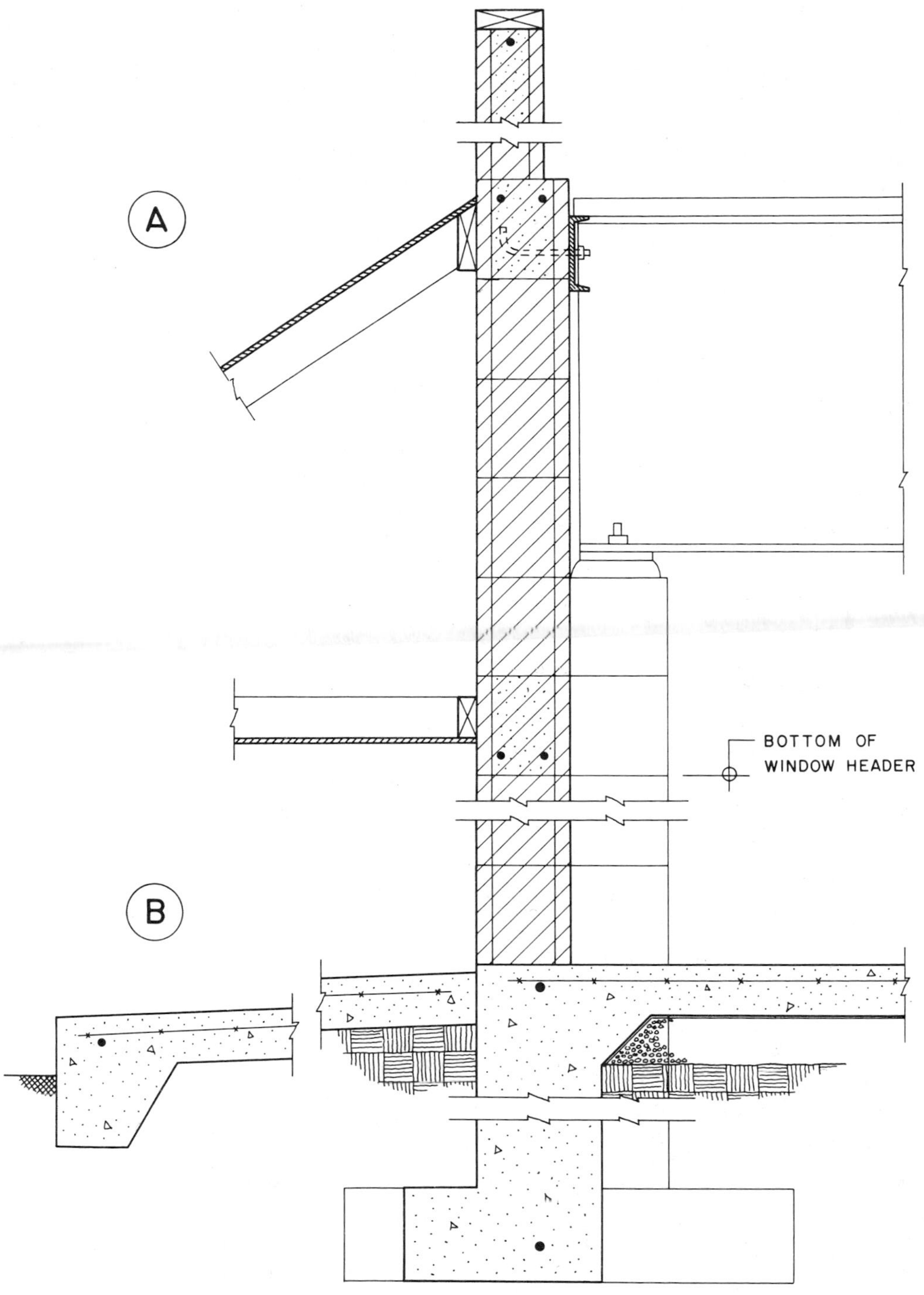

FIGURE 11.2. The front wall.

spacing, these would be used at the top of the wall, the bottom of the header, and the location of the canopy and roof edge bolting to the wall.

Detail B (Fig. 11.2)

This shows the foundation edge at the front, which is essentially similar to that for the wood structure. The sill bolts would be replaced by dowels for the masonry wall. The pier would be added below the pilaster to carry the load down to the widened footing.

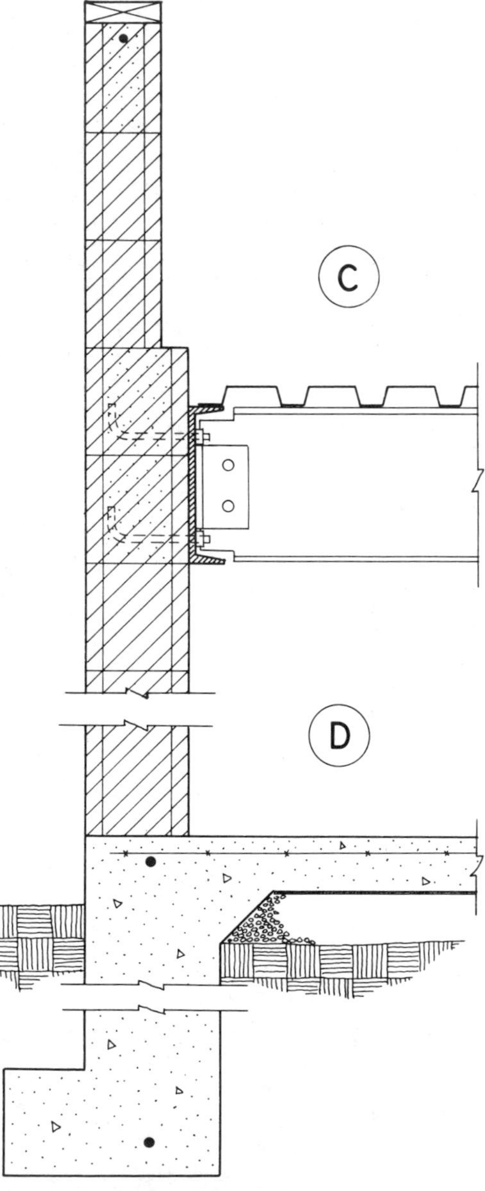

FIGURE 11.3. The end wall.

Detail C (Fig. 11.3)

This shows the roof edge condition at the building ends. The steel channel performs the dual task of providing vertical support for the ends of the purlins and transfers the lateral loads from the steel deck to the masonry wall. Because of the roof slope, the top of the steel channel varies 15 in. from front to rear of the building. The horizontal filled block courses and the cutoff to the narrower parapet would be staggered to accommodate this slope. A somewhat larger than usual cant would be used to cover the jog in the wall to the narrower parapet block.

Detail D (Fig. 11.3)

This detail is also essentially similar to that for the wood structure. If a footing of increased width is required, care should be taken to ensure that the centroid of the vertical loads is close to the center of the width of the footing.

Detail E (Fig. 11.4)

This shows the detail at the top of the interior bearing wall. The continuous steel beam on top of the wall is one means for providing the necessary support for the ends of the purlins and the transfer of lateral load from the steel deck to the wall. It is also possible to simply rest the ends of the purlins on a bearing plate and to provide some sort of blocking between the purlins so that the lateral loads are transferred through the welds between the deck and the purlins. The purlins would then be welded to the

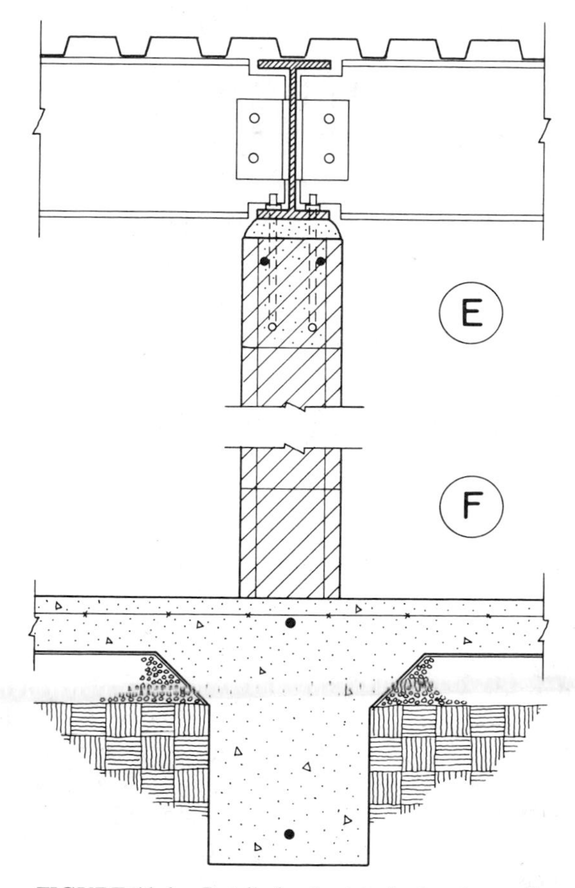

FIGURE 11.4. Details for the interior bearing wall.

plate and the plate bolted to the wall to complete the transfer.

Detail F (Fig. 11.4)

This shows the base of the interior wall with a typical trenched wall footing. Depending on the slab thickness and reinforcing and the nature of the subsoil, it may be advisable to provide this type of footing for all major interior walls. In the event that the slab is to be poured after the masonry walls are built, an optional footing would be as shown for the interior basement wall in Building One.

CHAPTER TWELVE

Alternate Solutions for the Steel and Masonry Structure

As with the wood structure for Building Two, there are many possible solutions and many variations for the details in each scheme.

12.1. STEEL TRUSS SYSTEM

Figure 12.1 shows a layout for a roof system utilizing open web steel joists. This system is essentially similar to the alternate system for the wood structure, although the spacing is free of the constraints of the plywood span and module limits.

With the joists on 6-ft centers the joist depth would be from 30 to 36 in., depending on the type of joist used. Assuming a design depth of 32 in., Figure 12.2 shows how the depth could be varied to facilitate roof drainage.

Figure 12.3 shows a modification of the front wall for this system. (For comparison see Fig. 11.2.) The parapet wall is reduced in thickness to 6 in. and the wall below the joist seat and down to the

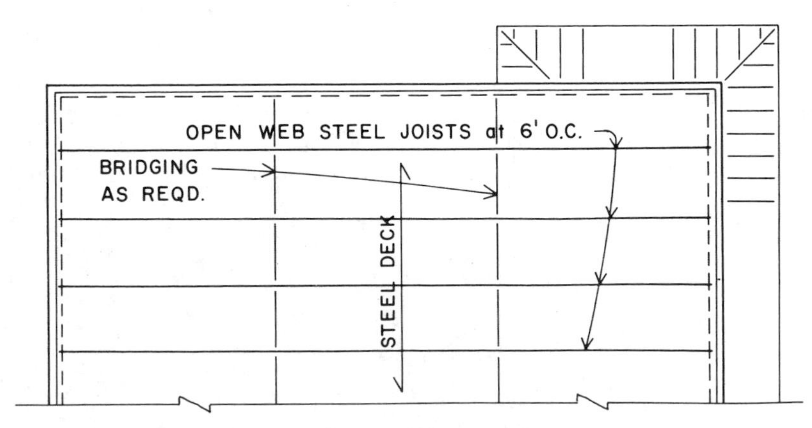

FIGURE 12.1. Structural plan—steel truss system.

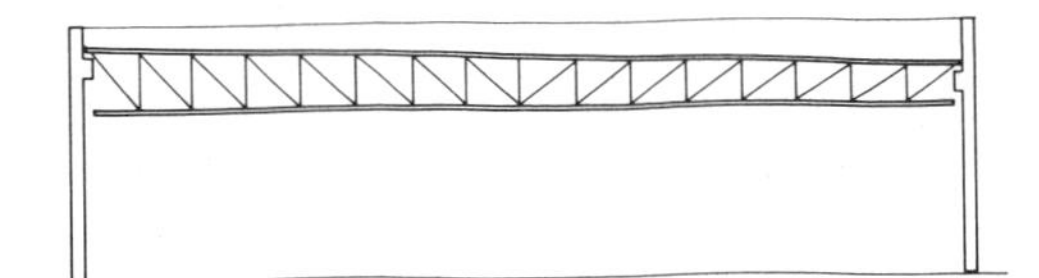

FIGURE 12.2. Optional depth variation for the steel truss.

bottom of the header is thickened to 12 in. This provides a 6-in. shelf to accommodate the steel ledger for the deck and the seat for the joist. The thickened portion produces a heavier header for the larger loading.

Below the header the wall could be maintained as a 12-in. wall or could con-

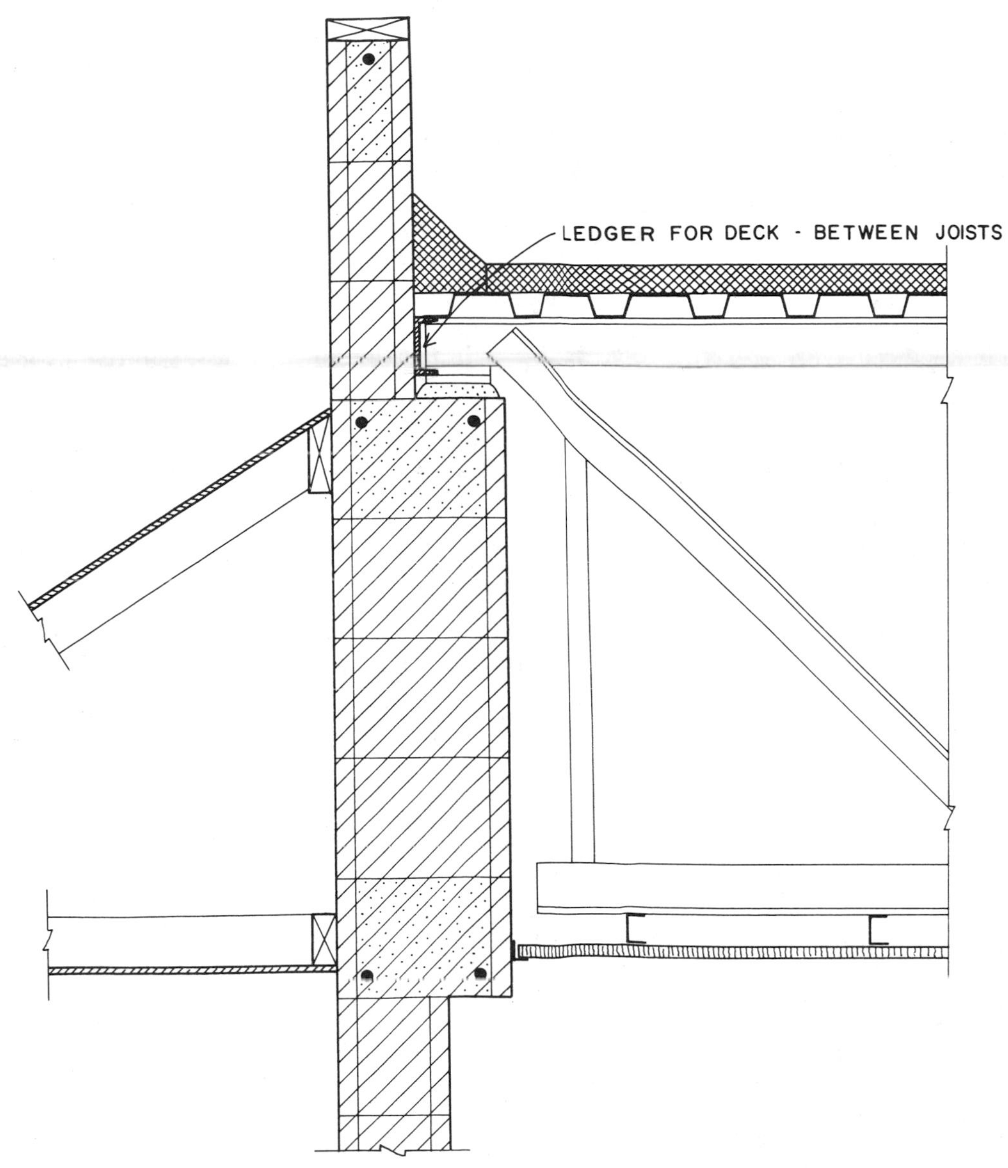

FIGURE 12.3. Front wall detail with the steel truss.

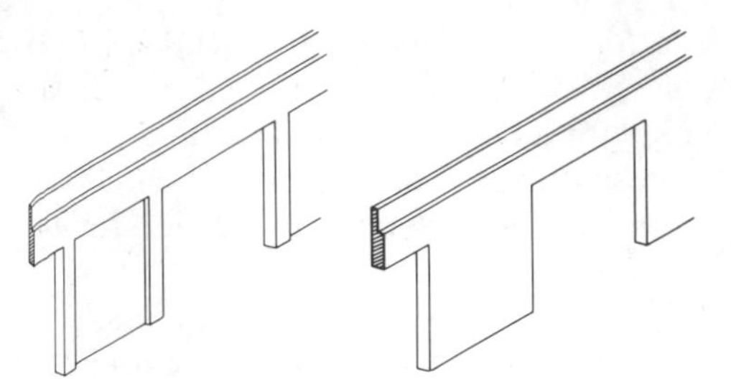

FIGURE 12.4. Options for the front wall.

sist of a series of 12-by-16-in. columns at the wall ends with an 8-in. wall between. These possibilities are shown in the sketches in Figure 12.4. At the north wall, without the header condition, the thickened portion could be reduced to two courses. If the wall is capable of the eccentric loading, it could be 8 in. below this point. If not, it could be increased to 12 in. or also use a series of pilaster columns.

12.2. SYSTEMS WITH INTERIOR SUPPORTS

Details for a system employing interior bearing walls were shown together with the clear span system in the construction drawings in Chapter 11. Figure 12.5 shows a structural plan for a system with a single row of interior columns. The 25-ft spacing of the columns in this case relates to the possible layout of interior walls. If the walls are used, the columns may thus be incorporated within the wall thickness and present less intrusion on the interior space. The interior columns support a series of steel beams.

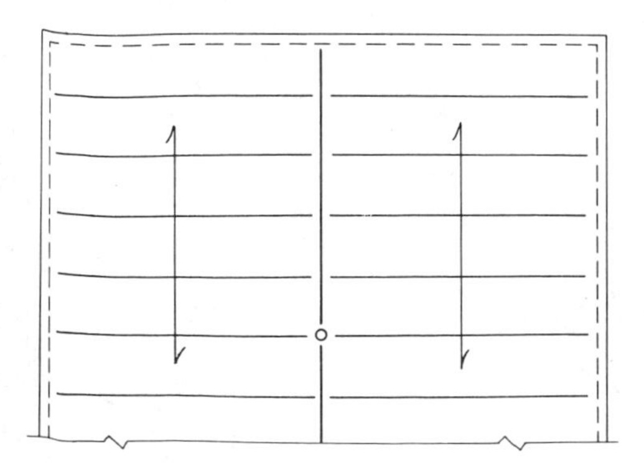

FIGURE 12.5. Structural plan—system with interior columns.

Spanning from the interior beams to the outside walls are open web steel joists that are similar to those used in the system illustrated in Section 12.1. In this system it is less likely that the joists would be tapered in profile to achieve roof drainage. More likely the top of the interior beam and the supports at the outside walls would be adjusted and the entire joist simply tilted to produce the necessary roof slope. This would require the use of a suspended structure for a flat ceiling surface.

12.3. ALTERNATE DECK SYSTEMS

With any of the steel framing system so far illustrated it would be possible to use a variety of decks. Steel decks are often used in combination with some insulating material such as preformed foam plastic planks or poured lightweight concrete. Plywood may be used with the nailing accommodated by wood strips attached to the tops of the joists.

Structural considerations include the magnitude of the roof live load, the need for diaphragm strength, and the means for support of a suspended ceiling, which in some cases is hung from the deck. Joist spacings must be established on the basis of the deck span limits and the plan size of deck units. Economy is usually achieved with a relatively light deck and closely spaced joists.

PART THREE

BUILDING THREE

Building Three is a small three-story office building. Assuming the building is to be built for investment, with a speculative rental occupancy or a sale for undetermined purpose, a feature typically desired is the adaptability of the building to change. With regard to the structure, this usually means an emphasis on minimizing permanent elements of the construction—notably on the interior of the building. Another feature of office buildings is the general use of some planning module, which may be used two dimensionally in developing the building plans, or simply be expressed in the size and spacing of the exterior windows. Other design constraints come from prevailing architectural styles, current building code requirements, and the currently popular means for construction—all of which change with time and place.

CHAPTER THIRTEEN

The Building and the Construction Alternatives

As with Building Two, there are innumerable possibilities for the construction of this building. However, in any given place at any given time, it will be found that the general planning and basic construction of such a building vary little from a limited set of choices. The purpose of this chapter is to present the design factors for the building and discuss the problem of choosing the appropriate construction system.

13.1. THE PROPOSED BUILDING DESIGN

Figure 13.1 presents the initial design scheme for the building in the form of plans and a full building section. In a reasonably rational process of design it is to be expected that the initial design may be modified somewhat in the process of developing the specific details of the building construction as well as the systems for lighting and environmental control. We assume that the plans shown are relatively firmly established, but will discuss some modifications that could improve the structure with certain alternate choices in the construction system.

Although the building exterior is obviously of great importance both architecturally and functionally, we will avoid dealing with the building elevations, except where the choice of certain structural features have a strong relation to the exterior form. We assume, however, that a fundamental requirement for the building is the provision of a significant amount of exterior window surface and the avoidance to long expanses of unbroken solid wall surface. Another assumption—which is reasonably evident in the plans—is that the building is freestanding on the site with all sides having a clear view.

13.2. DESIGN CRITERIA

The following will be assumed as criteria for the building design work:

Building Code:
1985 *Uniform Building Code* (Ref. 7)
Live loads:
Roof: UBC minimum, Table 23-C.
Floors: Office
= 50 psf [2.39 kPa]
Corridors
= 100 psf [4.79 kPa]
Partitions:
= 20 psf (UBC minimum, Sec. 2304(d)) [0.96 kPa]

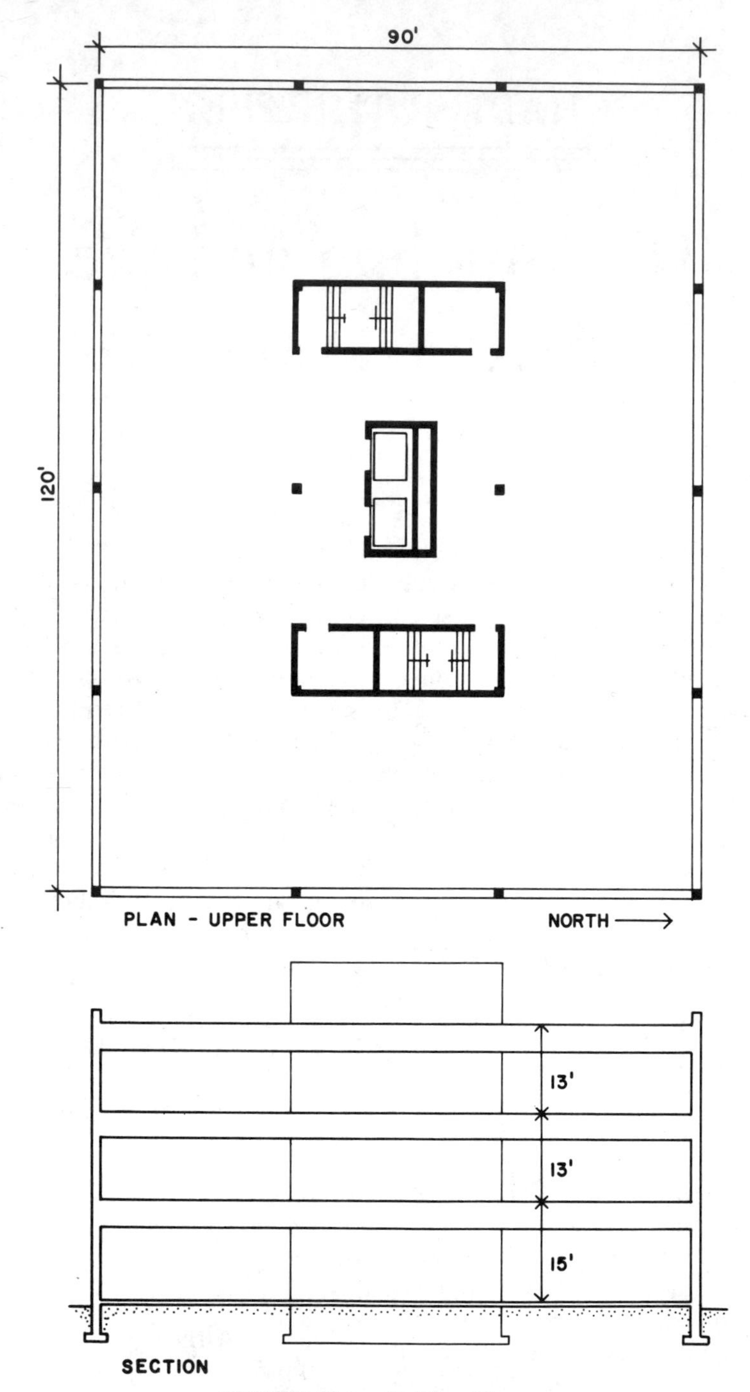

FIGURE 13.1. Building Three.

Wind: Map speed
= 80 mph; exposure B [129 km/h]
Seismic: Zone 3
Assumed construction loads:
Floor finish
= 5 psf [0.24 kPa]
Ceiling, lights, ducts
= 15 psf [0.72 kPa]
Walls (average surface weight):
Interior partitions
= 25 psf [1.20 kPa]
Exterior curtain wall
= 25 psf [1.20 kPa]

13.3. STRUCTURAL ALTERNATIVES

Fire codes permitting, the most economical structure for the building will be one that makes the most use of light wood frame construction. It is unlikely that the building would use all wood construction of the type illustrated in Building One, but a mixed system is quite possible. It is also possible to use steel, masonry, or concrete construction and eliminate wood, except for nonstructural uses. In addition to code requirements, consideration must be given to the building owners' preferences and to design criteria or standards for acoustic privacy, thermal control, and so on.

The plan as shown, with 30-ft square bays and a general open interior, is an ideal arrangement for a beam and column system in either steel or reinforced concrete. Other types of systems may be made more effective if some modifications of the basic plans are made. These changes may affect the planning of the building core, the plan dimensions for the column locations, the articulation of the exterior wall, or the vertical distances between the levels of the building.

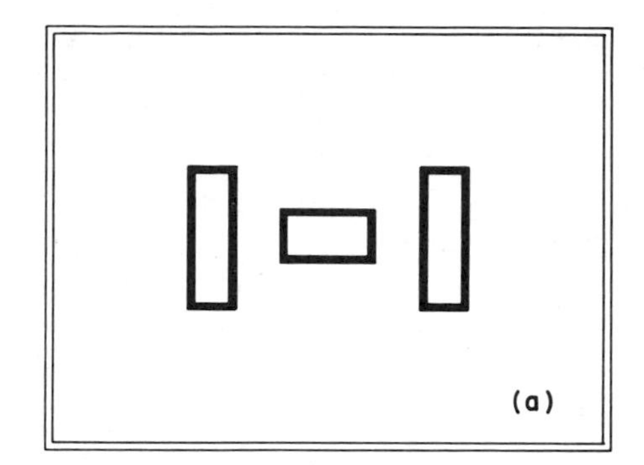

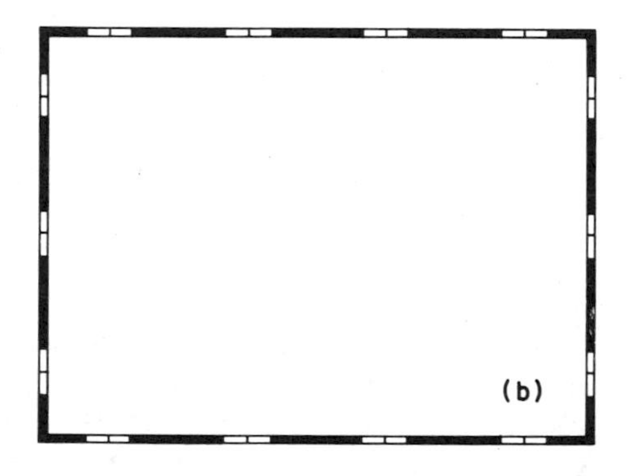

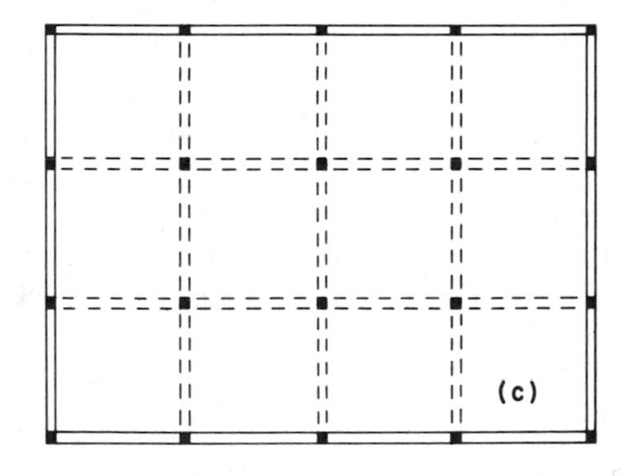

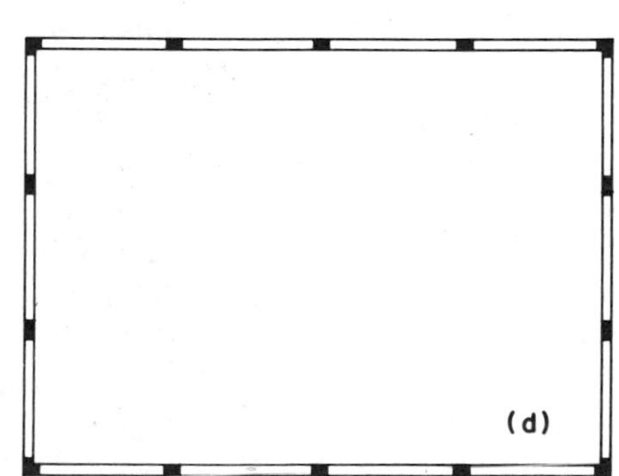

FIGURE 13.2. Options for the lateral bracing.

The general form and basic type of the structural system must relate to both the gravity and lateral force problems. Considerations for gravity require the development of the horizontal spanning systems for the roof and floors and the arrangement of the vertical elements (walls and columns) that provide support for the spanning structure. Vertical elements should be stacked, thus requiring coordinating the plans of the various levels.

The most common choices for the lateral bracing system would be the following (see Fig. 13.2):

1. *Core Shear Wall System* (Fig. 13.2a). This consists of using solid walls to produce a very rigid central core. The rest of the structure leans on this rigid interior portion, and the roof and floor constructions outside the core—as well as the exterior walls—are free of concerns for lateral forces as far as the structure as a whole is concerned.

2. *Truss-Braced Core*. This is similar in nature to the shear wall-braced core, and the planning considerations would be essentially similar. The solid walls would be replaced by bays of trussed framing (in vertical bents) using various possible patterns for the truss elements.

3. *Peripheral Shear Walls* (see Fig. 13.2b). This in essence makes the building into a tubelike structure. Because doors and windows must pierce the exterior, the peripheral shear walls usually consist of linked sets of individual walls (sometimes called piers).

4. *Mixed Exterior and Interior Shear Walls*. This is essentially a combination of the core and peripheral systems.

5. *Full Rigid Frame System* (see Fig. 13.2c). This is produced by using the vertical planes of columns and beams in each direction as a series of rigid bents. For this building there would thus be four bents for bracing in one direction and five for bracing in the other direction. This requires that the beam-to-column connections be moment resistive.

6. *Peripheral Rigid Frame System* (see Fig. 13.2d). This consists of using only the columns and beams in the exterior walls, resulting in only two bracing bents in each direction.

In the right circumstances any of these systems may be acceptable. Each has advantages and disadvantages from both structural design and architectural planning points of view. The braced core schemes were popular in the past, especially for buildings in which wind was the major concern. The core system allows for the greatest freedom in planning the exterior walls, which are obviously of major concern to the architect. The peripheral system, however, produces the most torsionally stiff building—an advantage for seismic resistance.

The rigid frame schemes permit the free planning of the interior and the greatest openness in the wall planes. The integrity of the bents must be maintained, however, which restricts column locations and planning of stairs, elevators, and duct shafts so as not to interrupt any of the column-line beams. If designed for lateral forces, columns are likely to be large, and thus offer more intrusion in the building plan.

Other solutions are also possible, limited only by the creative imagination of designers. In the chapters that follow we will illustrate the design of two possible structures for the building. We do not propose that these are ideal solutions, but merely that they are feasible alternatives. They have been chosen primarily to permit illustrating the design of the elements of the construction.

CHAPTER FOURTEEN

Design of the Masonry Wall Structure

A structural framing plan for one of the upper floors of Building Three is shown in Figure 14.1. The plan indicates major use of structural masonry walls for both bearing walls and shear walls. The lateral bracing scheme is a combination of the core and peripheral systems, as described in Section 13.3. The floor—and probably the roof—structures consist of deck–joist–beam systems that are supported mostly by the interior and exterior bearing walls. The material in this chapter consists of the design of the major elements of this system.

14.1. DESIGN OF THE DECK–JOIST–BEAM SYSTEM

There are a number of options for this system. The spans, the plan layout, the fire code requirements, the anticipated surfacing of floors and ceilings, the need for incorporating elements for wiring, piping, heat and cooling, fire sprinklers, and lighting are all influences on the choice of construction. We assume that the various considerations can be met with a system consisting of a plywood deck, wood joists, and steel beams.

The Deck

The plywood deck will be used for both spanning between the joists to resist gravity loads and as a horizontal diaphragm for distribution of wind and seismic forces to the shear walls. Choice of the deck material, the plan layout of the plywood panels, and the nailing must be done with both functions in mind. The spacing of the joists must also be coordinated with the panel layout. It is also necessary to anticipate floor surfacing in terms of both loading and construction accommodation.

We assume the use of the common size of plywood panel of 4 ft by 8 ft, and a joist spacing of 24 in., which is usually the maximum for floor systems of this type. Office floors will most likely be covered with carpet, thin tile, or wood flooring. The surface of the structural plywood deck, however, is too rough and uneven for these materials, thus resulting in needing some intermediate surfacing, which is most commonly either a second paneling with fiberboard or a thin coat of lightweight structural concrete.

The UBC Table No. 25-S-1 (see Appendix B) indicates that a $\frac{3}{4}$-in.-thick

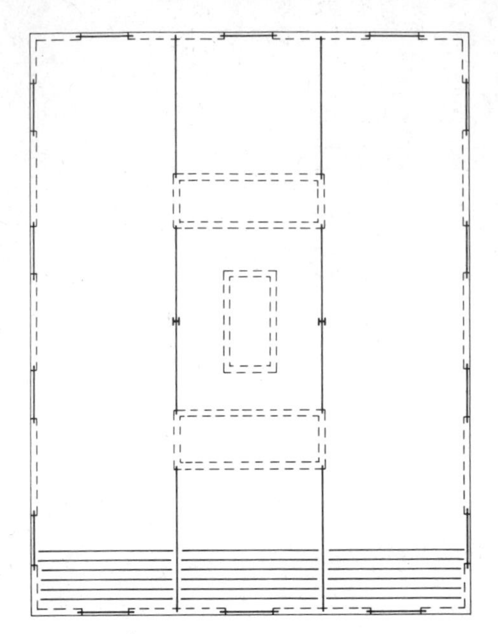

FIGURE 14.1. Structural plan—upper floor with the bearing walls.

deck may be used with the joist spacing of 24 in. A footnote to the UBC table states that the unsupported edges must be blocked or have tongue-and-groove joints if underlayment is not provided. Paneling for floors is quite commonly available with tongue-and-groove joints, but this is not actually required as we will allow for a 2-in. concrete fill.

The panel layout, edge nailing, and the need for blocking must be investigated as part of the design for diaphragm action.

The Joists

It is unlikely that the joists for this span and loading would be solid timber members. As the calculations will show, the required section is quite large and would be very expensive for joists at 24-in. centers. In most regions it is likely that fabricated joists are available that are both lighter and less expensive. For purpose of comparison, however, we will demonstrate the joist design procedure for a solid member.

Economy dictates that a relatively low grade of wood be used for the joist. We will try Douglas Fir–Larch, No. 1 grade, which is about as high a grade as is feasible. For this grade, the UBC yields the following data for sections 2- to 4-in. thick and 6 in. or more in width:

F_b = 1750 psi (repetitive use) [12.1 MPa]

F_v = 95 psi [0.66 MPa]

E = 1,800,000 psi [12.4 GPa]

Using these data we will design a joist for the 100 psf live load at the corridor. The design loads for a single 30 ft [9.14 m] span joist are thus as follows:

Live load:
100 psf × 2 ft = 200 plf [2.92 kN/m]
Dead load:
Carpet + pad at 5 psf
2-in. concrete fill at 10 lb/in.
= 20 psf
$\frac{3}{4}$-in. plywood at 3 psf
Ceiling, lights, ducts at 15 psf
Total unit DL: 43 psf
Load on joist: 43 × 2
= 86 plf
Estimate joist weight: 20 plf
Design DL for joist: 106 plf [1.55 kN/m]

The total design load for a joist is thus

$$\text{DL} + \text{LL} = 106 + 200$$
$$= 306 \text{ plf } [4.46 \text{ kN/m}]$$

and the maximum bending moment is

$$M = \frac{wL^2}{8} = \frac{306(30)^2}{8}$$
$$= 34{,}425 \text{ ft-lb } [46.7 \text{ kN-m}]$$

for which the required section modulus is

$$S = \frac{M}{F_b} = \frac{34{,}425 \times 12}{1750}$$
$$= 236 \text{ in.}^3 \; [3.87 \times 10^6 \text{ mm}^3]$$

There is no member in the size range that was assumed that has an S value this high. We must therefore find a new value for S that corresponds to the proper size range. For the "Beams and Stringers" category in the UBC the allowable bending stress for grade Dense No. 1 is 1550 psi [10.7 MPa]. The corresponding S is thus

$$S = \frac{34{,}425 \times 12}{1550}$$
$$= 267 \text{ in.}^3 \; [4.38 \times 10^6 \text{ mm}^3]$$

which may be satisfied with a 6 × 18 (S = 281 in.^{3}). This is a mammoth size member for joists at 24 in. on center. If the framing arrangement shown in Figure 14.1 is to be retained, a fabricated joist will have to be used. One possibility is to use an open web steel joist with provision for attachment of the plywood deck to its top flange. Using the joist loading determined previously, it will be found from Ref. 15 that a 22H6 joist at 9.7 lb/ft is adequate for both total load capacity and live load deflection of L/360. Other possibilities are a wood-trussed joist, a vertically laminated wood joist, or a wood plus plywood member (wood flanges and plywood web).

Another possibility is to alter the framing system slightly to produce a system similar to that used for the wood roof structure for Building Two. A layout of this type is shown in the partial framing plan in Figure 14.2. In this scheme the long span joists are replaced by purlins at 8-ft centers that support short span joists at 2 ft centers. As with the roof system for Building Two, an advantage with this scheme is that it can be used to provide full edge support for the plywood panels, producing a blocked diaphragm without the need for added blocking.

The purlins for the system in Figure 14.2 will carry approximately four times the joist load determined for the 2-ft-on-center joists. This would require the use of glue-laminated members or steel wide flange sections. We will not finish the design of this system, but will proceed with the design assuming the scheme in Figure 14.1 with fabricated joists.

The Beams

The large interior beams may be of wood glue-laminated construction, but are more likely to be steel wide flange sections. We assume the latter case, and assume the beam section to be as shown in Figure 14.3, with a 2-in. nominal member bolted to the top of the steel section and the joists carried by saddle-type hangers, with the joist tops level with the top of the wood nailer. If open web steel joists or wood-trussed joists are selected, their top chords would be supported on top of the beam, and the detail would be slightly different.

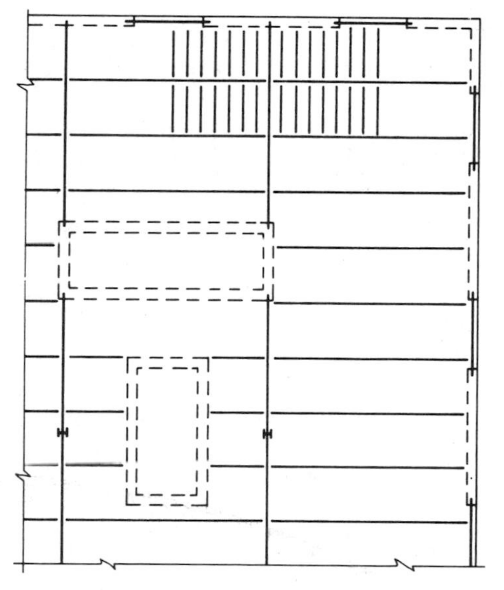

FIGURE 14.2. Alternate floor framing system.

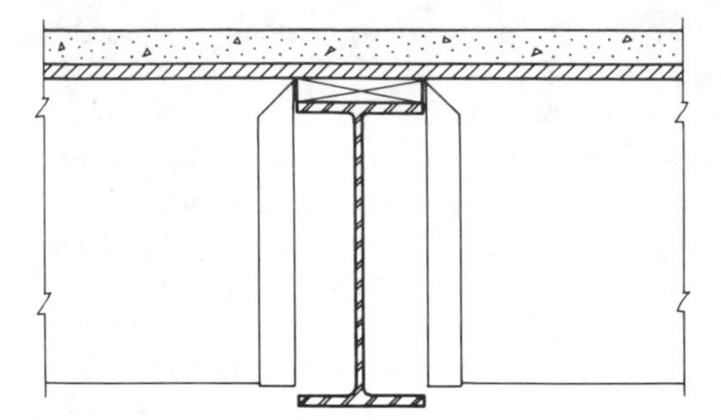

FIGURE 14.3. Detail at the steel beam.

We will design the beam for a live load of 100 psf and a unit dead load of 55 psf, which includes allowances for the weight of the joists, the beams, and any bridging, blocking, and so on. The heaviest loaded beam is the 30-ft beam that occurs at the building ends. This beam has a total load periphery of 30 ft by 30 ft or 900 ft^2. The UBC allows a reduction of the live load equal to

$$R = 0.08(A - 150)$$
$$= 0.08(900 - 150) = 60\%$$

The reduction is further limited to a maximum of 40% or

$$R = 23.1\left(1 + \frac{\text{DL}}{\text{LL}}\right)$$
$$= 23.1\left(1 + \frac{55}{100}\right) = 35.8\%$$

We will therefore use a 35% reduction, or a design live load of 65 psf.

Using the reduced live load, the total load carried by the 30-ft span beam is

$$W = (65 + 55) \times 30 \times 30$$
$$= 108{,}000 \text{ lb, or } 108 \text{ kips } [480 \text{ kN}]$$

and the maximum bending moment is

$$M = \frac{WL}{8} = \frac{108 \times 30}{8}$$
$$= 405 \text{ k-ft } [549 \text{ kN-m}]$$

From Table A.1, Appendix A, assuming full lateral support for the beam, possible choices are

W 27 × 84 $\quad M_R = 426$ k-ft

W 24 × 94 $\quad M_R = 444$ k-ft

W 18 × 106 $\quad M_R = 408$ k-ft

Deflection of the beam should be a minimum, for the total deflection of the joists must be added to that of the beam for the sag at the middle of the bay. An inspection of Figure A.2, Appendix A shows that either the 27- or 24-in. members will have quite low values of deflection. If a member shallower than 24 in. is desired to allow for passage of ducts, the deflection should be carefully investigated.

Design of the two-span beam at the core would be similar in procedure to that used for the basement beam in Building One. The shorter spans and reduced loading will result in a section considerably smaller than that required for the 30-ft span beam.

The short beams that span between the masonry walls at the building edge must be designed to carry the floor structure as well as the wall construction. Selection of the beam section and development of the connection details must be developed as part of the general design of the exterior walls. Some possible details are shown in Chapter 15.

The Steel Column

To provide for the continuous beam, the two columns at the core would be stacked with the beams as shown in Figure 14.4. Although this is not the way to make multistory columns in general, it can work with a relatively short building as long as the beams are strengthened with web stiffeners as demonstrated in the detail in Figure 14.4.

The type of column section used would depend mostly on desired archi-

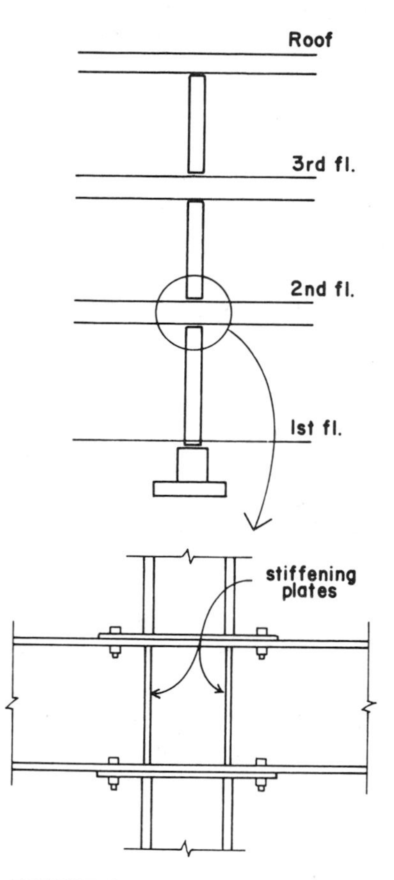

FIGURE 14.4. The interior column.

tectural detailing. Possibilities include a wide flange section, a round steel pipe, or a square tubular section. The column must be fireproofed, of course, which must be achieved with a jacket or by building it into the core wall construction.

The design of multistory steel columns is developed more thoroughly in the examples in Building Four.

The Masonry Walls

Because the walls are used for both gravity and lateral force resistance, we must investigate the lateral force distribution before proceeding with the wall design. This is done in the next section.

14.2. DESIGN FOR LATERAL FORCES

Design for lateral force effects includes consideration for wind and seismic forces. Design for wind is based on the requirements of the 1985 edition of the *Uniform Building Code* (Ref. 1) and an assumed basic wind speed of 80 mph. Seismic requirements are also based on the UBC and an assumed seismic zone 3 condition.

It is quite common, when designing for both wind and seismic forces, to have some parts of the structure designed for wind and others for seismic effects. In fact, what is necessary is to analyze for both effects and to design each element of the structure for the condition that produces the greater effect. Thus the shear walls may be designed for seismic effects, the exterior walls and window glazing for wind, and so on.

For wind it is necessary to establish the design wind pressure, defined by the code as

$$p = C_e C_q q_s I$$

where C_e is a combined factor including concerns for the height above grade, exposure conditions, and gusts. From UBC Table 23-G (see Appendix B), assuming exposure B:

$$C_e = 0.7 \text{ from 0 to 20 ft above grade}$$
$$= 0.8 \text{ from 20 to 40 ft}$$
$$= 1.0 \text{ from 40 to 60 ft}$$

and C_q is the pressure coefficient. Using the projected area method (method 2) we find from UBC Table 23-H (see Appendix B) the following.

For vertical projected area:

$$C_q = 1.3 \text{ up to 40 ft above grade}$$
$$= 1.4 \text{ over 40 ft}$$

For horizontal projected area (roof surface):

$$C_q = 0.7 \text{ upward}$$

The symbol q_s is the wind stagnation pressure at the standard measuring height of 30 ft. From UBC Table 23-F the q_s value for a speed of 80 mph is 17 psf.

For the importance factor I (UBC Table 23-K) we use a value of 1.0.

Table 14.1 summarizes the foregoing data for the determination of the wind pressures at the various height zones for Building Three. For the analysis of the horizontal wind effect on the building, the wind pressures are applied and translated into edge loadings for the horizontal diaphragms (roof and floors) as shown in Figure 14.5. Note that we have rounded off the wind pressures from Table 14.1 for use in Figure 14.5.

Figure 14.6 shows a plan of the building with an indication of the masonry walls that offer potential as shear walls for resistance to north–south lateral force. The numbers on the plan are the approximate plan lengths of the walls. Note that although the core construction actually produces vertical tubular-shaped elements, we have considered only the walls parallel to the load direction. The walls shown in Figure 14.6 will share the total wind load delivered by the diaphragms at the roof, third-floor, and second-floor levels (H_1, H_2, and H_3, respectively, as shown in Fig. 14.5). Assuming the building to be a total of 122-ft wide in the east–west direction, the forces at the three levels are

$$H_1 = 195 \times 122 = 23{,}790 \text{ lb [106 kN]}$$

$$H_2 = 234 \times 122 = 28{,}548 \text{ lb [127 kN]}$$

$$H_3 = 227 \times 122 = 27{,}694 \text{ lb [123 kN]}$$

and the total wind force at the base of the shear walls is the sum of these loads, or 80,032 lb [356 kN].

Although the distribution of shared load to masonry walls is usually done on the basis of a more sophisticated analysis for relative stiffness (as was done for Building Two), if we assume for the moment that the walls are stiff in proportion to their plan lengths (as is done with plywood walls), we may divide the maximum shear load at the base of the walls by the total of the wall plan lengths to obtain an approximate value for the maximum shear stress. Thus

$$\text{maximum shear:} \quad v = \frac{80{,}032}{260}$$

$$= 308 \text{ lb/ft of wall length [4.49 kN/m]}$$

This is quite a low force for a reinforced masonry wall, which tells us that if wind alone is of concern we have considerable overkill in terms of total shear walls. However, because we will find that the seismic forces are considerably greater for our example, we will reserve final judgment on the structural scheme.

Table 14.2 presents the analysis for determining the building weight to be used for computation of the seismic effects in the north–south direction. In the tabulation we have included the weights of all the walls, which eliminates the necessity for adding the weights of the shear walls in any subsequent analysis of individual walls. Tabulations are done separately for the determination of loads to the three upper diaphragms (eventually producing three forces similar to H_1,

TABLE 14.1. Design Wind Pressures for Building Three

Height above Average Level of Adjoining Ground (ft)	C_e	C_q	Pressure[a] p (psf)
0–20	0.7	1.3	15.47
20–40	0.8	1.3	17.68
40–60	1.0	1.4	23.80

[a] Horizontally directed pressure on vertical projected area: $p = C_e \times C_q \times 17$ psf.

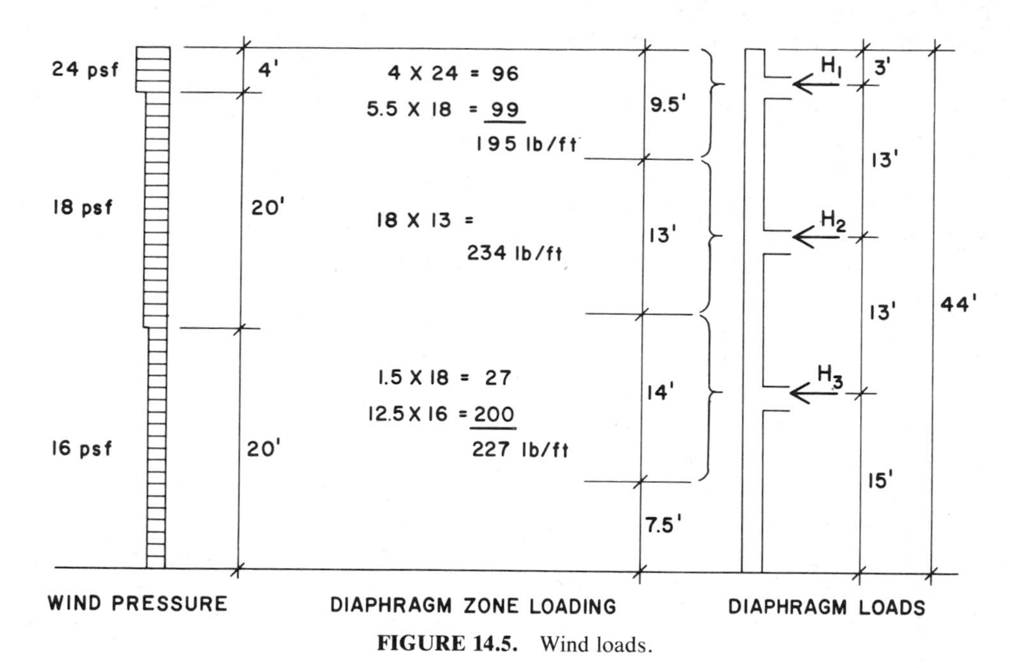

FIGURE 14.5. Wind loads.

H_2, and H_3, as determined for the wind loading). Except for the shear walls, the weight of the lower half of the first-story walls is assumed to be resisted by the first-floor-level construction (assumed to be a concrete structure poured directly on the ground) and is thus not part of the distribution to the shear wall system.

For the determination of the base shear we now use UBC Formula 12-1:

$$V = ZIKCSW$$

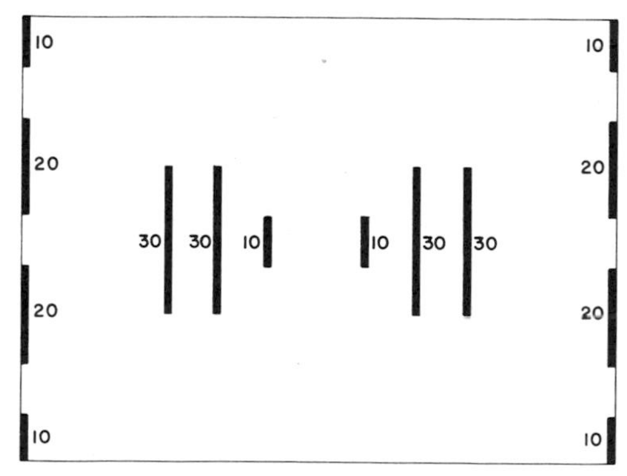

FIGURE 14.6. Shear walls—north–south load.

For seismic zone 3: $Z = \frac{3}{4}$.

For our building as for wind: $I = 1.0$.

For the masonry shear wall structure: $K = 1.33$.

For the three-story building with S not determined, we use the maximum value of the product of $CS = 0.14$.

Thus

$$V = \tfrac{3}{4}(1.0)(1.33)(0.14)(2775.9)$$
$$= 388 \text{ k } [1724 \text{ kN}]$$

This total force must be distributed to the roof and second floor in accordance with the requirements of Section 2312(e) of the UBC. The force at each level, F_x, is determined from Formula 12–7 as

$$F_x = (V)(w_x h_x)/\sum_{i=1}^{n} w_i h_i$$

where F_x = the force to be applied at each level x

w_x = the total dead load at level x

h_x = the height of level x above the base of the structure

(Notice that F_t has been omitted from the formula because T is less than 0.7 sec.)

The determination of the F_x values is shown in Table 14.3. Distribution of the total forces at each level to the individual shear walls requires two considerations. The primary concern is for the functioning of the horizontal diaphragms. First, if these are considered to be infinitely stiff, then the distribution to the individual walls will be strictly in terms of their relative stiffness or deflection. Second, if the horizontal diaphragms are considered to be quite flexible (in their diaphragm spanning actions), then the distribution to the shear walls will be on a peripheral basis.

Figure 14.7 shows the building plan with the north–south shear walls and a breakdown of peripheral distribution assuming the flexible horizontal diaphragm. On this basis, the end shear walls each carry one eighth of the total shear and the core walls carry three fourths of the shear. In this approach the next step would be to consider the relative stiffness of the group of walls in each of the zones and to distribute forces to the individual walls.

In truth, the nature of the diaphragms is most likely somewhere between the two extremes described (just as most structural connections are neither pinned nor fully fixed, but actually partially fixed). It is thus not uncommon in practice for designers to investigate both conditions and to incorporate data from both analyses into their designs.

TABLE 14.2. Dead Load for the North–South Seismic Force

Level	Source of Load	Unit Load (psf)		Load (kips)
Roof	Roof and ceiling	25	120 × 90 × 25 =	270
	Masonry walls	60	480 × 9.5 × 60 =	273.6
	Window walls	15	140 × 9.5 × 15 =	20.0
	Interior walls	10	200 × 5 × 10 =	10.0
	Penthouse + equipment (estimate total)		=	25.0
	Subtotal			598.6
Third floor	Floor	55	120 × 90 × 55 =	594.0
	Masonry walls	60	480 × 13 × 60 =	374.4
	Window walls	15	140 × 13 × 15 =	27.3
	Interior walls	10	200 × 9 × 10 =	18.0
	Subtotal			1013.7
Second floor	Floor	55	120 × 90 × 55 =	594.0
	Masonry walls	60	480 × 14 × 60 =	403.2
	Window walls	15	140 × 14 × 15 =	29.4
	Interior walls	10	200 × 10 × 10 =	20.0
	Subtotal			1046.6
First floor	Shear walls	60	260 × 7.5 × 60 =	117.0
	(Remainder of first floor direct to ground)			
Total dead load for base shear			=	2775.9 [12.347 × 10^3 kN]

TABLE 14.3. Seismic Loads: Building Three

Level	w_x (kips)	h_x (ft)	$w_x h_x$	F_x[a] (kips)
Roof	598.6	41	24,543	138.8
Third floor	1013.7	28	28,384	160.5
Second floor	1046.6	15	15,699	88.7
			68,626	

[a] $F_x = \frac{388}{68,626}(w_x h_x)$

For either approach it is necessary to consider the relative stiffness of the walls of various plan length. The most common means for doing this is the method illustrated in the design of Building Two (see Fig. 10.1). Figure 14.8 shows an analysis for the relative stiffness of the walls with the three plan lengths of 10 ft, 20 ft, and 30 ft. The walls are assumed to be cantilevered from fixed bases and the distributions shown are for the roof load, for which the wall height (h in the table and figure) is 41 ft. For a precise analysis separate distributions should be made for the distribution of the floor diaphragm loads using the shorter wall heights. If this is done, it will be found that the percent of load carried by the shorter walls will be considerably increased.

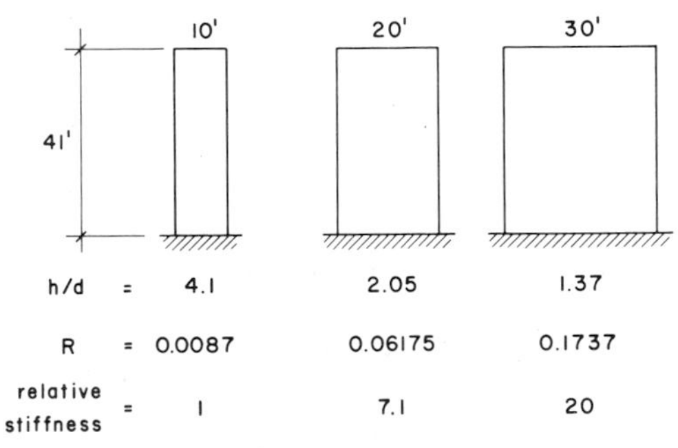

FIGURE 14.8. Relative stiffness of the masonry walls.

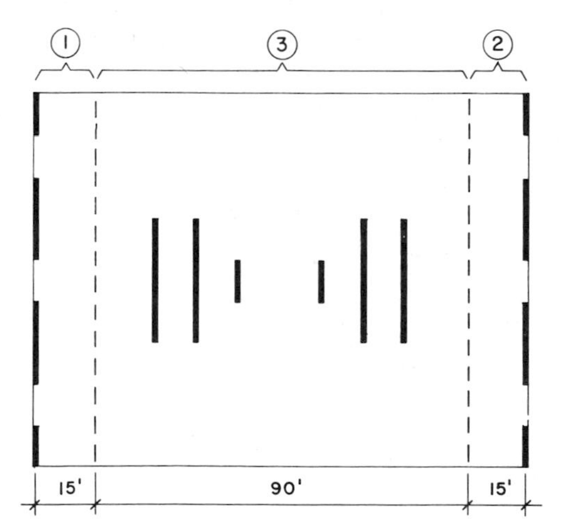

FIGURE 14.7. Peripheral distribution—north–south load.

Referring to Figure 14.7, if we consider the group of core walls in peripheral zone 3, their total combined stiffness is

$$4 \times 0.1737 = 0.6948$$
$$2 \times 0.0087 = \underline{0.0174}$$
$$\text{Total} = 0.7122$$

The portion of load carried by a single 10-ft-long wall will thus be

$$\frac{0.0087}{0.7122} = 0.0122 \text{ or barely more than } 1\%$$

It is therefore reasonable to assume that the 30-ft walls carry the entire load to the core zone. For a single pair of walls constituting one stair plus a rest room tower, the portion of the full lateral load will thus be

$$\tfrac{1}{2} \times \tfrac{3}{4} \times F_x = \tfrac{3}{8} \times F_x$$

Referring to Table 14.3 and Figure 14.9, the loads for a single tower are

$$H_1 = \tfrac{3}{8} \times 138.8 = 52.05 \text{ k } [232 \text{ kN}]$$

$$H_2 = \tfrac{3}{8} \times 160.5 = 60.2 \text{ k } [268 \text{ kN}]$$

$$H_3 = \tfrac{3}{8} \times 88.7 = 33.3 \text{ k } [148 \text{ kN}]$$

and the total overturning moment about the base of the wall at the first floor level is

$H_1 \times 41 = 52.05 \times 41$		
	$= 2134$ k-ft	2894 kN-m
$H_2 \times 28 = 60.2 \times 28$		
	$= 1686$	2286
$H_3 \times 15 = 33.3 \times 15$		
	$= 500$	678
Total	$= 4320$ k-ft	5858 kN-m

For the dead load moment that resists this overturn effect we make the following assumptions:

1. The walls are 8-in. concrete block weighing 60 psf of wall surface. For the entire tower the total weight is thus approximately

$$80 \times 41 \times 60 = 196.8 \text{ k } [875 \text{ kN}]$$

2. As bearing walls, the tower walls carry approximately 1800 ft^2 of roof or floor periphery, which results in a supported dead load of

55×1800

$= 99$ k/floor or 198 k total [881 kN]

25×1800

$= 45$ k of roof load [200 kN]

This results in a total load (G in Fig. 14.9) of 439.8 k [1956 kN] and a restoring dead load moment of

$$439.8 \times 15 = 6597 \text{ k-ft } [8946 \text{ kN}]$$

The safety factor against overturn is thus

$$\frac{6597}{4320} = 1.53$$

which indicates that there is no real need for a tiedown force.

There will, of course, be a considerable tiedown developed in the form of the dowelling of the wall reinforcing into the foundation. If the tower foundation is as shown in Figure 14.10, the construction results in an additional dead weight of approximately 150 k, not including the earth fill. For an investigation of the footing, the total dead load is thus approximately 690 k, and the unit soil pressure for dead load is

$$q = \frac{690{,}000}{11.7 \times 31.7} = 1860 \text{ psf } [89 \text{ kPa}]$$

The equivalent eccentricity of the dead load caused by the seismic overturn force is

$$e = \frac{4320}{690} = 6.26 \text{ ft } [1.91 \text{ m}]$$

which is slightly less than the kern limit of $D/6$ ($41.7/6 = 6.95$ ft); thus the maxi-

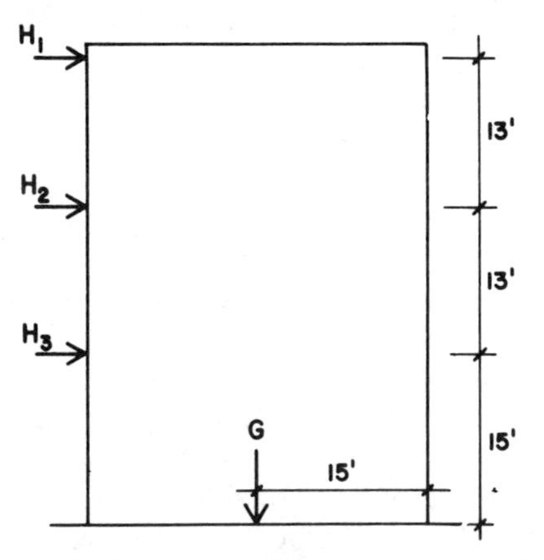

FIGURE 14.9. Stability of the 30-ft wall.

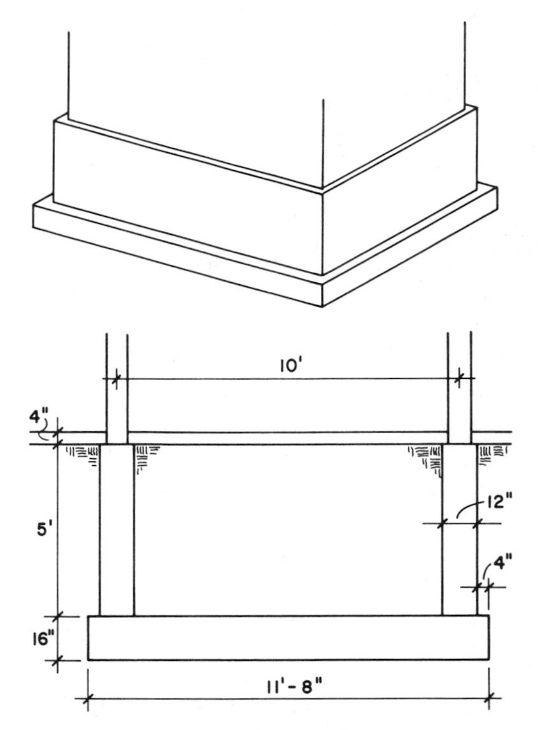

FIGURE 14.10. Foundation for the stair tower.

mum soil stress due to dead load plus seismic load will be approximately 3500 psf. This should not be a problem, unless the soil is highly compressible.

At the base of the tower the total lateral shear ($H_1 + H_2 + H_3$) is 145.5 kips. This will be shared by the two walls, although the shear stress will be slightly higher in the wall with the door openings. If we assume the 30-ft wall to be reduced to a net width of 22 ft by the openings and an 8-in. wall with 40% voided blocks, the shear analysis is:

$$\text{unit shear force: } \frac{1.5 \times 145{,}500}{2 \times 22}$$

$$= 4960 \text{ lb/ft [72.4 kN/m]}$$

$$\text{unit stress: } f_v = \frac{4960}{7.5 \times 12 \times 0.60}$$

$$= 92 \text{ psi [845 kPa]}$$

This is a bit high, but not beyond the capacity of the reinforced masonry wall. (Note the increase of shear force by 50% per UBC 2407(h)4F(i).)

There are, of course, many additional elements that must be designed for a complete design of the structure for Building Three. We intend, however, to illustrate only some of the major planning issues and the design of some of the major structural elements. Some additional concerns are discussed in the next chapter in conjunction with the development of the construction drawings.

An alternative structure for Building Three is presented in Chapter 16. Some additional issues regarding structures for multistory buildings are discussed in the chapters that present the design of Building Four.

CHAPTER FIFTEEN

Construction Drawings for the Masonry Wall Structure

Construction of this building is subject to a wide range of variations in terms of the selection of materials, structural components, and the specific details for fabrication and assembly. A complete presentation of all the necessary details for construction would be quite exhaustive, and therefore we will limit the discussion to the general cases and the most typical details.

15.1. CONSTRUCTION ILLUSTRATED IN PREVIOUS CHAPTERS

Many construction drawings shown in the illustrations in previous chapters are applicable to the situations in this building. The following is a description of some of these drawings:

1. Illustrations for the roof construction of Building Two are shown in Chapters 7 and 11. The wood construction in Chapter 7 and the steel framing in Chapter 11 are essentially the same as might be utilized for this building. Except for the possible requirements for more stringent fire resistance, the roof of a single-story building is not essentially different from that for a multistory building. The details shown for Building Two are thus in general applicable to the similar situations that occur in Building Three.

2. Illustrations for the masonry wall construction in Building Two are shown in Chapters 9, 10, and 11. Many of the general situations described for the development of the reinforced concrete masonry construction apply equally to the construction for Building Three. Some additional details for the exterior walls are presented in Section 15.2.

3. Illustrations for the design work are included in Chapter 14. The following drawings show some of the details for the construction as they were presented and discussed in connection with the development of the design.

Figure 14.1 shows the framing plan for the typical upper floor. Framing of the roof would probably be similar, although different sizes and spacing of components may be used.

In Figure 14.2 is an alternate framing layout for a system using beams, purlins, and joists. This is most likely to be used for the roof structure if wood construction is selected.

4. Figure 14.3 shows a section through a typical interior beam carrying wood joists. If trussed joists of either wood or steel are used, this detail would be modified to permit the seating of the extended ends of the truss top chords on the top of the carrying beam.

5. Figure 14.4 demonstrates the general arrangement for the stacking of the columns and beams at the building core. This detail is recommended only if the continuity of the beams is considered to be a top priority in the design.

6. Figure 14.10 shows the general form of the base of the tower structure that encloses the stairs and rest rooms in the core.

15.2. DETAILS OF THE EXTERIOR WALLS

The following drawings illustrate some of the possibilities for the construction of the exterior walls.

Figure 15.1 shows the situation at the joint between the roof and the exterior wall. For the transfer of both gravity and lateral forces, a steel channel section is bolted to the face of the masonry wall to form a ledger. Gravity loads are transferred to the ledger through the metal hangers that support the ends of the joists. If trussed joists are used, the detail is modified to permit the seating of the extended top chords of the joists on top of the ledger. Lateral forces are transferred from the roof deck by nailing the deck edges to the wood member that is bolted to the top of the channel. Waterproofing of the roof-to-wall joint and the back of the parapet and forming of the parapet top are subject to variation in detail. Figure 15.1 shows the use of a multilayered, hot-mopped roofing that is applied continuously over the roof surface, up the 45° cant at the wall joint and up the back of the parapet wall to its top, where it is sealed by the one-piece metal top. No provision is indicated for drainage of the roof, although it is, of course, a major concern in the general development of the roof layout and detailing.

Also shown in this drawing are the forming for the ceiling and the inside surface of the wall, which indicate the use of wood framing with gypsum drywall surfacing. If fire codes do not permit the use of wood, the construction may be developed in a similar fashion using framing elements of light gauge steel. The wall cavity created by the studs may be used for installation of insulation materials.

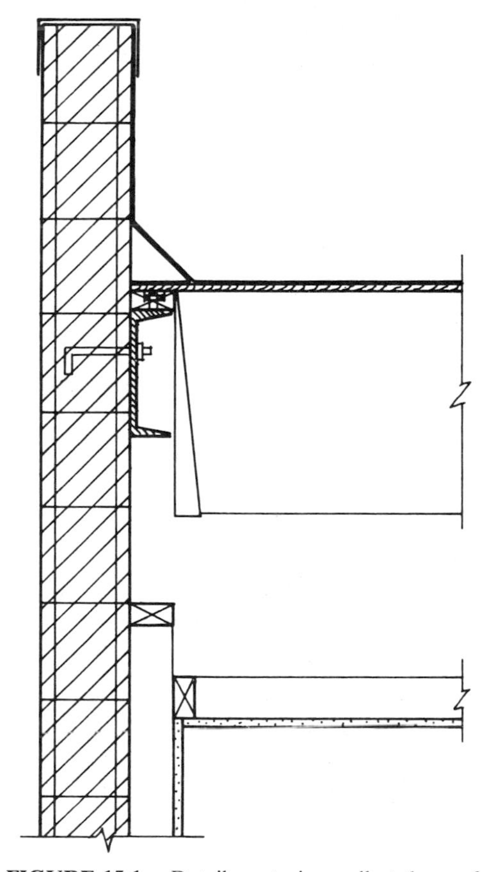

FIGURE 15.1. Detail—exterior wall at the roof.

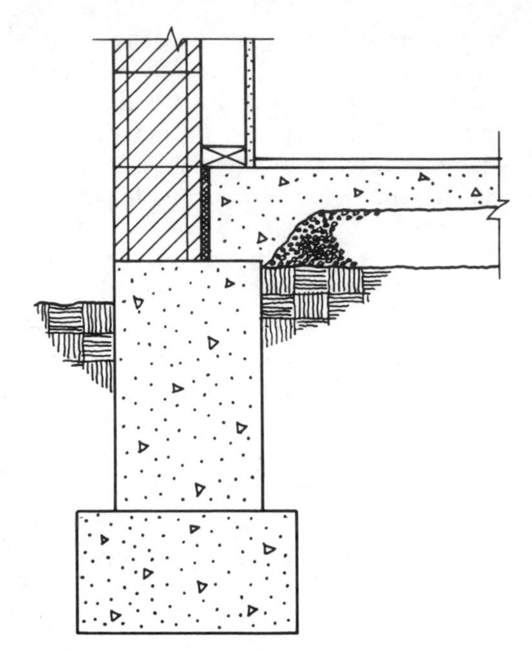

FIGURE 15.2. Detail—exterior wall at the first floor.

Figure 15.2 shows the situation at the base of the exterior wall, indicating the use of a poured concrete slab floor for the first story and a simple, shallow-bearing footing and short grade wall. Again, many variations are possible for this detail, depending on the soil conditions, the need for frost protection, provision for various wall and floor finishes, extent of thermal protection required, and so on. The situations discussed for the isolated piers and their footings in Building Two may also be applied here. The considerable overturning moments generated by the lateral forces will probably require some significant bending and shear resistance from the grade wall, and may require that it have a greater height than that required for soil-bearing considerations alone. The latter situation is quite similar to that which occurs for

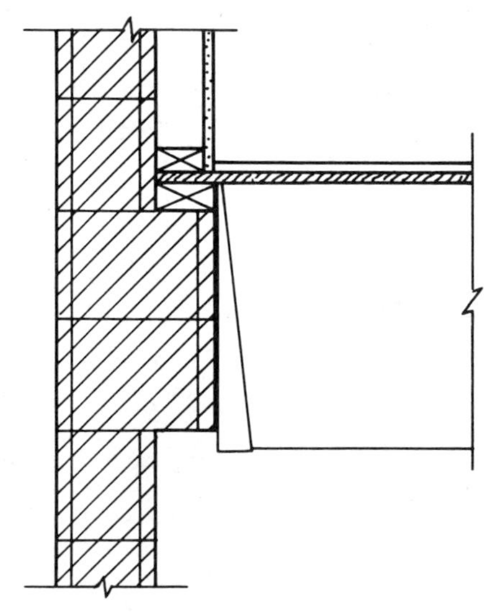

FIGURE 15.3. Support of the floor structure.

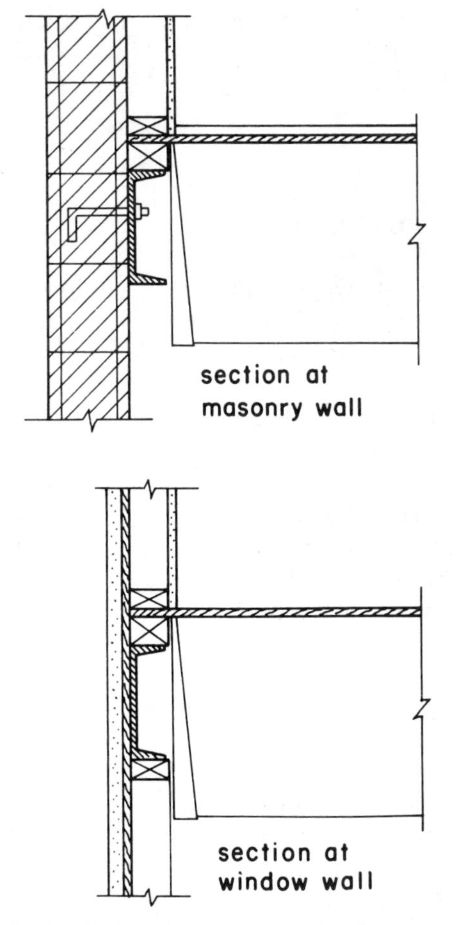

FIGURE 15.4. Alternate details for the support of the floor structure.

the concrete frame structure discussed in the next chapter, to which the reader is referred.

Figure 15.3 demonstrates a possible detail for the joint that occurs at the intersection of the upper floors and the exterior wall. In this case a ledger is formed for the support of the floor joists by widening the masonry wall. This widened member would be filled with concrete and reinforced to form a continuous beam within the wall. The joists and deck are then attached in a manner not essentially different from that shown for the roof structure in Figure 15.1.

For purposes of resisting both vertical and lateral loads, the masonry piers in the exterior walls would consist only of the solid wall portions between the windows, as shown in the plan in Figures 13.2b and 14.1. However, at the level of the floor construction there will also be solid portions of the exterior wall. The details for construction of this spandrel strip between the masonry piers must be developed as part of the general design of the building exterior. One possibility is that the entire exterior wall be made of masonry, and the detail between the windows be the same as that at the piers, as in Figure 15.3. In this case, the masonry beam in the drawing would be the element that spans between the piers, carrying the wall above as well as the edge of the floor.

Figure 15.4 shows another alternative for the exterior wall in which the masonry construction is limited to the piers and a lighter, stucco-covered wall between the piers. In this example the ledger at the piers consists of a steel channel bolted to the wall face, as was used at the roof in Figure 15.1. As shown in the lower drawing, the steel channel is then used to span across the space between the piers to provide support for the wall and the edge of the floor.

If the soil and general site conditions do not favor the use of the concrete slab on grade as the structure for the first floor, it may be necessary to use a concrete-framed system at this location. Details for such a system are shown in the construction drawings for Building Four.

CHAPTER SIXTEEN

Design of the Concrete Structure

A structural framing plan for the upper floors in Building Three is presented in Figure 16.1, where the use of a poured-in-place slab and beam system of reinforced concrete is indicated. Support for the spanning structure is provided by concrete columns. The system for lateral load resistance is that shown in Figure 13.2d, which utilizes the exterior columns and spandrel beams as rigid frame bents. This is a highly indeterminate structure for both gravity and lateral force design, and its precise engineering design would undoubtedly be done with a computer-aided system. We will discuss the major design considerations and illustrate the use of some simplified techniques for an approximate analysis and design of the structure.

16.1. DESIGN OF THE SLAB AND BEAM FLOOR STRUCTURE

As shown in Figure 16.1, the basic floor framing system consists of a series of beams at 10-ft centers that support a continuous, one-way spanning slab and are supported by column line girders or directly by the columns. We will discuss the design of three elements of this system: the continuous slab, the four-span beam, and the three-span spandrel girder.

The design conditions for slab, beam, and girder are indicated in Figure 16.2. Shown on the diagrams are the positive and negative moment coefficients as given in Chapter 8 of the ACI Code (Ref. 5). Use of these coefficients is quite reasonable for the design of the slab and beam. For the girder, however, the presence of the concentrated loads makes the use of the coefficients improper according to the ACI Code. But for an approximate design of the girder, their use will produce some reasonable results.

Figure 16.3 shows a section of the exterior wall that demonstrates the general nature of the construction. On the basis of the criteria given in Chapter 13 and the construction shown, we determine the loadings for the slab design as follows:

Floor live load:
  100 psf (at the corridor) [4.79 kPa]
Floor dead load:
  Carpet and pad at 5 psf
  Ceiling, lights, and ducts at 15 psf
  2-in. lightweight concrete fill at 18 psf
  Assumed 5-in. thick slab at 62 psf
    Total dead load: 100 psf [4.79 kPa]

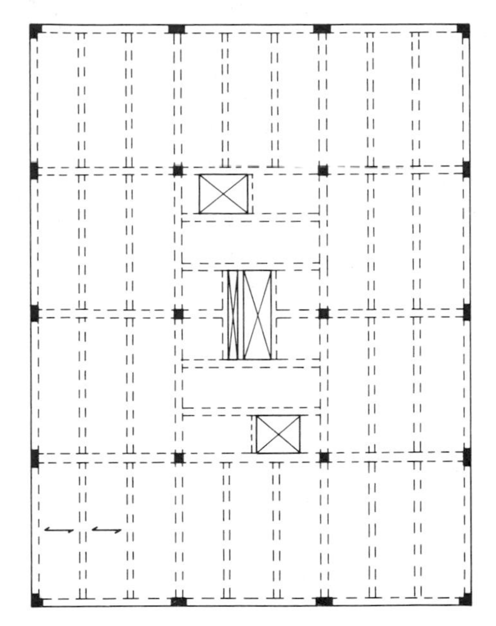

FIGURE 16.1. Structural plan—upper floor with the concrete slab and beam system.

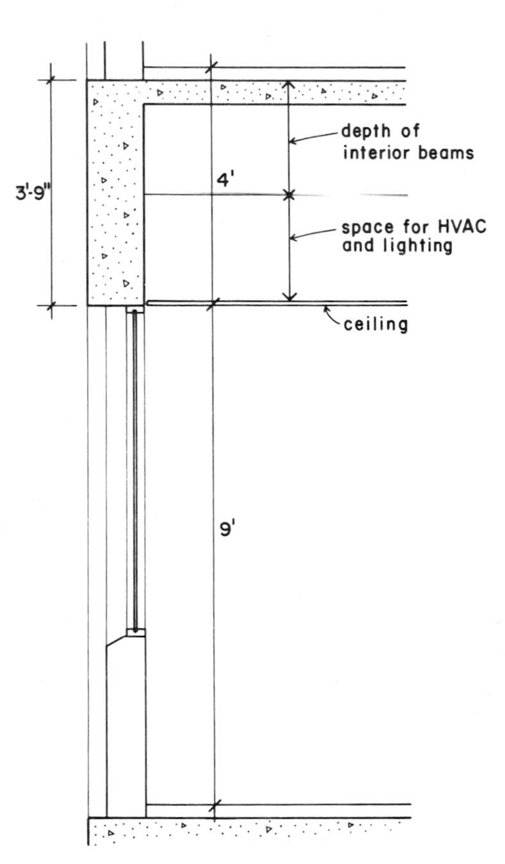

FIGURE 16.3. Typical exterior wall at the upper floor.

Bending moments in the slab will be determined as

$$M = CwL^2 = C(200)(9)^2$$
$$= C(16{,}200) \quad \text{in ft-lb}$$

In this determination C is the moment coefficient shown in Figure 16.2 for the appropriate location in the slab. The unit load w is for a 1-ft wide strip of slab; L is the clear span in feet assuming a 12-in.-wide beam. We have chosen the 5-in. slab for an estimate primarily because this is probably the minimum thicknesses that will provide the necessary fire resistance to satisfy the building code. Assuming $\frac{3}{4}$-in. cover for the reinforcing, the effective depth will be approximately 4 in. (with a No. 4 bar). Using the working stress method, the area of steel required per foot of slab width will thus be

$$A_s = \frac{M}{f_s \times jd} = \frac{C(16{,}200)(12)}{24{,}000(0.9)(4)} = 2.25\ C$$

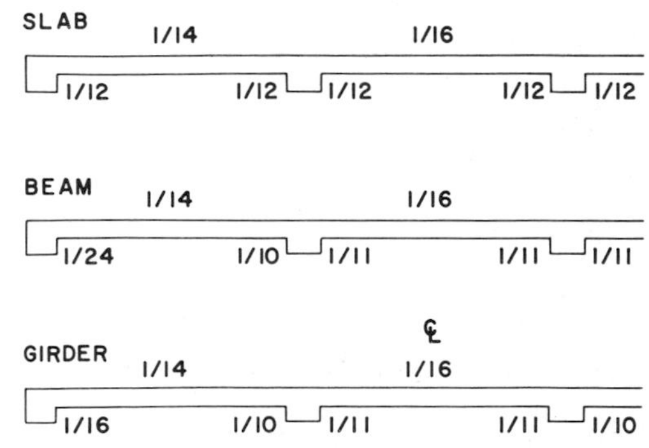

FIGURE 16.2. Moment coefficients for the approximate analysis.

The slab design is illustrated in Figure 16.4. Various possible choices are shown

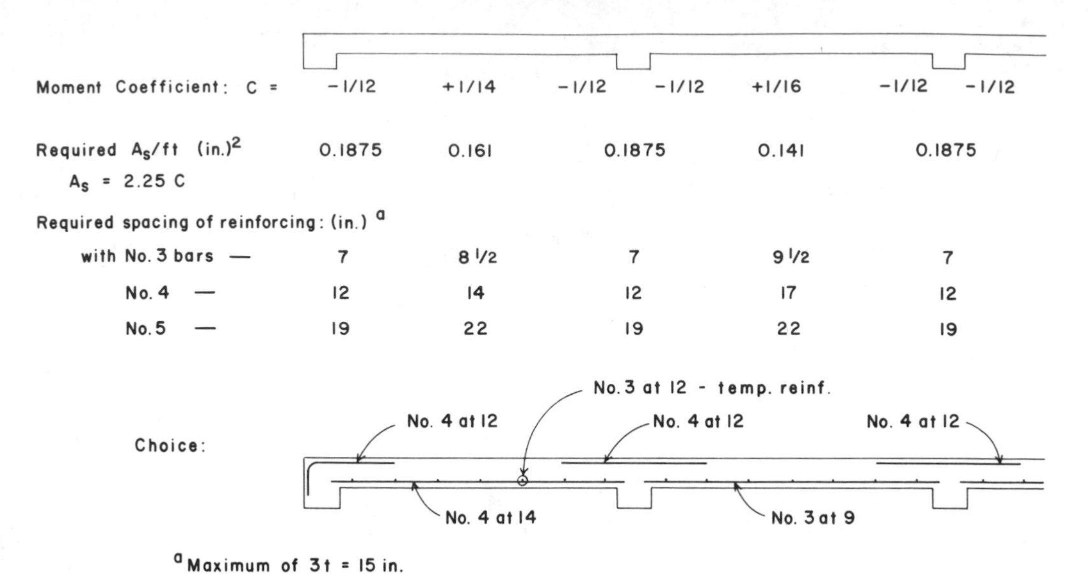

FIGURE 16.4. Design of the continuous slab.

for the reinforcing. The most common grade of reinforcing would be that with a yield strength of 60 ksi, which is what we have used for the design. We have thus far not indicated the concrete strength, but will use a fairly low grade with f'_c of 3000 psi.

Flexural and shear stresses in the concrete should be investigated, although the 5-in. slab will be found to be more than adequate for this span and load. It is also well above the recommended minimum thickness for deflection control in the ACI Code.

Inspection of the framing plan in Figure 16.1 reveals that there is a large number of different beams in the structure for the floor with regard to individual loadings and span conditions. Two general types are the beams that carry only uniformly distributed loads as opposed to those that also provide some support for other beams; the latter produce a load condition consisting of a combination of concentrated and distributed loading. We now consider the design of one of the uniformly loaded beams.

The beam that occurs most often in the plan is the one that carries a 10-ft-wide strip of the slab as a uniformly distributed loading, spanning between columns or supporting beams that are 30 ft on center. Assuming the supports to be approximately 12 in. wide, the beam has a clear span of 29 ft and a total load periphery of $29 \times 10 = 290$ ft². Using the UBC provisions for reduction of live load,

$$R = 0.08(A - 150)$$

$$= 0.08(290 - 150) = 11.2\%$$

We round this off to a 10% reduction, and, using the loads tabulated previously for the design of the slab, determine the beam loading as follows:

Live load per foot of beam span (with 10% reduction):

$$0.90 \times 100 \times 10$$

$$= 900 \text{ lb/ft} \quad \text{or} \quad 0.90 \text{ k/ft } [13.1 \text{ kN/m}]$$

Slab and superimposed dead load:

$$100 \times 10$$

$$= 1000 \text{ lb/ft} \quad \text{or} \quad 1.0 \text{ k/ft } [14.6 \text{ kN/m}]$$

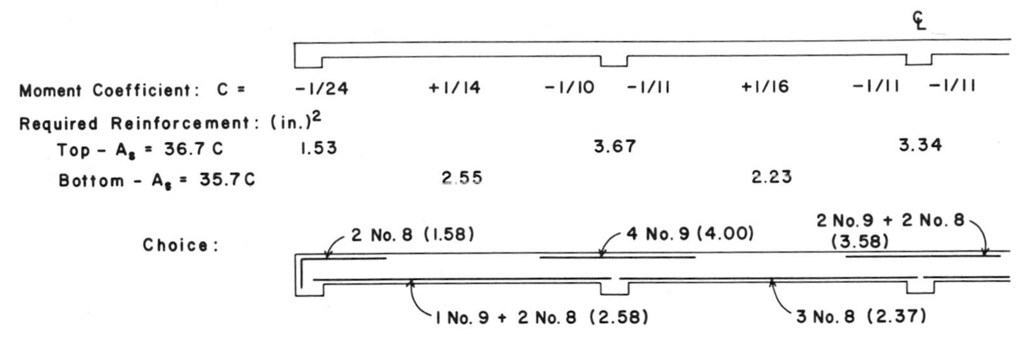

FIGURE 16.5. Design of the four-span beam.

The beam stem weight, estimating a size of 12 by 20 in. for the beam stem extending below the slab, is

$$\frac{12 \times 20}{144} \times 150 \text{ lb/ft}^3$$

$$= 250 \text{ lb} \quad \text{or} \quad 0.25 \text{ k/ft [3.65 kN/m]}$$

The total uniformly distributed load is thus

$$0.90 + 1.0 + 0.25$$

$$= 2.15 \text{ k/ft [31.35 kN/m]}$$

Let us now consider the design of the four-span continuous beam that occurs in the bays on the north and south sides of the building and is supported by the north–south spanning column-line beams that we will refer to as the girders. The approximation factors for design moments for this beam are given in Figure 16.2, and a summary of the design is presented in Figure 16.5. Note that the design provides for tension reinforcing only, thus indicating that the beam dimensions are adequate to prevent a critical condition with regard to flexural stress in the concrete. Using the working stress method, the basis for this is as follows.

Maximum bending moment in the beam is

$$M = \frac{wL^2}{10}$$

$$= \frac{(2.15)(29)^2}{10}$$

$$= 181 \text{ k-ft [245 kN-m]}$$

Then, for a balanced section, using factors from Table A.7, Appendix A,

required bd^2

$$= \frac{M}{R} = \frac{181 \times 12}{0.204}$$

$$= 10{,}647 \text{ in.}^3 \ [175 \times 10^6 \text{ mm}^3]$$

If $b = 12$ in.,

$$d = \sqrt{\frac{10{,}647}{12}} = 29.8 \text{ in. [757 mm]}$$

With minimum concrete cover of 1.5 in. on the bars, No. 3 U-stirrups, and moderate-sized flexural reinforcing, this d can be approximately attained with an overall depth of 32 in. This produces a beam stem that extends 27 in. below the slab, and is thus slightly heavier than that assumed previously. Based on this size, we will increase the design load to 2.25 k/ft for the subsequent work.

Before proceeding with the design of the flexural reinforcing, it is best to investigate the situation with regard to shear to make sure that the beam dimensions are adequate. Using the approximations given in Chapter 8 of the ACI Code, the maximum shear is considered to be 15% more than the simple span shear and to occur at the inside end of the exterior spans. We thus consider the following.

The maximum design shear force is

$$V = 1.15 \times \frac{wL}{2} = 1.15 \times \frac{2.25 \times 29}{2}$$

$$= 37.5 \text{ k [167 kN]}$$

For the critical shear stress this may be reduced by the shear between the support and the distance of d from the support; thus

$$\text{critical } V = 37.5 - \frac{29}{12} \times 2.25$$

$$= 32.1 \text{ k [143 kN]}$$

Using a d of 29 in., the critical shear stress is

$$v = \frac{V}{bd} = \frac{32{,}100}{29 \times 12} = 92 \text{ psi [634 kPa]}$$

With the concrete strength of 3000 psi, this results in an excess shear stress of 32 psi that must be accounted for by the stirrups. The closest stirrup spacing would thus be

$$s = \frac{A_v f_s}{v'b} = \frac{0.22 \times 24{,}000}{32 \times 12}$$

$$= 13.75 \text{ in. [348 mm]}$$

Because this results in quite a modest amount of shear reinforcing, the section may be considered to be adequate.

For the approximate design shown in Figure 16.5, the required area of tension reinforcing at each section is determined as

$$A_s = \frac{M}{f_s jd} = \frac{C \times 2.25 \times (29)^2 \times 12}{24 \times 0.89 \times 29}$$

$$= 36.7C$$

Based on the various assumptions and the computations we assume the beam section to be as shown in Figure 16.6. For the beams the flexural reinforcing in the top that is required at the supports must pass either over or under the bars in the tops of the girders. Because the girders will carry heavier loadings, it is probably wise to give the girder bars the favored position (nearer the outside for greater value of d) and thus to assume the positions as indicated in Figure 16.6.

At the beam midspans the maximum positive moments will be resisted by the combined beam and slab section acting as a T-section. For this condition we assume an approximate internal moment arm of $d - t/2$ and may approximate the required steel areas as

$$A_s = \frac{M}{f_s(d - t/2)}$$

$$= \frac{C \times 2.25 \times (29)^2 \times 12}{24 \times (29 - 2.5)} = 35.7C$$

The beams that occur on the column lines are involved in the lateral force resistance actions and are discussed in Section 16.3.

Inspection of the framing plan in Figure 16.1 reveals that the girders on the north–south column lines carry the ends of the beams as concentrated loads at the third points of the girder spans. Let us consider the spandrel girder that occurs at the east and west sides of the building. This member carries the outer ends of the first beams in the four span rows and

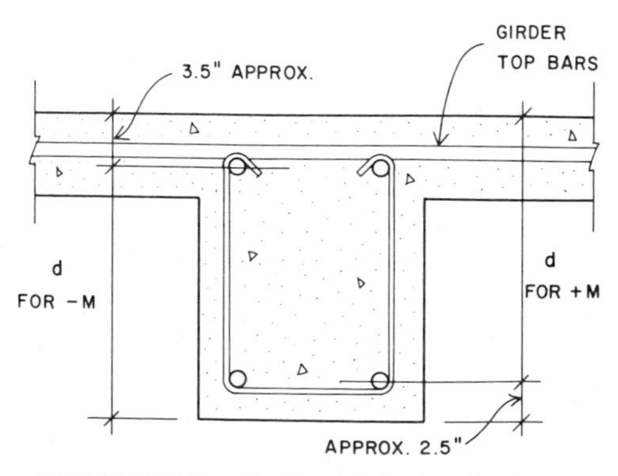

FIGURE 16.6. Section of the interior beam.

in addition carries a uniformly distributed load consisting of its own weight and that of the supported exterior wall. The form of the girder and the wall was shown in Figure 16.3. From the framing plan note that the exterior columns are widened in the plane of the wall. This is done to develop the peripheral bent system, as will be discussed later.

For the spandrel girder we determine the following:

Assumed clear span: 28 ft [8.53 m].

Floor load periphery based on the carrying of two beams and half the beam span load, is

$$15 \times 20 = 300 \text{ ft}^2 \; [27.9 \text{ m}^2]$$

Note: This is approximately the same total load area as that carried by a single beam, so we will use the live load reduction of 10% as determined for the beam.

Loading from the beams:

Dead load: 1.35 k/ft × 15 ft
= 20.35 k

Live load: 0.90 k/ft × 15 ft
= 13.50 k

Total 33.85 k say 34 k [151 kN]

Uniformly distributed load:

$$\text{Spandrel beam weight: } \frac{12 \times 45}{144 \times 150}$$

= 560 lb/ft

Wall assumed at 25 psf: 25 × 9
= 225 lb/ft

Total
= 785 lb/ft say 0.8 k/ft [11.7 kN/m]

For the uniformly distributed load approximate design moments may be found using the moment coefficients as was done for the slab and beam. Values for this procedure are given in Figure 16.2. The ACI Code does not permit the use of this procedure for concentrated loads, but we may adapt some values for an approximate design using moments for a beam with the third point loading. Values of positive and negative moments for the third point loading may be obtained from various references, including Refs. 7, 11, and 12.

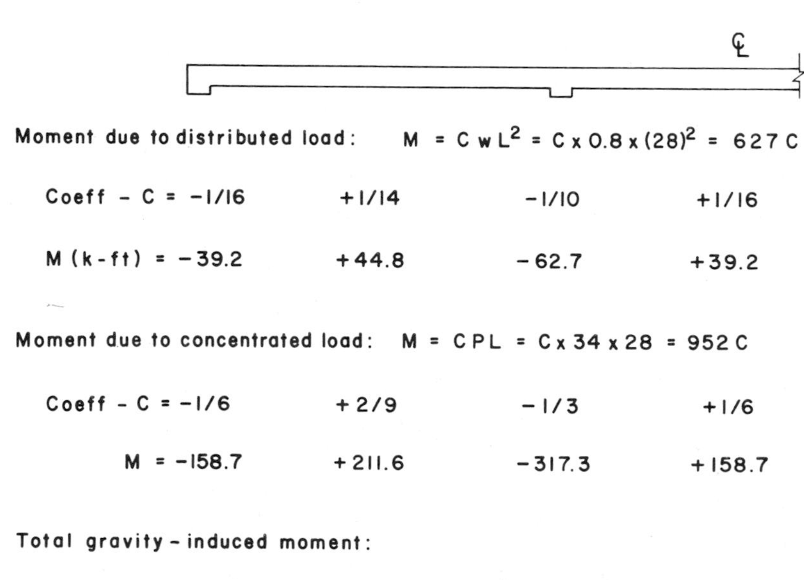

FIGURE 16.7. Gravity moments for the girder.

Figure 16.7 presents a summary of the work for determining the design moments for the spandrel girder under gravity loading. Moment values are determined separately for the two types of load and then added for the total design moment.

We will not proceed further with the girder design at this point, for the effects of lateral loading must also be considered. The moments determined here for the gravity loading will be combined with those from the lateral loading in the discussion in Section 16.3.

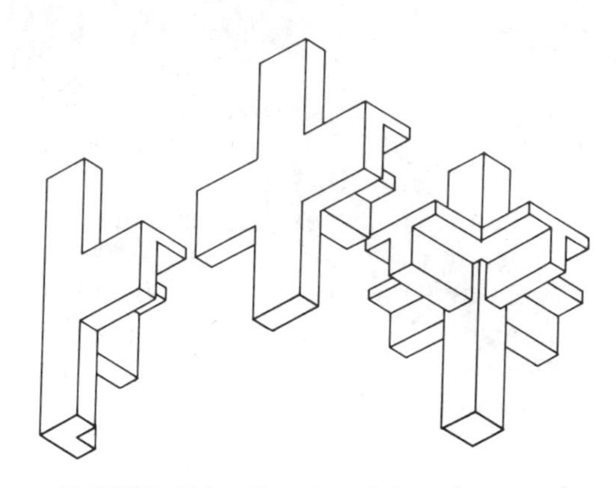

FIGURE 16.8. Framing of the columns and beams.

TABLE 16.1. Column Axial Loads due to Gravity (kips)

		Corner Column 225 ft²			Intermediate Exterior Column 450 ft²			Interior Column 900 ft²		
Level	Load Source	DL	LL	Total	DL	LL	Total	DL	LL	Total
PR	Roof							8	5	
	Wall							5		
	Column									
	Total/level							13	5	
	LL reduction									
	Design load									18
R	Roof	32	5		63	9		126	18	
	Wall	10			10			10		
	Column	8			8			8		
	Total/level	50	5		81	9		157	23	
	LL reduction									
	Design load			55			90			180
3	Floor	32	17		63	34		126	68	
	Wall	10			10			10		
	Column	8			8			8		
	Total/level	100	22		162	43		301	91	
	LL reduction	24%	17		60%	17		60%	36	
	Design load			117			179			337
2	Floor	32	17		63	34		126	68	
	Wall	11			11			11		
	Column	10			10			10		
	Total/level	153	39		246	77		448	159	
	LL reduction	42%	23		60%	31		60%	64	
	Design load			176			277			512

16.2. DESIGN OF THE CONCRETE COLUMNS

The four general cases for the columns are (see Fig. 16.8):

The interior column carrying primarily only axial gravity loads.

The intermediate exterior columns on the north and south sides carrying the ends of the interior girders and functioning as members of the peripheral bents for lateral resistance.

The intermediate exterior columns on the east and west sides carrying the ends of the column line beams and functioning as members of the peripheral bents.

The corner columns carrying the ends of the spandrel beams and functioning as the end members in both peripheral bents.

A summation of the gravity loads for the columns is given in Table 16.1. In the table we have assumed the existence of a penthouse structure of light metal construction. For the roof we have assumed a dead load equal to that for the floors and a live load of 20 psf. Wall loads for the exterior columns are based on the construction shown in Figure 16.3. Live load reduction is based on the UBC method as illustrated previously for the design of horizontal framing.

For the interior column it is assumed that a single square size will be used for the three-story-high column. The size would be one that works with minimum reinforcing for the low load at the top story and within the limit for maximum reinforcing for the load at the bottom story. It is necessary, of course, to develop the layout of this column so that it fits into the walls of the building core. The latter may make some shape other than a square more desirable; we have nevertheless assumed a square shape and present a possible design as shown in Figure 16.9. The design is based on the

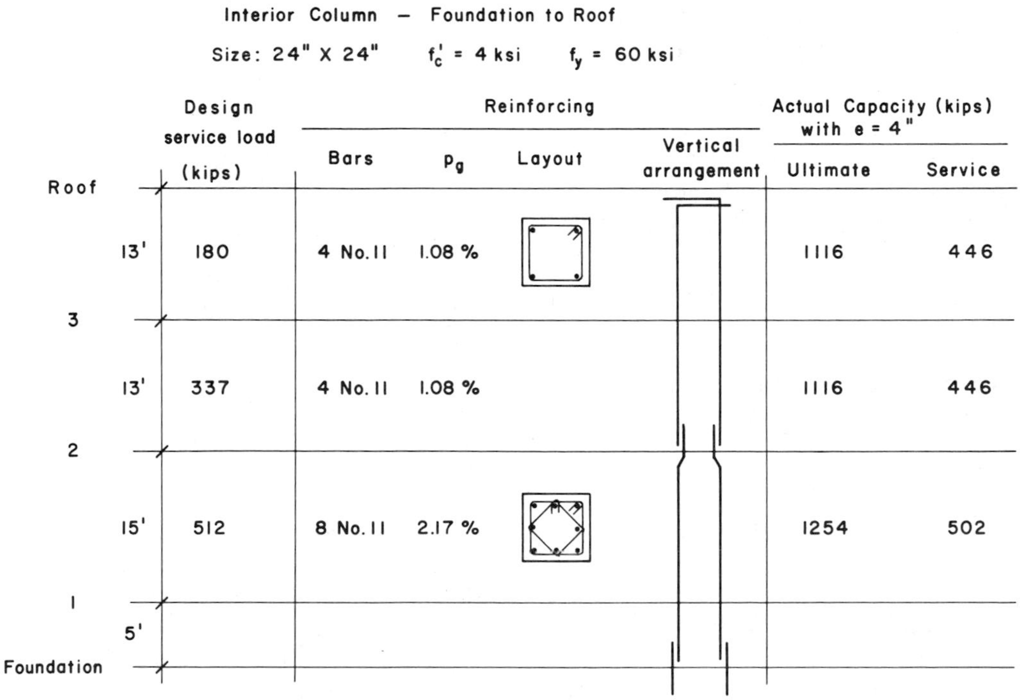

Interior Column — Foundation to Roof

Size: 24" X 24" f'_c = 4 ksi f_y = 60 ksi

Level	Story height	Design service load (kips)	Reinforcing: Bars	p_g	Layout	Vertical arrangement	Actual Capacity (kips) with e = 4": Ultimate	Service
Roof								
	13'	180	4 No. 11	1.08 %			1116	446
3								
	13'	337	4 No. 11	1.08 %			1116	446
2								
	15'	512	8 No. 11	2.17 %			1254	502
1								
	5'							
Foundation								

FIGURE 16.9. Design of the interior column.

use of concrete with f'_c of 4000 psi and Grade 60 reinforcing. Load capacities have been derived from the load tables in the *CRSI Handbook* (Ref. 16) using 40% of the capacities as determined by strength design; the procedure required for the working stress method is described as the Alternate Design Method in Appendix B of the ACI Code (Ref. 5). A minimum eccentricity of 15% of the column dimension has been assumed to allow for some bending caused by the floor and roof framing.

For the intermediate exterior column there are four actions to consider:

1. The vertical compression induced by gravity, as determined in Table 16.1.
2. Bending moment induced by the interior framing that intersects the wall column; the columns are what provides the end moments shown in Figures 16.4 and 16.5.
3. Bending moments in the plane of the wall induced by unbalanced conditions in the spandrel beams and girders.
4. Bending moments induced by the actions of the peripheral bents in resisting lateral loads.

For the corner column the situation is similar to that for the intermediate exterior column, that is, bending on both axes. The forms of the exterior columns as shown on the plan in Figure 16.1 have been established in anticipation of the major effects described. Further discussion of these columns will be deferred, however, until after we have investigated the situations of lateral loading.

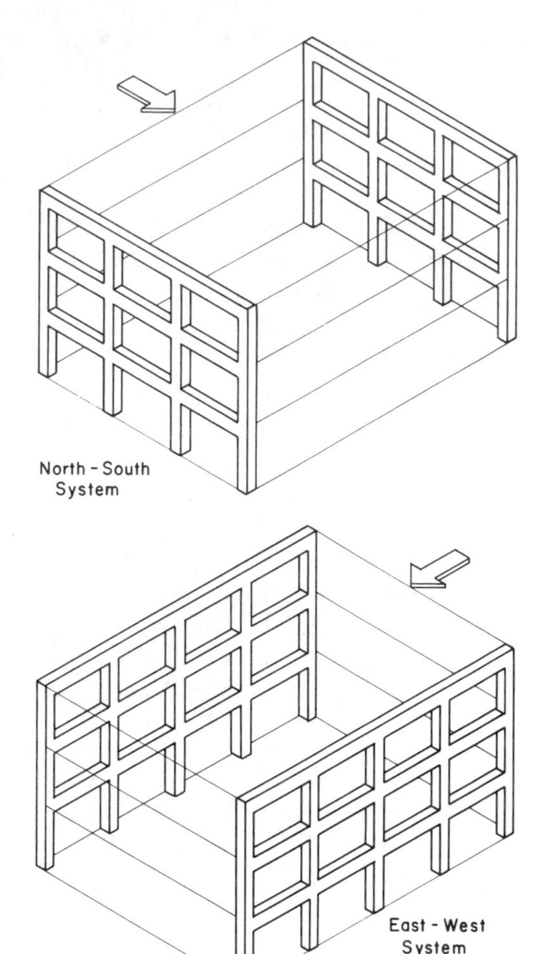

FIGURE 16.10. The peripheral bent system.

16.3. DESIGN FOR LATERAL FORCES

The lateral force resisting systems for the concrete structure are shown in Figure 16.10. For force in the east–west direction the resistive system consists of the horizontal roof and floor slabs and the exterior bents (columns and spandrel beams) on the north and south sides. For force in the north–south direction the system utilizes the bents on the east and west sides.

If the lateral load is the same in both directions, the stress in the slab (shear in the horizontal diaphragm) is critical for the north–south loading because the slab has less width for resistance to this loading. The loads are not equal, however. Wind force will be greater in the north and south directions because the building has a greater profile in this direction. This makes it even more obvious that this will be the loading critical for the slab in design for wind. However, for seismic load, a true dynamic analysis reveals that the load effect is greater in the east–west direction because the resistive bents are slightly stiffer in this direction. In any event, it is unlikely that the 5-in.-thick slab with properly anchored edge reinforcing at the spandrels will be critically stressed for any loading.

Our considerations for lateral load will be limited to the seismic loading in the north–south direction and to investigations of the effects on the columns and spandrel beams on the east and west sides. Both the wind and seismic loadings were determined for the masonry and steel structure in Chapter 14. The wind loading is the same for the concrete structure, but the seismic effects will be modified by the difference in building weight caused by the heavier concrete structure and the different value to be used for the K factor.

Review of the designs for the floor systems indicate that the total weight of the horizontal construction in the concrete building is more than twice that in the masonry and steel building. Assuming other elements of the construction to be approximately the same in both buildings, inspection of the weight tabulation in Table 14.2 indicates that the total weights are approximately 25% higher for the concrete building. Thus, for use in the UBC formula for seismic load, the values for W should be increased by 25%.

On the other hand, the UBC gives a value of only 0.67 for K for the building with a rigid frame system. This is one half of that for the shear wall (box system). Thus a 50% reduction is effected. Note that the K of 0.67 is to be used only for a building with a "ductile moment-

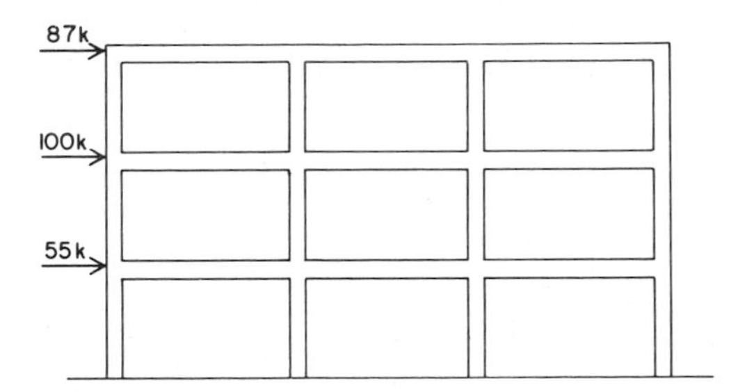

FIGURE 16.11. Loading of the north–south bent.

resisting space frame" that requires a great number of special considerations in both the design and the construction detailing for the concrete frame.

With the approximate adjustments for weight and different bracing, we may use the values determined in Table 14.3 with an adjustment of $1.25 \times 0.5 = 0.625$. Figure 16.11 shows the typical total loading in the north–south direction. As there are two bents, the values from Fig. 16.11 are divided by two for the design of the bent. Note that even with the reductions the seismic loading is still higher than the wind loading determined in Chapter 14.

For an approximate analysis we consider the individual stories of the bent to behave as shown in Figure 16.12, with the columns developing an inflection point at their midheight points. Because the columns all move the same distance, the shear load in a single column may be assumed to be equal to the cantilever deflecting load and the individual shears to be proportionate to the stiffnesses of the

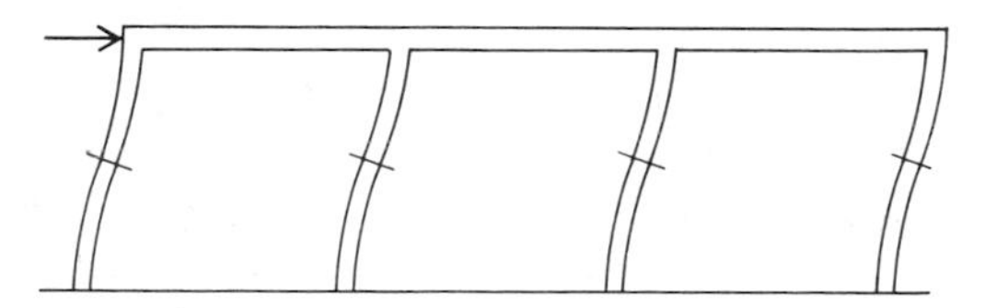

FIGURE 16.12. Deformation of the bent columns.

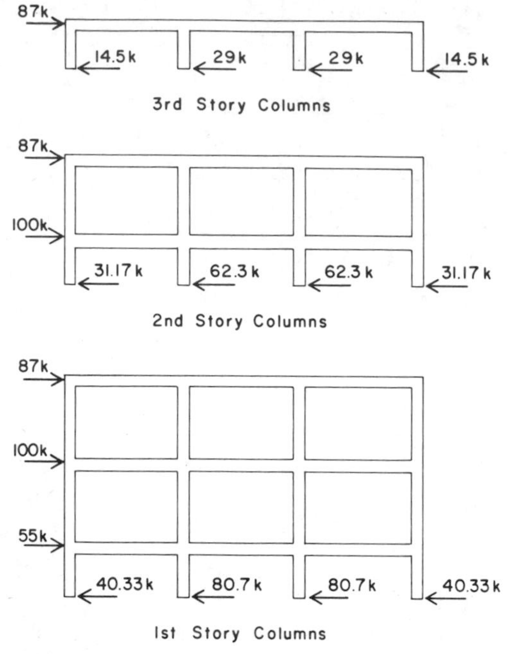

FIGURE 16.13. Analysis for the column shears.

columns. If the columns are all of equal stiffness in this case, the total load would be simply divided by four. However, the end columns are slightly less restrained as there is a beam on only one side. We will assume the net stiffness of the end columns to be one half that of the interior columns. Thus the shear force in the end columns will be one sixth of the load and that in the interior columns one third of the load. The column shears for each of the three stories is thus as shown in Figure 16.13.

The column shear forces produce moments in the columns. With the column inflection points assumed at midheight, the moment produced by a single shear force is simply the product of the force and half the column height. These moments must be resisted by the end moments in the rigidly attached beams, and the actions are as shown in Figure 16.14. These effects due to the lateral loads may now be combined with the previously determined effects of gravity loads for an approximate design of the columns and beams.

For the columns, we combine the axial compression forces with any gravity-induced moments and first determine that the load condition without lateral effects is not critical. We may then add the effects of the moments caused by lateral loading and investigate the combined loading condition, for which we may use the one third increase in allowable stress. Gravity-induced beam moments are taken from Figure 16.7 and are assumed to induce column moments as shown in Figure 16.15. The summary of design conditions for the corner and interior column is shown in Table 16.2. The design values for axial load and moment and approximate sizes and reinforcing are shown in Figure 16.16. Column sizes and reinforcing were obtained from the tables in the *CRSI Handbook* (Ref. 16) using concrete with f'_c = 4 ksi and Grade 60 reinforcing.

The spandrel beams (or girders) must be designed for the combined shears and moments due to gravity and lateral effects. Using the values for gravity-induced moments from Figure 16.7 and the values for lateral load moments from Figure 16.14, the combined moment conditions are shown in Figure 16.17. For design we must consider both the gravity only moment and the combined effect. For the combined effect we use three fourths of the total combined values to reflect the allowable stress increase of one third.

Figure 16.18 presents a summary of the design of the reinforcing for the spandrel beam at the third floor. If the construction that was shown in Figure 16.3 is retained with the exposed spandrel beams, the beam is quite deep. Its width should be approximately the same as that of the column, without producing too massive a section. The section

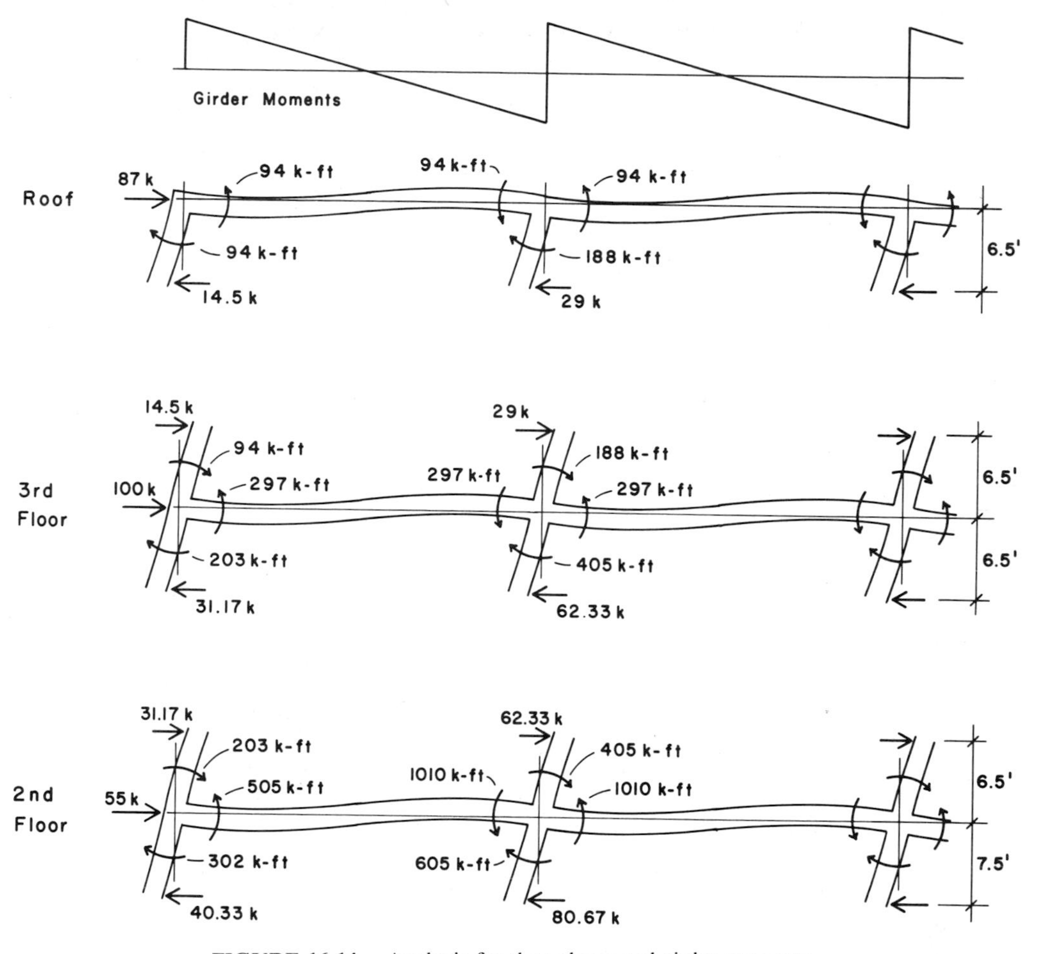

FIGURE 16.14. Analysis for the column and girder moments.

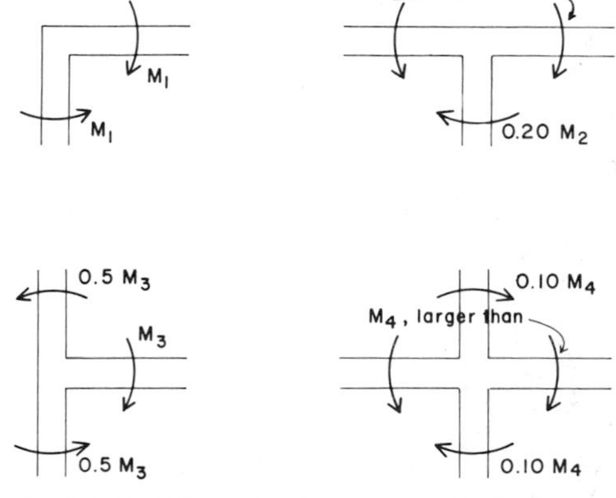

FIGURE 16.15. Gravity moments in the bents.

shown is probably adequate, but several additional considerations must be made as will be discussed later.

For computation of the required steel areas we assume an effective depth of approximately 40 in. and use

$$A_s = \frac{M}{f_s jd} = \frac{M \times 12}{(24)(0.9)(40)} = 0.0139\ M$$

Because the beam is so deep, it is advisable to use some longitudinal reinforcing at an intermediate height in the section—especially on the exposed face.

Shear design for the beams should

TABLE 16.2. Summary of Design Data for the Bent Columns

	Column	
	Intermediate	Corner
Axial gravity design load from Table 16.1 (kips)		
Third story	90	55
Second story	179	117
First story	277	176
Assumed gravity moment on bent axis (k-ft) from Figures 16.7 and 16.15		
Third story	60	120
Second story	40	100
First story	40	100
Moment from lateral force (k-ft) from Figure 16.14		
Third story	94	47
Second story	203	102
First story	302	151

Level	Intermediate Column					Corner Column				
	Axial Load (kips)	Moment (k-ft)	e (in.)	Column Dimensions (in.)	Reinforcing No. – Size	Axial Load (kips)	Moment (k-ft)	e (in.)	Column Dimensions (in.)	Reinforcing No. – Size
Roof–3	90 X 3/4 = 68	154 X 3/4 = 115	20	20 X 28	6 - 9	55 X 3/4 = 41	167 X 3/4 = 125	36	20 X 24	6 - 10
3–2	179 X 3/4 = 134	243 X 3/4 = 182	16	20 X 28	6 - 9	117 X 3/4 = 88	202 X 3/4 = 152	21	20 X 24	6 - 10
2–1	277 X 3/4 = 208	342 X 3/4 = 257	15	20 X 28	6 - 11	176 X 3/4 = 132	251 X 3/4 = 188	17	20 X 24	6 - 11

FIGURE 16.16. Design of the bent columns.

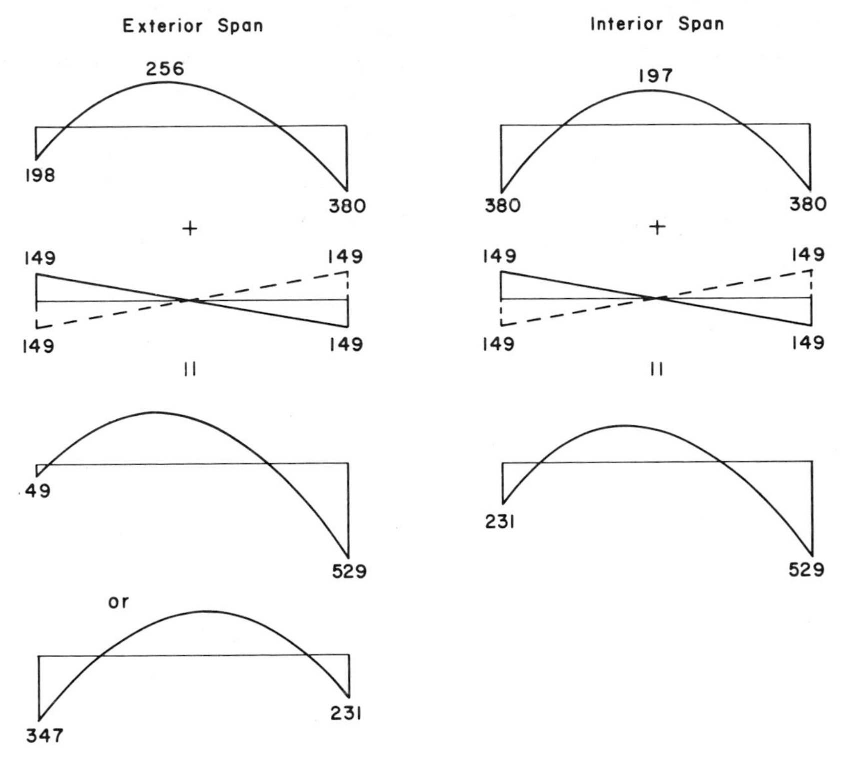

FIGURE 16.17. Combined moment for the girder.

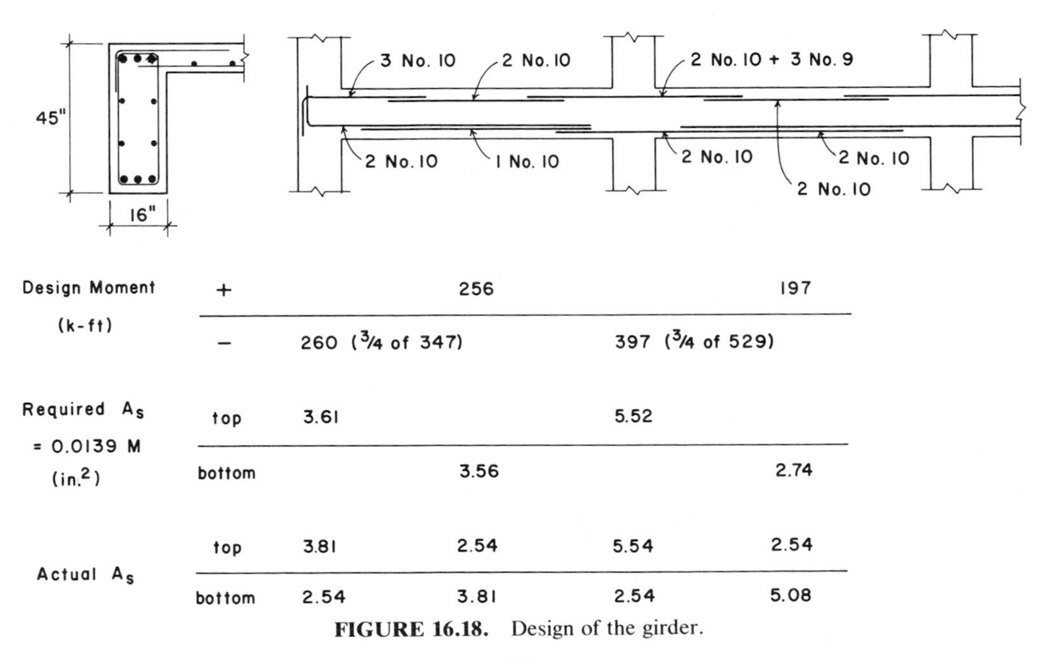

Design Moment (k-ft)	+		256		197
	−	260 (3/4 of 347)		397 (3/4 of 529)	
Required A_s = 0.0139 M ($in.^2$)	top	3.61		5.52	
	bottom		3.56		2.74
Actual A_s	top	3.81	2.54	5.54	2.54
	bottom	2.54	3.81	2.54	5.08

FIGURE 16.18. Design of the girder.

also be done for the combined loading effects. The closed tie form for the shear reinforcing—as shown in Figure 16.18—is used for considerations of torsion as well as the necessity for tying the compressive reinforcing.

With all of the approximations made, this should still be considered to be a very preliminary design for the beam. It should, however, be adequate for use in preliminary architectural studies and for sizing the members for a dynamic seismic analysis and a general analysis of the actions of the indeterminate structure.

CHAPTER SEVENTEEN

Construction Drawings for the Concrete Structure

Development of the construction drawings for this example is based on the use of an exposed concrete structure on the building exterior. It is also possible to use the same basic structural scheme with a completely covered structure. The principal difference would be in the form and detail of the exterior columns and the spandrel beams.

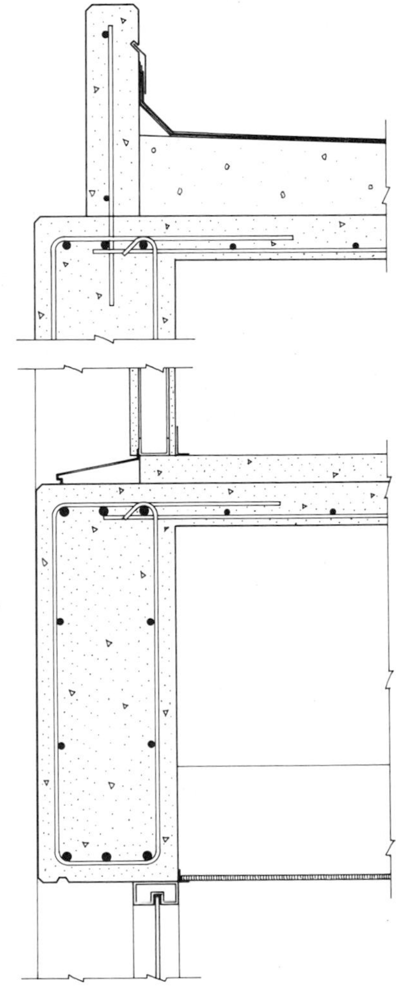

FIGURE 17.1. Details of the exposed concrete structure at the roof and upper floor.

17.1. CONSTRUCTION ILLUSTRATED IN OTHER CHAPTERS

Some of the drawings shown in the illustration of the work in other chapters are applicable to similar situations that may occur in this building. The following is a description of some of these drawings:

1. Figure 16.1 presents the general form of the framing system for the typical floor. This would also probably be used for the roof structure.
2. Figure 22.1 demonstrates the foundation plan for Building Four. The form of this construction would be similar for Building Three.
3. Figure 20.5 shows a section through a framed slab and beam system poured directly on fill. If it is not feasible to use a simple paving slab for the first floor (as has been shown for this build-

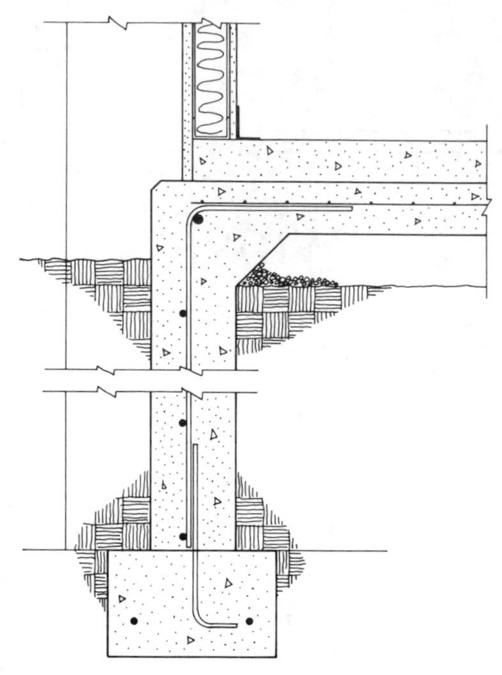

FIGURE 17.2. Details of the concrete structure at the first floor.

ing), this form of construction may be used for Building Three.

4. Figure 16.6 shows the form of the typical interior beam and its reinforcing.

5. Figure 22.8 shows a section through a beam at the edge of one of the large openings for a stair, elevator, or vertical duct.

If the concrete structure is not exposed at the building edges, the details illustrated for the steel structure—or any of a number of schemes—could be used for this building.

17.2. DETAILS OF THE EXTERIOR WALLS

The general form of the wall is shown in Figure 16.3. Enlarged details of the concrete structure at the roof and typical upper floor are presented in Figure 17.1. The detail at the roof illustrates the use of a short parapet wall of concrete construction and a lightweight concrete insulating fill with a conventional multilayered membrane roofing.

Figure 17.2 shows a section at the edge of the first floor. The floor paving slab is supported by a short grade wall that is carried down to the level of the top of the column footings. Problems of frost as well as the level at which the column footings must be placed would generally dictate the necessary height for the grade wall.

PART FOUR

BUILDING FOUR

Building Four is a multistory office building. Complete structural calculations for this building would be quite voluminous. We will therefore show only the design of some of the typical elements of the structural system. The building will be designed first as a steel frame and then as a reinforced concrete frame, with some options shown for each material.

CHAPTER EIGHTEEN

The Building and the Construction Alternatives

18.1. THE BUILDING

The basic form of the building is shown in Figures 18.1–18.3. As with Building Three, the intent is to have minimum permanent construction to allow for rearrangement of interior partitioning. This will be achieved by using a basic planning module, in this case 4 ft, within which the partitioning, modular ceiling system, and exterior wall mullions and columns will be coordinated. Partitioning may then be accomplished with various patented demountable wall systems, although masonry or plastered walls could also be used.

The building must be conceived as a vertical, superimposed stack of four separate plans: the basement, ground floor, typical office floor, and roof-penthouse. Location of columns, elevator shafts, stairs, duct shafts, and risers for the plumbing, power, and communications systems must be coordinated from level to level. The key plan is that of the typical office floor, since its functioning is the reason for the building's existence.

We assume the exterior wall to have the configuration shown in the section in Figure 18.3. This is a major architectural design concern and subject to considerable variation in terms of choice of materials, size and shape of windows, location of the columns and spandrel beams with respect to the wall, and so on. The wall is quite thick, permitting the total incorporation of the columns. This results in a smooth surface for the wall on both the inside and the outside. As we have stated, this is one of countless variations.

Distribution of wiring for power and communication will be accomplished through wiring conduits built into the floor structure. This system was widely used in the past, although alternatives now exist that include ceiling distribution, floor surface and wall surface distribution, and raised floors with a void space between the structure and the raised floor.

A peripheral hot water heating system will be incorporated in the exterior wall, as shown in the drawings. Ventilation, cooling, and supplementary heating will be achieved through a system incorporated in the ceiling space, using supply ducts from the major vertical risers in the core.

Major equipment for the HVAC, power, communication, and elevator systems will be housed in the roof penthouse and in the basement. The building management offices and equipment and

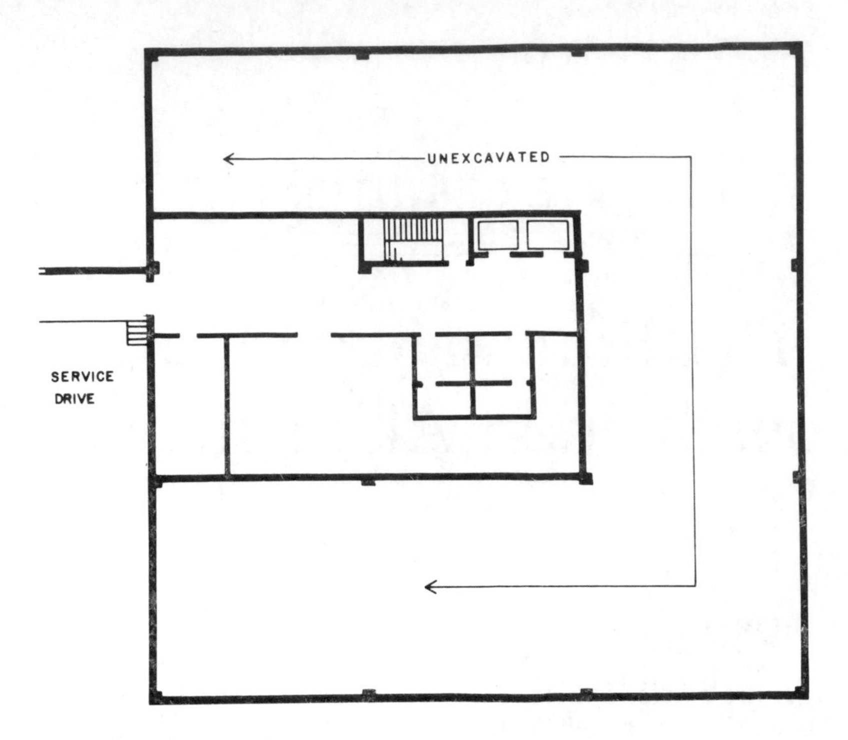

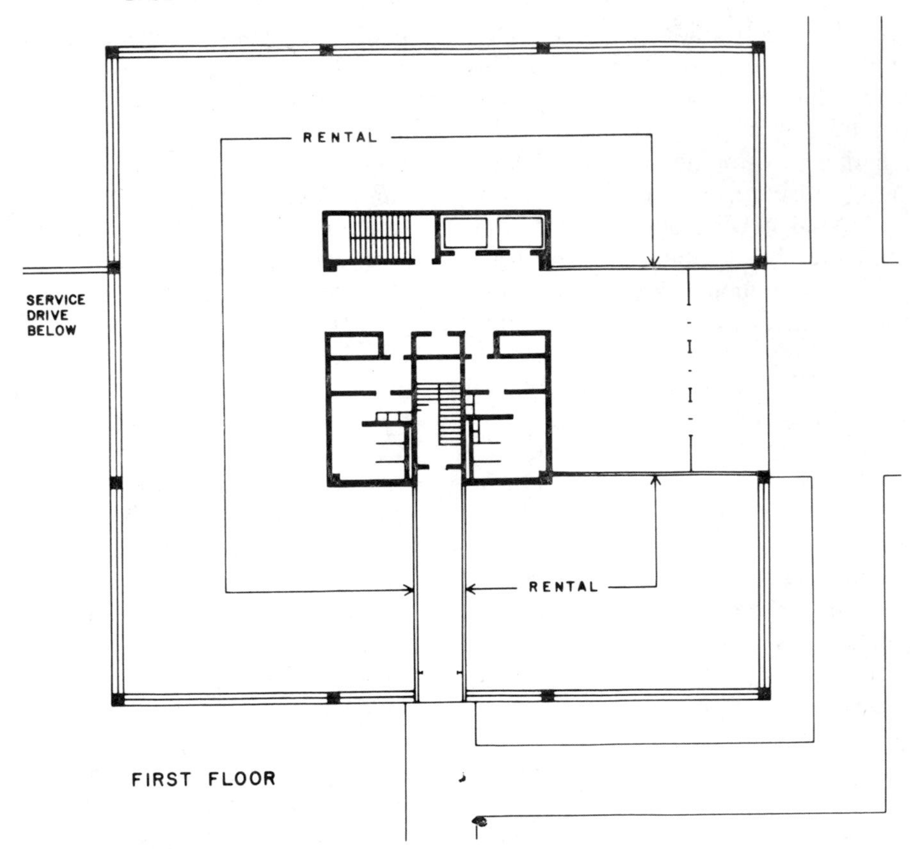

FIGURE 18.1. Plans for Building Four.

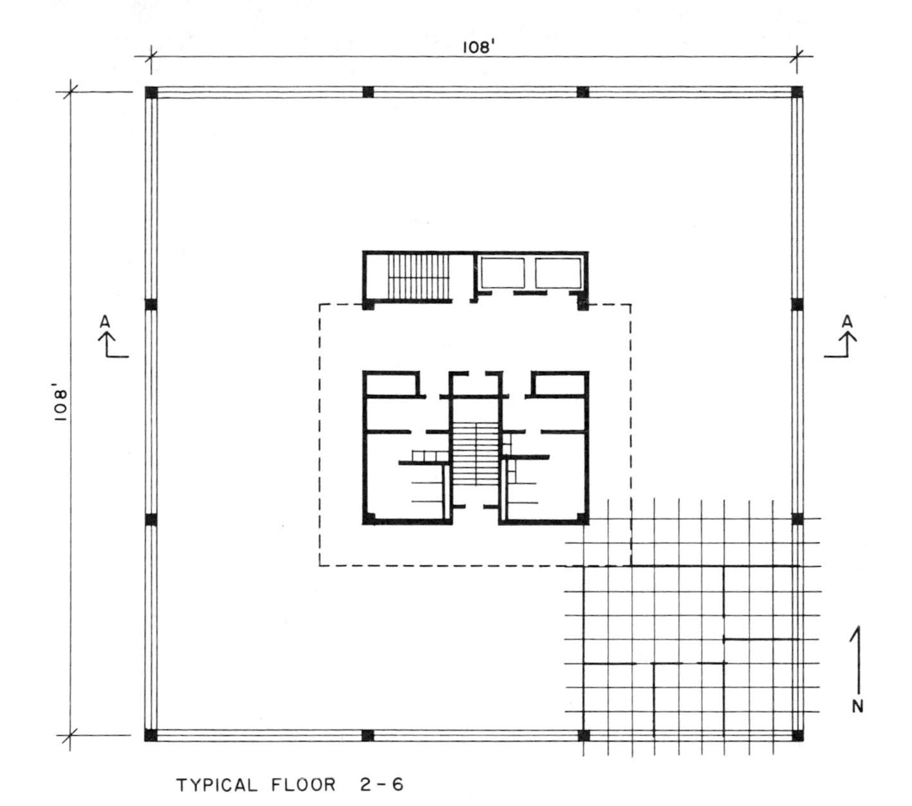

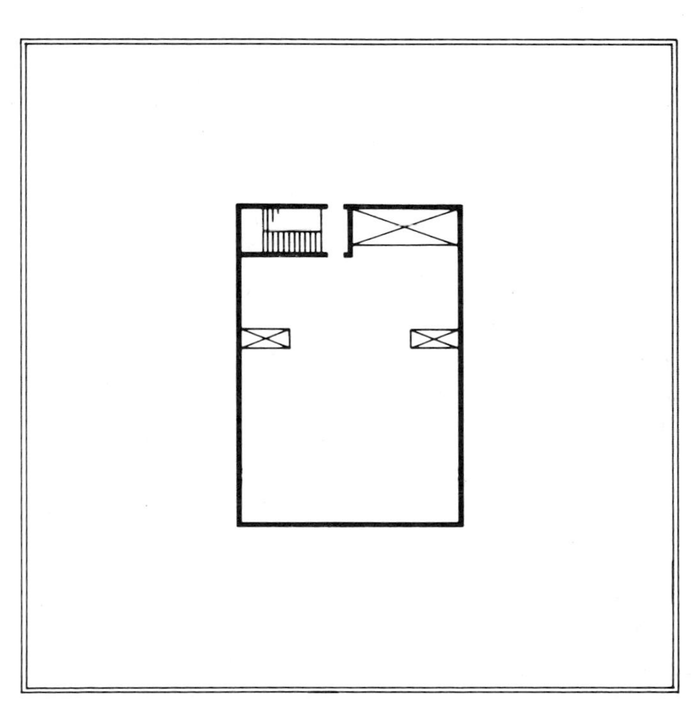

FIGURE 18.1. *(Continued)*

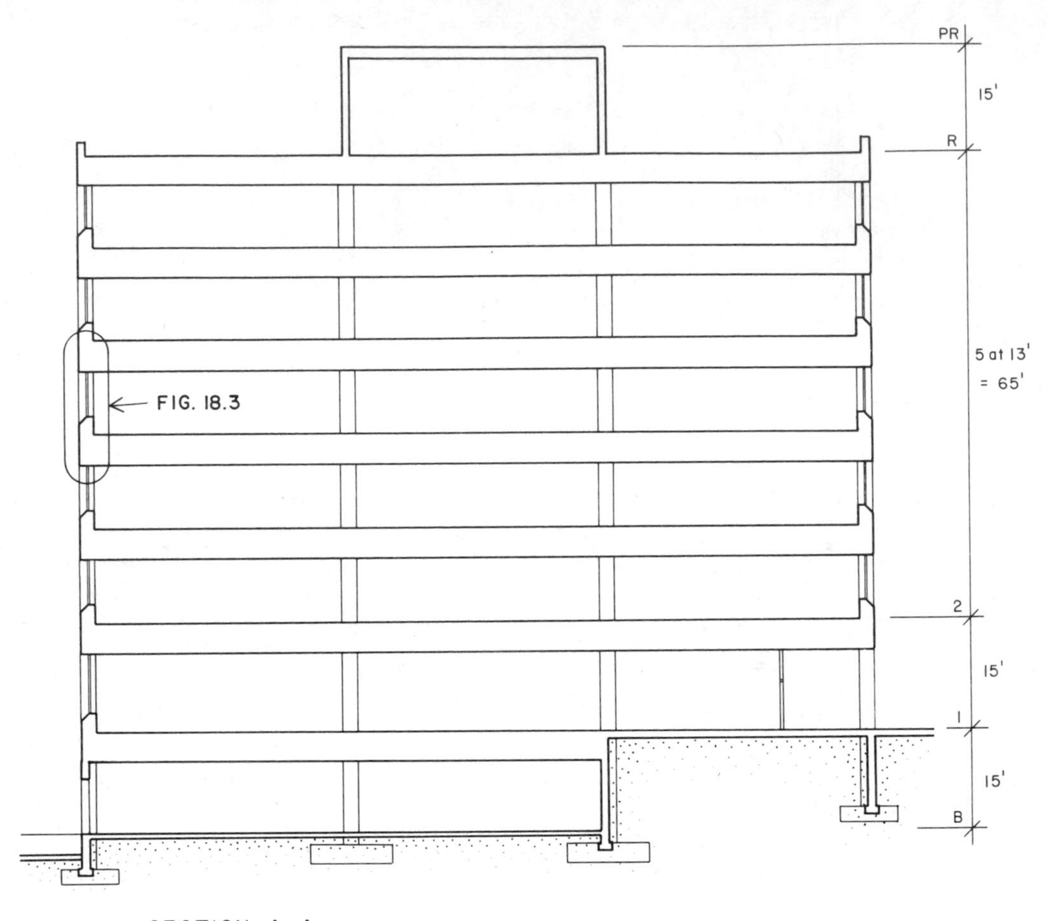

FIGURE 18.2. Section of Building Four.

facilities for the maintenance staff will be in the basement. The service entrance is also at the basement level.

The following will be assumed as criteria for the building design work.

Building Code: 1985 *Uniform Building Code* (Ref. 7)

Live loads:

Roof: UBC minimum, Table 23-C

Floor: Offices
= 50 psf [2.39 kPa]
Corridors
= 100 psf [4.79 kPa]

Partitions:
= 20 psf (UBC minimum, section 2304d) [0.96 kPa]

Wind: Map speed
= 80 mph, exposure B [129 km/h]

Soil capacity:
= 8 ksf maximum [383 kPa]

18.2. THE STRUCTURE

As with Building Three, there are many options for the structure of this building. At six stories it is not a high rise building, but it is nevertheless sufficiently higher than Building Three to eliminate some options—notably those using bearing walls or wood structural elements.

For the basic frame the choices are almost exclusively either one of steel or

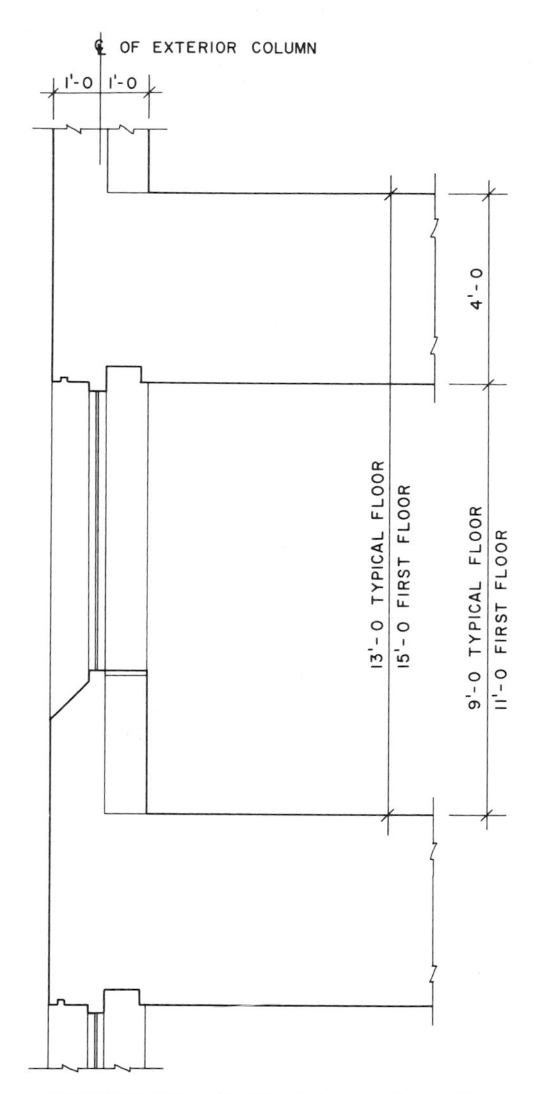

FIGURE 18.3. Profile of the exterior wall.

of reinforced concrete. For the roof and floor framing the choice of elements depends on the plan arrangements and the required clear spans. With the 36-ft spans in this building the major spanning elements will be significantly larger than those used in Building Three. Larger spans exist, but most office buildings have spans in the range of those used here.

As with Building Three, we will illustrate two designs. The first will use steel wide flange columns and beams to develop a full rigid frame, using all the column line bents for lateral resistance. The infilling structure for the floors and roof will use wide flange beams and a formed steel deck with a concrete fill. Some other options for the deck are discussed in Chapter 23.

For the second design we will use a full rigid frame of reinforced concrete columns and beams and a horizontal structure with slab and beams similar to that used in Building Three. Some other options for the horizontal structure are discussed in Chapter 23.

18.3. THE BUILDING SURFACE

General configuration of the exterior walls was discussed in Section 18.1. This is, of course, one of many options and is a major concern for the architect. Some considerations regarding alternatives for the walls are discussed in Chapter 23.

From a strictly structural design point of view, some practical considerations must be made for the exterior building surface. The exterior columns and the edge framing members must be integrated into the wall and the relative position and form of all the elements must be developed. The weight of the wall construction must be determined for the design of the columns and the edge framing.

In addition to the data and illustrations given thus far, the assumptions regarding the exterior walls that are necessary for the structural design are developed throughout the work in the following chapters. In general, however, we have avoided dealing with how the building looks as much as possible.

CHAPTER NINETEEN

Design of the Steel Structure

Figure 19.1 shows the framing system for the typical floor, using steel H-shaped columns and steel beams and girders. The basic floor system consists of the 9-ft-on-center beams supporting a one-way spanning deck. Every fourth beam frames directly to the columns; the remaining beams are supported by the girders. Four vertical bents are described in each direction and constitute the lateral load resisting system, together with the floor diaphragms.

Within this basic system scheme there are some variables to be considered. The spacing of the beam system is one. The type of deck used and the depth restriction for the beams would influence this choice. Within the 36-ft column module logical possibilities are 6-, 9-, 12-, and 18-ft spacings. The 9-ft spacing chosen seems reasonable with the steel formed deck we are using.

Orientation of the steel columns is another consideration. Because we are using the frame for lateral load resistance, the system shown was chosen so that there are eight columns in each direction turned to present their major stiffness. Although not exactly geometrically symmetrical, this does give reasonable biaxial stiffness to the frame.

As shown, location of major openings for duct shafts, elevators, and stairs should be developed so as not to interfere with any of the girders or beams on the column lines. The major plumbing and power risers should also not be on the column lines.

Similar framing systems would likely be used for the roof and the portion of the first floor over the basement. A concrete slab or a concrete framing system poured on fill would be used over the unexcavated areas of the ground floor.

Two options exist for the steel column bases. These may occur at the ground-floor level, with concrete piers or columns extending down to the footings. The other option is to extend the steel columns down to the footings and simply encase them in concrete up to the ground-floor level. For a building of this height either option is feasible.

Some consideration must be given to the fireproofing of the steel frame and deck. We will assume this to be accomplished as follows:

1. Top of the steel deck and exposed faces of the spandrel beams and beams at openings: poured concrete (probably of lightweight aggregate).

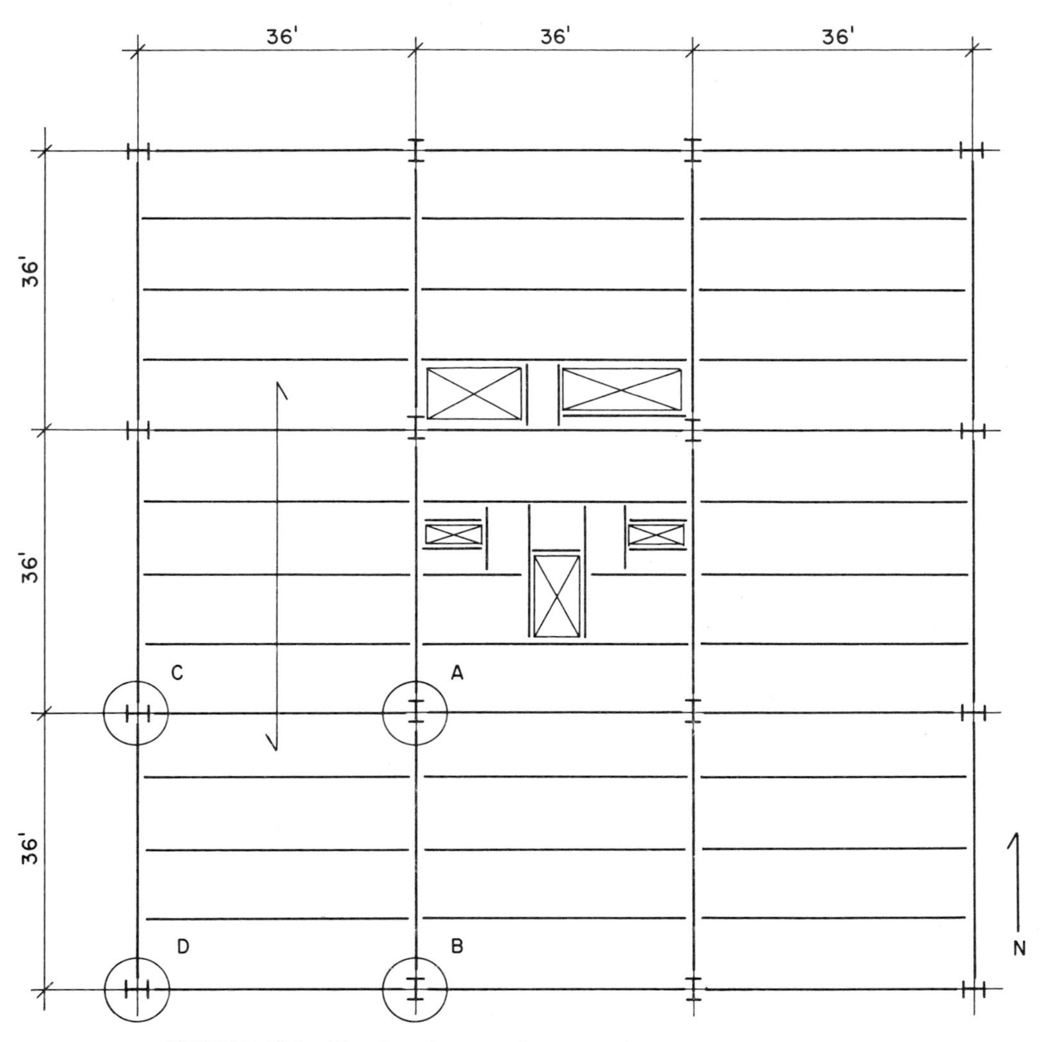

FIGURE 19.1. Framing plan—steel structure for the typical upper floor.

2. Exterior columns, interior sides of beams and girders, and the underside of the steel deck: sprayed-on fireproofing.
3. Interior columns: metal lath and plaster.

A36 steel will be used for all steel frame members. Elastic analysis will generally be used in the design.

19.1. DESIGN OF THE TYPICAL FLOOR

Loads:

Live loads:

50 psf for offices [2.39 kPa]

100 psf for lobby, corridors [4.29 kPa]

2000-lb concentrated loads (per UBC 2304c) [8.90 kN]

Dead loads:

Deck (steel + concrete fill)
= 35 psf [1.68 kPa]

Ceiling, lights, ducts, and so on
= 15 [0.72 kPa]

Partitions (per UBC 2304d)
= 20 [0.96 kPa]

Total dead load on deck
= 70 psf [3.35 kPa]

Several options are possible for the steel deck. Since a two-way distribution system for the power and communication systems is to be incorporated in the deck and fill, the choice must obviously be done in cooperation with the design of these systems and with the office layout design. Figure 19.2 shows some details and options for this type of system. We will assume a system using a $1\frac{1}{2}$-in.-deep corrugation and $2\frac{1}{2}$-in. minimum fill over the deck for a total depth of 4 in.

The typical beam is a 36-ft span, simple beam carrying the following load:

Dead load:

$9 \times 36 \times 70$ psf = 22,680 lb [101 kN]

Beam at 50 lb/ft = 1,800 [8]

Total = 24,480 lb [109 kN]

Live load:

$9 \times 36 \times (50 \times 0.75)$ = 12,150 lb [54 kN]

(with 25% reduction per UBC 2306)

Total load on beam
= 36,630 lb [163 kN]

The steel deck will normally be attached in a manner that will provide continuous

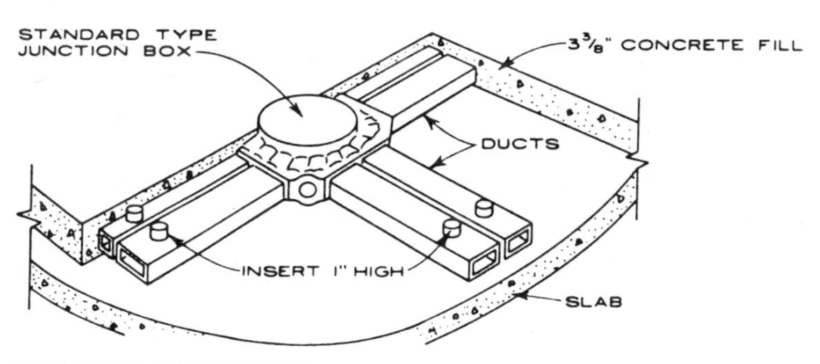

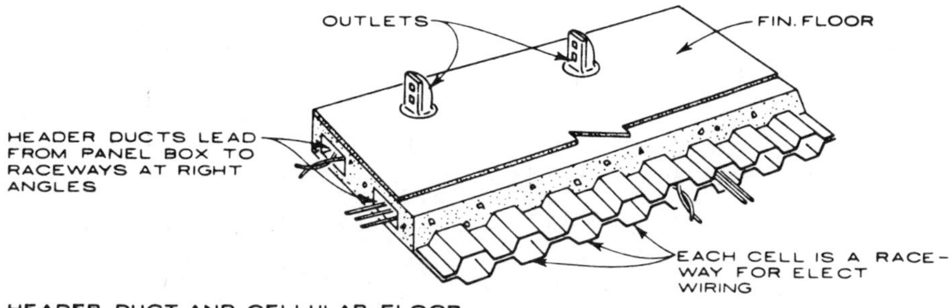

FIGURE 19.2. Details of wiring systems incorporated in floors. (From *Architectural Graphic Standards*, Ref. 8.)

lateral support for the beams. We may thus select a section from the tables based only on maximum bending stress from which we find the following options (see Appendix A, Table A.4):

$W\ 18 \times 50 \qquad W\ 21 \times 49$

Because these two choices are approximately the same weight, the selection is arbitrary and would involve other considerations, some of which are:

1. *Deflection.* The deeper beam will have less deflection, although both are within design limits.

2. *Depth.* The shallower beam will allow more headroom within the ceiling space for ducts and lights. The floor-to-floor height may thus be reduced; a savings of a few inches per floor adds up quickly in the multistory building.

3. *Flange Width.* The 21-in. beam is 1 in. narrower—not a major difference, but it relates to clearance at floor openings.

4. *Bent Members.* If deeper members are required for the beams and girders in the bents, the depth savings previously mentioned may become less meaningful.

In view of the last comment, the selection of the typical floor beam would probably be delayed until some analysis of the bents has been done.

19.2. THE VERTICAL COLUMN–BEAM BENTS

The seven-story-high, three-bay-wide bents are considerably indeterminate. In addition, loading conditions are numerous. The three basic loading conditions are:

1. Gravity dead load plus live load. This includes the consideration of skip loading for the live load.

2. Wind load plus gravity load for maximum combinations. This condition, using the allowable one third increase for stresses, would be compared with the gravity only condition for critical design.

3. Wind load plus dead load only for possible reversal effects, tiedown requirements, and so on.

There are essentially four different column–beam bents: the interior and the exterior bents in each of the directions. The lack of symmetry of the core, the basement, and the penthouse causes some minor variations on these as well. In present professional practice it is hard to conceive of the final analysis of these bents being done without a computer program in view of the relative availability of such programs and the facilities for their use. The approximate analysis shown here should be used only for preliminary sizing of members and connections to be used for cost estimates, feasibility studies, and development of architectural details.

The East–West Exterior Bent

This consists of the four exterior columns and the spandrel beams. Loading on the beam consists of approximately one half of the floor load on the typical beam plus the wall load:

One half beam DL:		
4.5 × 36 × 70	= 11,340 lb	[50.4 kN]
One half beam LL:		
4.5 × 36 × 50	= 8,100	[36.0]
Wall at 30 psf:		
13 × 36 × 30	= 14,040	[62.4]
Beam + fireproofing estimate at		
200 lb/ft × 36 ft	= 7,200	[32.0]
Total load	= 40,680 lb	[180.86 kN]

For an approximate design we can assume a critical maximum moment with a value somewhat less than that for a simple span beam. We assume the maximum moment to be 85% of that for the simple span and can thus use a total load of 85% of that just determined. (The value of the moment is proportional to that of the total load.) Thus

Design $W = 0.85 \times 40.7 = 34.6$ k [154 kN]

This load can now be used to find a section from the load–span tables (Appendix A, Table A.4) as

> W 21 × 44 carries 36 kips on the 36-ft span.
>
> W 18 × 45 carries 35 kips.

At the roof the design load will be lower because of less wall load, no partitions, and the lower live load. For preliminary purposes we may reduce the beam by a few sizes.

The East–West Interior Bent

The loading on these beams is the same as for the typical beam. The total load, previously determined, is close to that for the spandrel, so that we could use the same preliminary sizes. The loading of the middle span at the core should be determined once the exact core layout and the materials of the core walls are known.

The North–South Interior Bent

This consists of the interior girders, the two interior columns, and the two exterior columns labeled B. The principal loading on the girders consists of the end reactions of the beams. Added to this is a uniform load consisting of the weight of the girder and a strip of the floor directly over the girder. For simplification we will consider a concentrated load equal to the full 36-ft span load on the beam and will ignore the uniform load. This produces the quarter point loading shown in Figure 19.3. Note that this is slightly less than the beam load because of the higher reduction factor for live load since the girder carries more floor area.

The quarter point loading produces a moment diagram sufficiently similar to the parabolic diagram for uniform load, so that we may use an equivalent uniform load, as shown in the figure. Assuming, as before, that the rigid frame continuity reduces the actual critical moment to approximately 85% of the simple beam effect, we use a design load of

$$W = 0.85 \times 136.8 = 116 \text{ k [516 kN]}$$

And from the load–span tables (Appendix A, Table A.4):

> W 24 × 100 carries 123 k [547 kN].
>
> W 27 × 102 carries 119 k [529 kN].
>
> W 30 × 99 carries 120 k [534 kN].

The North–South, Exterior Bent

For the spandrel girders the floor loading is approximately one half of that for the interior girders and the wall load is the same as for the east–west spandrel beams.

One half of girder EUL:	$\frac{136.8}{2}$	
	= 68.4 k	[304 kN]
Wall load	= 14.0	[62]
Total load	= 82.4 k	[366 kN]
Design W: 0.85 × 82.4		
	= 70 k	[311 kN]

And from the load–span tables (Appendix A, Table A.4):

> W 24 × 76 carries 78 k [347 kN].
>
> W 27 × 82 carries 82 k [334 kN].

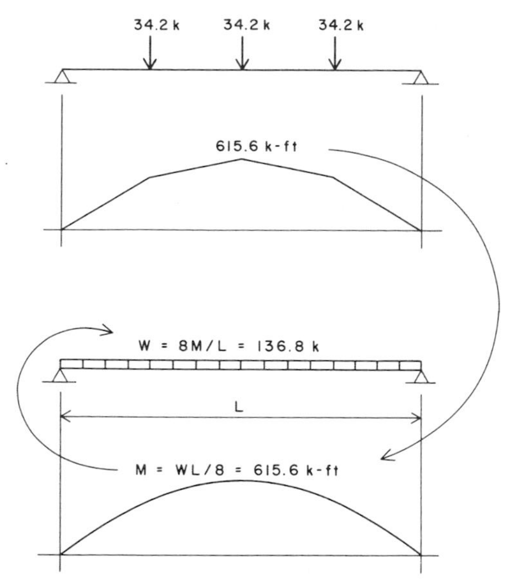

FIGURE 19.3. Equivalent uniform load for the girder.

On the basis of this first approximation a preliminary sizing of the beams and girders is made. These initial assumptions are shown in Table 19.1.

19.3. DESIGN OF THE COLUMNS FOR GRAVITY LOAD

The next step is to determine approximate column sizes. This will be done using the gravity loads only as a first trial. To do this we will make some assumptions for the moments induced in the columns by the rigid frame continuity. Then the axial loads will be tabulated so that a design can be made for the axial load plus moment conditions.

Figure 19.4 shows the assumptions for the relations of moment between the horizontal and vertical members of the bents. On the basis of these assumptions and the previous load calculations, we make the first approximation of the column moments as shown in Table 19.2. In Table 19.2 the fractions used to obtain the column moments are estimated from the conditions shown in Figure 19.4. Thus for column A at the roof in the north–south bent the column moment is assumed to be $\frac{1}{10}$ of the simple beam moment ($\frac{1}{8}$ wL^2) for the girder.

The gravity load calculation for the axial column loads is shown in Table 19.3. We assume the loads on columns B and C to be essentially the same. The live loads and dead loads are tabulated separately so that they may be combined for various purposes in the foundation design as well as the column design.

With the moment assumptions and axial loads we may proceed to pick approximate column sizes. Before doing so we must make some assumptions about the location of column splices. These are usually made a short distance above the beam level and are commonly not made at every floor, since the connections are quite expensive and time consuming in the field erection process. Figure 19.5 shows two possible schemes for the location of the splices. We will select scheme one and omit the sixth floor splice for the purpose of column size selection.

TABLE 19.1. First Approximation—Bent Beams and Girders

Level	North–South Interior Girders	North–South Exterior Girders	East–West Interior Beams	East–West Exterior Beams
Roof	*W* 24 × 94	*W* 21 × 62	*W* 18 × 35	*W* 18 × 35
2–6	*W* 24 × 117	*W* 24 × 76	*W* 21 × 44	*W* 21 × 44

TABLE 19.2. First Approximation—Column Moments (k-ft) due to Gravity

	North–South				East–West			
	A	B	C	D	A	B	C	D
Framing member causing moment	Interior girder	Interior girder	Spandrel girder	Spandrel girder	Interior beam	Spandrel beam	Interior beam	Spandrel beam
At Roof								
Simple beam M	493	493	297	297	132	146	132	146
Column moment condition[a]	2	1	2	1	2	2	1	1
Portion of M assumed for column	1/10	1/3	1/10	1/3	1/10	1/10	1/3	1/3
Column M	49	164	30	99	13	15	44	49
At Floor								
Simple beam M	616	616	371	371	165	183	165	183
Column moment condition[a]	4	3	4	3	4	4	3	3
Portion of M assumed for column	1/10	1/4	1/10	1/6	1/10	1/10	1/3	1/3
Column M	62	154	37	61	16	18	55	61

[a] See Figure 19.4.

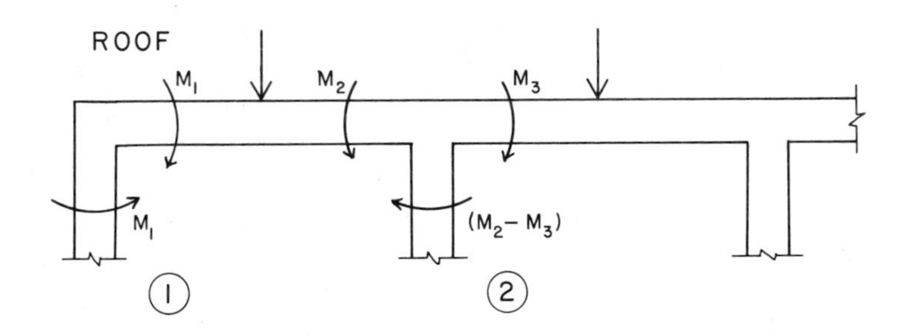

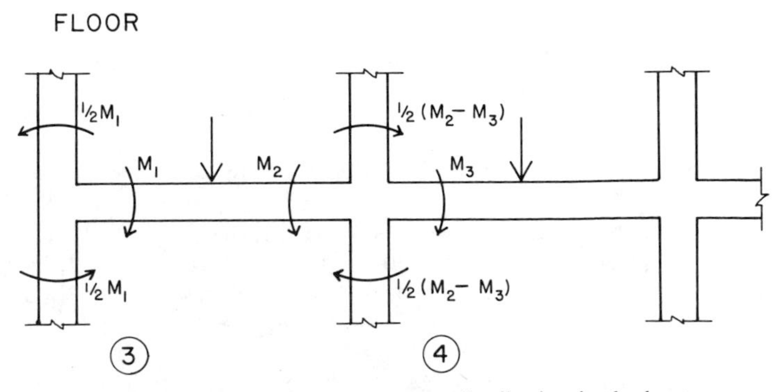

FIGURE 19.4. Assumed moment distribution in the bents.

TABLE 19.3. Column Axial Loads due to Gravity

Level		Column A, 1296 ft^2			Columns B and C, 648 ft^2			Column D, 324 ft^2		
		DL	LL	Total						
PR	Roof	15	10							
	Wall	20								
	Column	—								
	Total/level	35	10							
	LL reduction		—							
	Design load			45						
R	Roof	78	80		39	13		20	7	
	Wall	—			4			4		
	Column	2			2			2		
	Total/level	115	90		45	13		26	7	
	LL reduction		—			—			—	
	Design load			205			58			33
6	Floor	104	65		52	33		26	17	
	Wall	—			13			13		
	Column	2			2			2		
	Total/level	221	155		112	46		67	24	
	LL reduction	(60%)	62		(60%)	18		(50%)	12	
	Design load			283			130			79
5	Floor	104	65		52	33		26	17	
	Wall	—			13			13		
	Column	3			3			2		
	Total/level	328	220		180	79		108	41	
	LL reduction	(60%)	88		(60%)	32		(60%)	16	
	Design load			416			212			124
4	Floor	104	65		52	33		26	17	
	Wall	—			13			13		
	Column	3			3			3		
	Total/level	435	285		248	112		150	58	
	LL reduction	(60%)	114		(60%)	45		(60%)	23	
	Design load			549			293			173
3	Floor	104	65		52	33		26	17	
	Wall	—			13			13		
	Column	3			3			3		
	Total/level	542	350		316	145		192	75	
	LL reduction	(60%)	140		(60%)	58		(60%)	30	
	Design load			682			374			222
2	Floor	104	65		52	33		26	17	
	Wall	—			13			16		
	Column	4			3			3		
	Total/level	640	415		384	178		237	92	
	LL reduction	(60%)	166		(60%)	71		(60%)	37	
	Design load			816			455			274
1	Floor	104	65		52	33		26	17	
	Wall	60			20			20		
	Column	6			5			5		
	Total/level	820	580		461	211		288	109	
	LL reduction	(60%)	192		(60%)	85		(60%)	44	
	Design load			1012			546			332

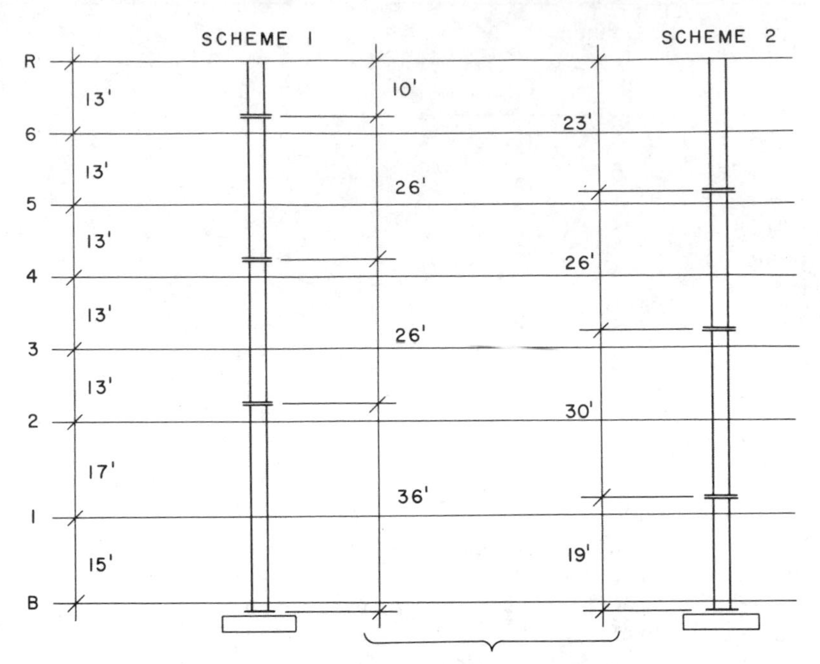

FIGURE 19.5. Options for the column splices.

TABLE 19.4. First Approximation of Column Sizes

		Column A		Column B		Column C		Column D	
Level	Assumed Critical KL	Design Load[a]	Choice	Design Load	Choice	Design Load	Choice	Design Load	Choice
R	19 ft	416	*W* 14	212	*W* 14	212	*W* 14	124	*W* 14
6		138	× 132	342	× 132	122	× 109	135	× 120
5		96		108		222		366	
		650		662		556		625	
4	19 ft	682	*W* 14	374	*W* 14	374	*W* 14	222	*W* 14
		138	× 176	342	× 159	122	× 145	135	× 145
3		96		108		222		366	
		916		824		718		723	
2	22 ft	1012	*W* 14	455	*W* 14	546	*W* 14	274	*W* 14
1		138	× 257	342	× 193	122	× 193	135	× 159
		96		108		222		366	
B		1246		905		890		775	

[a] Design load = $P + P'_x \times P'_y$.

The column selection is thus reduced to determining three sizes for each column in the plan. The floor load moments will thus be added to the axial loads at the basement, the second story, and the sixth story. For this approximation we assume a K factor for slenderness of 1.5 for the unbraced rigid frame. Column interaction is considered by using the B_x and B_y factors to convert the moments to additional axial load. With a final design load and the appropriate KL we then pick a column size from the AISC column load tables; 14-in. wide flange members will be used throughout.

The following is an example of this process, using the data for column A. Table 19.4 summarizes the design loads, KL assumptions, and column choices for the four types of columns.

Column A—First Approximation

At the fourth story:
Axial load = 416 k [1850 kN], KL = 1.5 × 13 = 19.5.
Use 19 ft [5.79 m]. (We guess at B_x and B_y for a try; then verify them after making a selection.) Try $B_x = 0.185$, $B_y = 0.50$.
$M_x = 62$ k-ft $M_y = 16$ k-ft [84, 22 kN-m]

$$\begin{aligned} P + P'_x + P'_y &= 416 + 0.185(62 \times 12) \\ &\quad + 0.50(16 \times 12) \\ &= 416 + 138 + 96 \\ &= 650 \text{ k } [2891 \text{ kN}] \end{aligned}$$

Pick: W 14 × 132 (see Appendix A, Table A.2).
Check: $B_x = 0.186$, $B_y = 0.521$.

At the second story:
Axial load = 682 k [3034 kN], KL = 19 ft [5.79 m], with M_x and M_y as before. Try $B_x = 0.185$, $B_y = 0.50$.
$P + P'_x + P'_y = 682 + 138 + 96 = 916$ k [4074 kN]
Pick: W 14 × 176, $B_x = 0.184$, $B_y = 0.484$.

At the basement:
Axial load = 1012 k [4501 kN], KL = 22 ft [6.71 m], with moments and bending factors as before.
$P + P'_x + P'_y = 1012 + 139 + 96 = 1246$ k [5542 kN]
Pick: W 14 × 257, $B_x = 0.182$, $B_y = 0.470$.

Inspection of the basement plan in Figure 18.1 will show that the majority of the first-floor structure consists of construction over unexcavated ground. We will assume this to be of reinforced concrete poured directly onto backfill. The load from this portion of the floor will be transmitted directly to the basement and grade walls and the foundations. Thus the only first-floor loadings transmitted to the steel columns will be from the first-floor construction over the basement area. Illustrations of this construction are shown in the construction drawings in Chapter 20.

With this assumption, the majority of the steel columns would be designed for the first-story loading. Only the four interior columns and the two exterior columns on the west side would be designed for the basement load condition. In a final, more exact analysis, however, it may be found that the first-story condition is actually more critical for these columns as well, because of a higher K factor, greater story height, and so on. Thus, although the basement loads tabulated in Table 19.3 may be useful for sizing the column base plates and the foundations, the column design would probably be for the first-story loads.

With approximate sizes for the columns established, we may now do a slightly more accurate analysis for the gravity moments on the bents. This will consist of an analysis of the three-span beams, assuming the columns to be fixed at the levels above and below. The two loading conditions considered are those of full live load on all spans and live load on the exterior spans only. The first will produce the maximum beam moment and the second the maximum column moments. We will do the analysis for the beams at the roof, sixth floor, and second floor and interpolate for values between these levels.

Figures 19.6 through 19.8 show the analysis of the east–west exterior bent using the method of moment distribution. Results of this analysis and similar ones for the other bents, not shown for sake of space, are summarized in Table 19.5. These slightly more accurate values could now be used to select a second set of approximate member sizes. However, we have not so far considered the effects of wind on the bents. It is therefore probably a better procedure to do an approximate analysis for wind effects and then to combine the results with the gravity analysis for the next approximation of sizes.

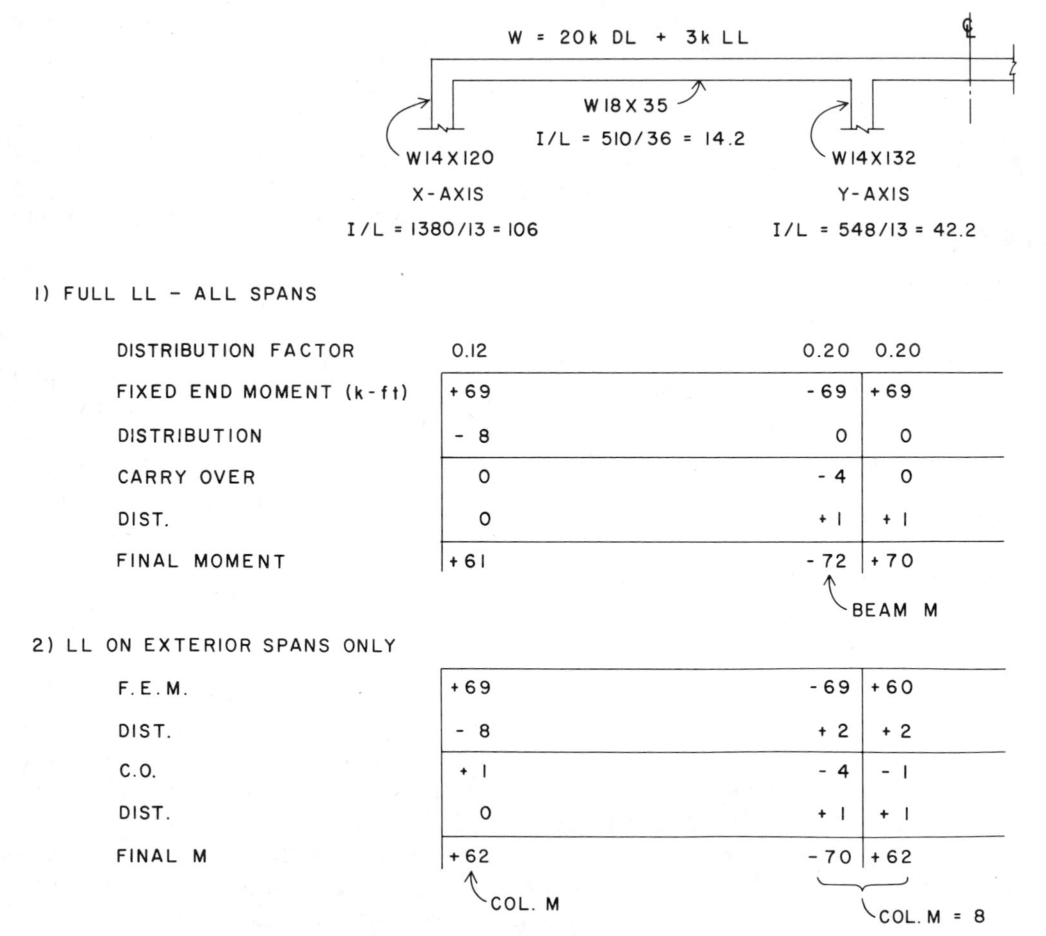

FIGURE 19.6. Moment distribution: roof gravity load.

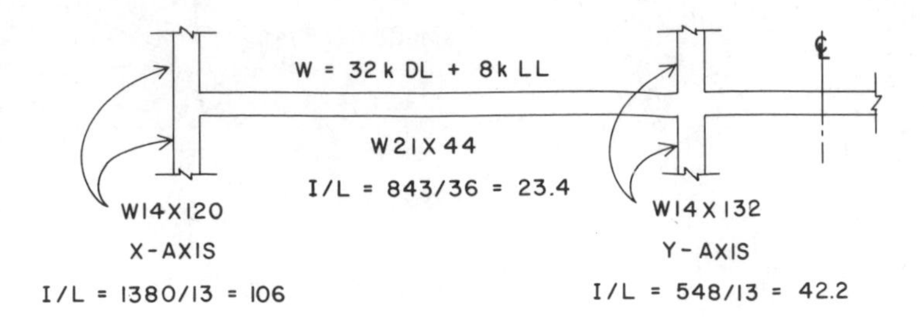

1) FULL LL - ALL SPANS

DIST. FACTOR	0.10	0.18	0.18
F.E.M.	+120	-120	+120
DIST.	-12	0	0
C.O.	0	-5	0
DIST.	0	+1	+1
FINAL M	+108	-124	+121

BEAM M (at -124)

2) LL ON EXTERIOR SPANS ONLY

F.E.M.	+120	-120	+96
DIST.	-12	+4	+4
C.O.	+2	-6	-2
DIST.	0	+1	+1
FINAL M	+110	-121	+99

COL. M = 110/2 = 55 COL. M = 22/2 = 11

FIGURE 19.7. Moment distribution: sixth-floor gravity load.

TABLE 19.5. Second Approximation—Bent Moments due to Gravity

	East–West Exterior			East–West Interior			North–South Exterior			North–South Interior		
Level	Column D, *x* Axis	Column B, *y* Axis	Beam	Column C, *x* Axis	Column A, *y* Axis	Beam	Column D, *y* Axis	Column C, *y* Axis	Girder	Column B, *x* Axis	Column A, *x* Axis	Girder
R	62	8	72	84	7	98	110	16	225	219	36	385
6	56	11	124	51	15	115	66	19	241	158	63	451
5	56	11	124	51	15	115	66	19	241	158	63	451
4	56	11	124	52	16	114	69	21	241	161	67	449
3	56	11	124	52	16	114	69	21	241	161	67	449
2	56	11	124	52	16	113	72	22	240	164	71	446

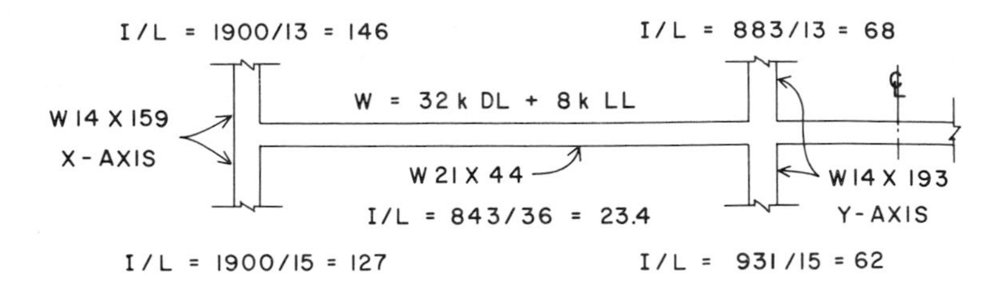

1) FULL LL - ALL SPANS

DIST. FACTOR	0.08	0.13	0.13
F.E.M.	+120	-120	+120
DIST.	-10	0	0
C.O.	0	-5	0
DIST.	0	+1	+1
FINAL M	+110	-124	+121

BEAM M (−124)

2) LL ON EXTERIOR SPANS ONLY

F.E.M.	+120	-120	+96
DIST.	-10	+3	+3
C.O.	+1	-5	-1
DIST.	0	+1	+1
FINAL M	+111	-121	+99

COL. M = 111/2 = 56 COL. M = 22/2 = 11

FIGURE 19.8. Moment distribution: second-floor gravity load.

19.4. DESIGN FOR WIND

Complete design for wind effects on this building would include the following:

- Inward and outward pressure on the exterior walls, involving the sizing of window glazing, structural mullions, attachment of wall elements to the structure, and so on.
- Inward pressure and uplift on the roof.
- Diaphragm action of the roof and floor decks.
- Lateral rigid frame action of the column–beam bents.
- Overturn, sliding, and lateral earth pressures at the building-to-ground interface.

Because we are not detailing the window wall, we will not consider its wind resistance other than to assume that it acts to transfer the wind load to the edges of the horizontal structure at all levels. The metal deck and concrete fill will be considered to be adequate to transfer these forces at all levels to the rigid bents. Our

main concern will be with the wind effects on the steel column–beam bents and the foundations.

We assume that the horizontal diaphragms are sufficiently stiff so that the force in shear at each story is distributed to the columns in proportion to their individual stiffnesses. Thus in each direction the shear on an individual column will be calculated as the total story shear times the ratio of the individual column stiffness to the sum of all column stiffness values at that story. For the first approximate analysis we ignore the effect of rotation at the joints and assume the column stiffness to be simply the I/L value.

Figure 19.9 shows the method for approximation of the shear and moment in the columns due to wind load. The total shear force at each story is taken as the wind load on the portion of the building above the midheight of the story. This total shear force is distributed between the columns at each story as previously discussed.

Figure 19.10 presents a summary of the basis for determining the wind forces on the building as a function of wind in a north–south direction. The wind pressures are derived essentially from the criteria in the UBC for a basic wind speed of 80 mph. The standard wind stagnation pressure (q_s) for this speed is 17 psf, and the design wind pressures for the various zones of height are as shown in Table 19.6. Because some sheltering is likely in the lowest zone (0–20 ft), we have determined the design pressure for both exposure conditions at this level.

The values from Table 19.6 have been rounded off, as shown in Figure 19.10, to obtain some simpler values for the wind pressures for each of the building levels. The total force applied to a single horizontal diaphragm (roof or floor deck) is the product of the pressure at that level times the distance between midstory

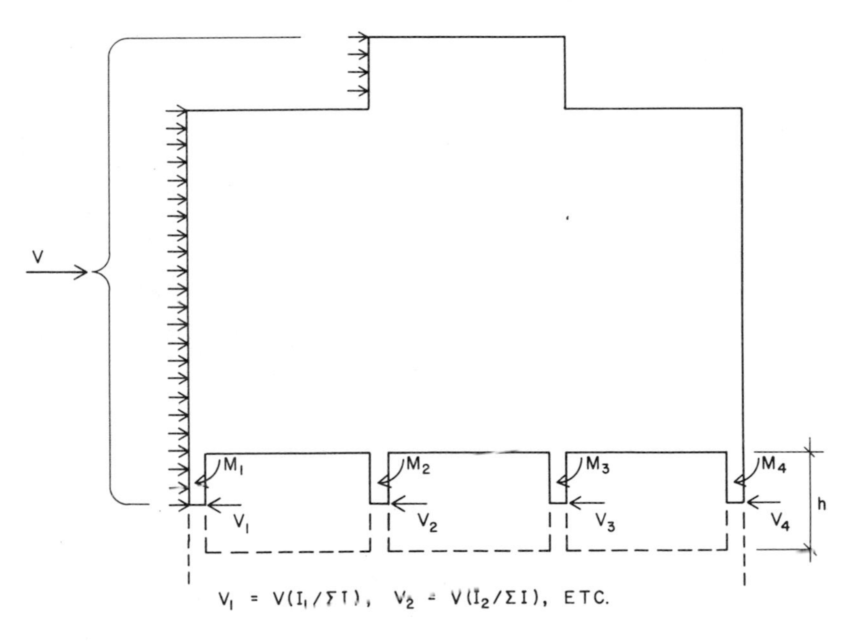

FIGURE 19.9. Assumed wind shear and moments in the columns.

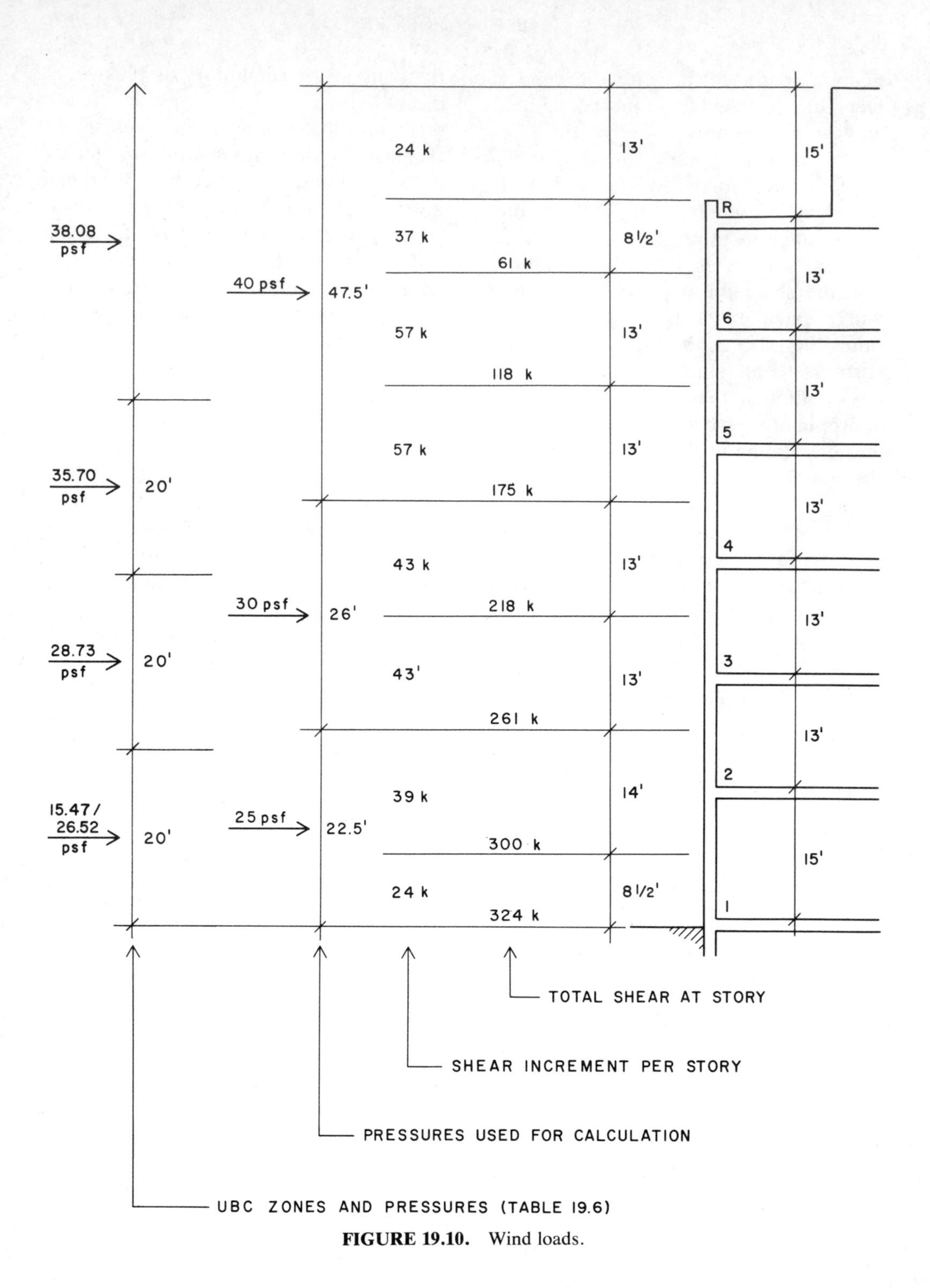

FIGURE 19.10. Wind loads.

TABLE 19.6. Design Wind Pressures[a]

Height Zone (ft)	Assumed Exposure Condition	C_e	C_q	$p = C_e \times C_q \times q_s$ (psf)
0–20	B/C	0.7/1.2	1.3	15.47/26.52
20–40	C	1.3	1.3	28.73
40–60	C	1.5	1.4	35.70
60+	C	1.6	1.4	38.08

[a] Based on UBC method 2 (projected area) with basic wind speed of 80 mph and q_s of 17 psf.

heights above and below and the building width. Thus for the fourth floor

$$H = 30 \text{ psf} \times 110 \text{ ft} \times 13 \text{ ft} = 42{,}900 \text{ lb}$$

This is rounded off to 43 kips in Figure 19.10.

Note that with the scheme used for column splices there are only three column sizes in the seven-story columns. The critical column design locations are thus at the basement or first story, at the upper portion of the second story, and at the upper portion of the fourth story. In addition, assuming the structure to be reasonably biaxially symmetrical in plan, we may reduce the design at each story to the four typical columns of A, B, C, and D. The result of these simplifications means that we must design only 12 columns, although there are actually a

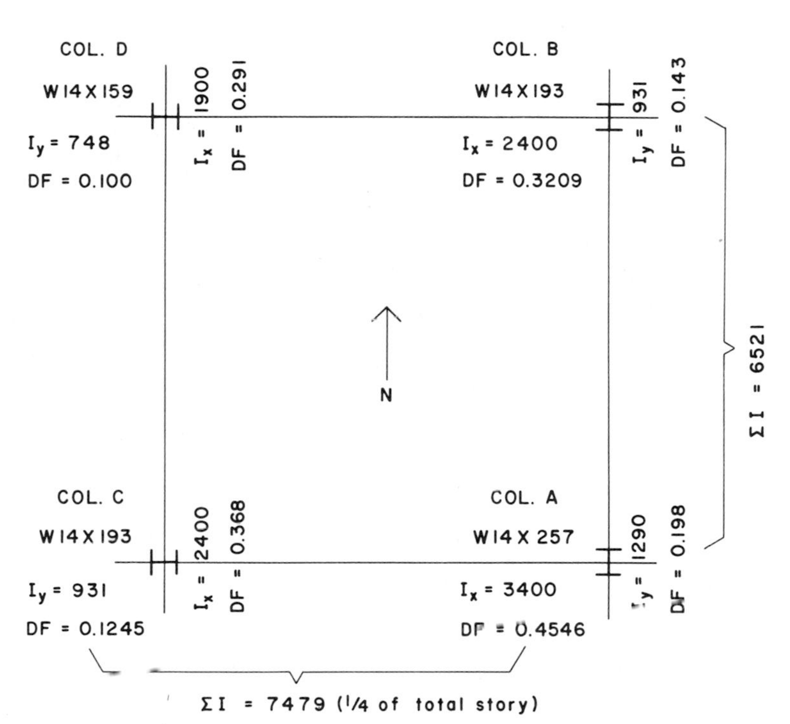

FIGURE 19.11. Relative stiffness of columns: lower tier.

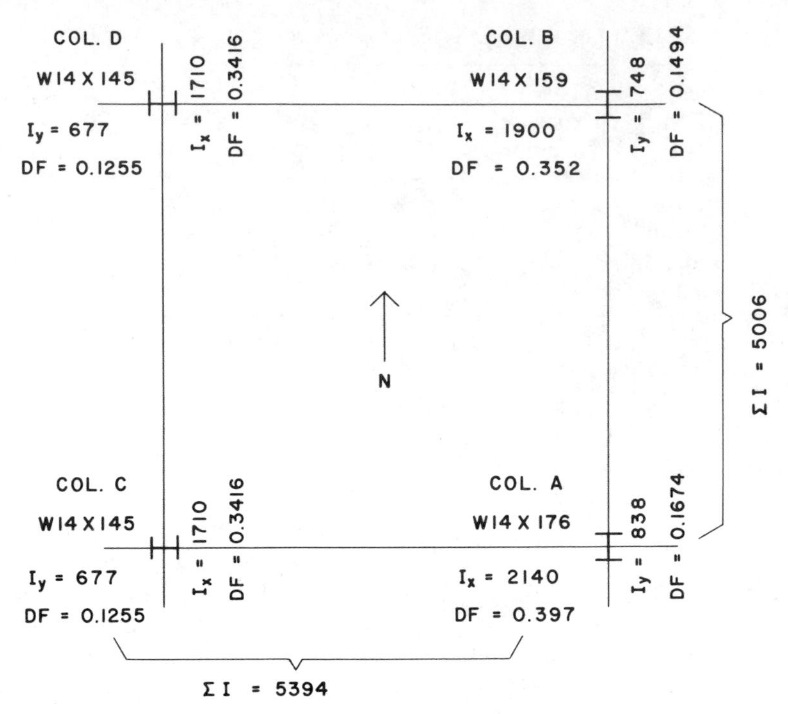

FIGURE 19.12. Relative stiffness of columns: middle tier.

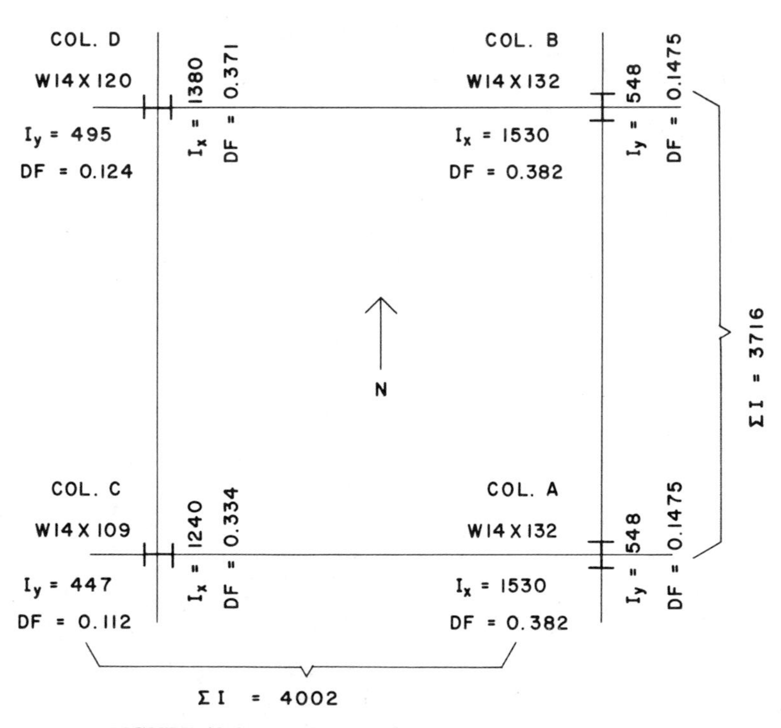

FIGURE 19.13. Relative stiffness of columns: upper tier.

total of 16 in the plan which, when multiplied by seven stories, equals 112 columns.

Figures 19.11 through 19.13 show the basis for determining the distribution of shears at the three critical stories. The column sizes assumed for this distribution are those approximated from the gravity load analysis, as summarized in Table 19.4. A summary of the shears and moments in the columns and the moments in the bent beams and girders are shown in Table 19.7. Figure 19.14 illustrates the basis for determining the moments in the horizontal members of the bents.

In addition to the horizontal shear effect on the building, the wind produces a bending effect on the whole building, resulting in axial loads on the columns: tension on the windward side, compression on the leeward side. In a very tall, relatively slender building these may be of considerable magnitude. With this building—which is quite squat in profile, being wider than it is tall—the axial loads thus produced are relatively small.

The basis for determination of these axial loads is shown in Figure 19.15. The structure is assumed to flex about its axis of symmetry, and a moment of inertia is determined on the basis of the column areas. For calculation the areas are assumed as relative values; one for the smallest column (D) and proportionate numbers for the others. Table 19.8 shows the calculation of this I value, consisting of a summation of the products of the column areas times the square of their distance from the axis of bending.

The axial load on an individual column is thus

$$P = \frac{(\text{relative } A)(\text{total story } M) \times (\text{distance of column from axis})}{(\text{total } I \text{ for the story})}$$

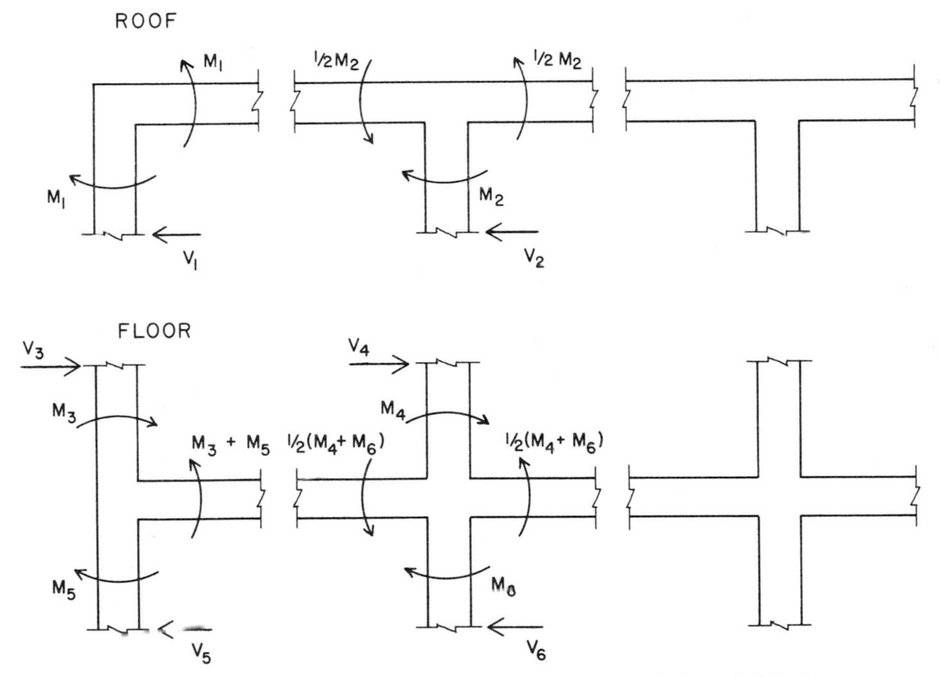

FIGURE 19.14. Assumed moment distribution in the bents: wind load.

TABLE 19.7. Wind Moments in the Bent Members

Level	Total Story Shear (kips)	Column:	A		B		C		D	
		Axis:	*x*	*y*	*x*	*y*	*x*	*y*	*x*	*y*
R	——	Beam *M*	19	7	38	7	33	6	38	12
		Column DF	0.382	0.148	0.382	0.148	0.334	0.112	0.371	0.124
	61	Column *V*[a]	5.8	2.3	5.8	2.3	5.1	1.7	5.7	1.9
		Column *M*[b]	38	15	38	15	33	11	37	12
6	——	Beam *M*	56	24	111	21	95	16	111	36
		Column DF	0.382	0.148	0.382	0.148	0.334	0.112	0.371	0.124
	118	Column *V*	11.3	4.4	11.3	4.4	9.9	3.3	10.9	3.7
		Column *M*	73	29	73	29	64	21	71	24
5	——	Beam *M*	91	35	182	35	155	26	182	60
		Column DF	0.382	0.148	0.382	0.148	0.334	0.112	0.371	0.124
	175	Column *V*	16.7	6.5	16.7	6.5	14.6	4.9	16.2	5.4
		Column *M*	109	42	109	42	95	32	105	35
4	——	Beam *M*	126	51	234	48	213	38	229	80
		Column DF	0.397	0.167	0.352	0.150	0.342	0.126	0.342	0.126
	218	Column *V*	21.6	9.1	19.2	8.2	18.6	6.9	18.6	6.9
		Column *M*	140	59	125	53	121	45	121	45
3	——	Beam *M*	156	66	275	59	264	48	264	48
		Column DF	0.397	0.167	0.352	0.150	0.342	0.126	0.342	0.126
	261	Column *V*	25.9	10.9	23.0	9.8	22.3	8.2	22.3	8.2
		Column *M*	168	71	150	64	145	53	145	53
2	——	Beam *M*	212	91	329	72	347	61	314	111
		Column DF	0.455	0.198	0.321	0.143	0.368	0.124	0.291	0.100
	300	Column *V*	34.1	14.9	24.1	10.7	27.6	9.3	21.8	7.5
		Column *M*	256	112	181	80	207	70	164	56
1	——									

[a] Column V = (DF)(total story shear/4).
[b] Moments in kip-ft.

For the first-story columns:

Column D:

$$P = \frac{(1.0)(15{,}833)(54)}{28{,}719} = 29.8 \text{ kips}$$

Column C:

$$P = \frac{(1.16)(15{,}833)(54)}{28{,}719} = 34.5 \text{ kips}$$

Column B:

$$P = \frac{(1.16)(15{,}833)(18)}{28{,}719} = 11.5 \text{ kips}$$

Column A:

$$P = \frac{(1.56)(15{,}833)(18)}{28{,}719} = 15.5 \text{ kips}$$

Because these are small values in comparison to the axial loads and moments due to gravity and the moments due to wind, we will not use them for the approximate design.

19.5. SECOND APPROXIMATION OF THE COLUMNS AND BENTS

We will now consider the effects of wind plus gravity on the columns and beams of the bents. Our procedure is to determine the critical load for gravity only and compare it with three fourths of the load for wind plus gravity. We then use the higher of the two for design.

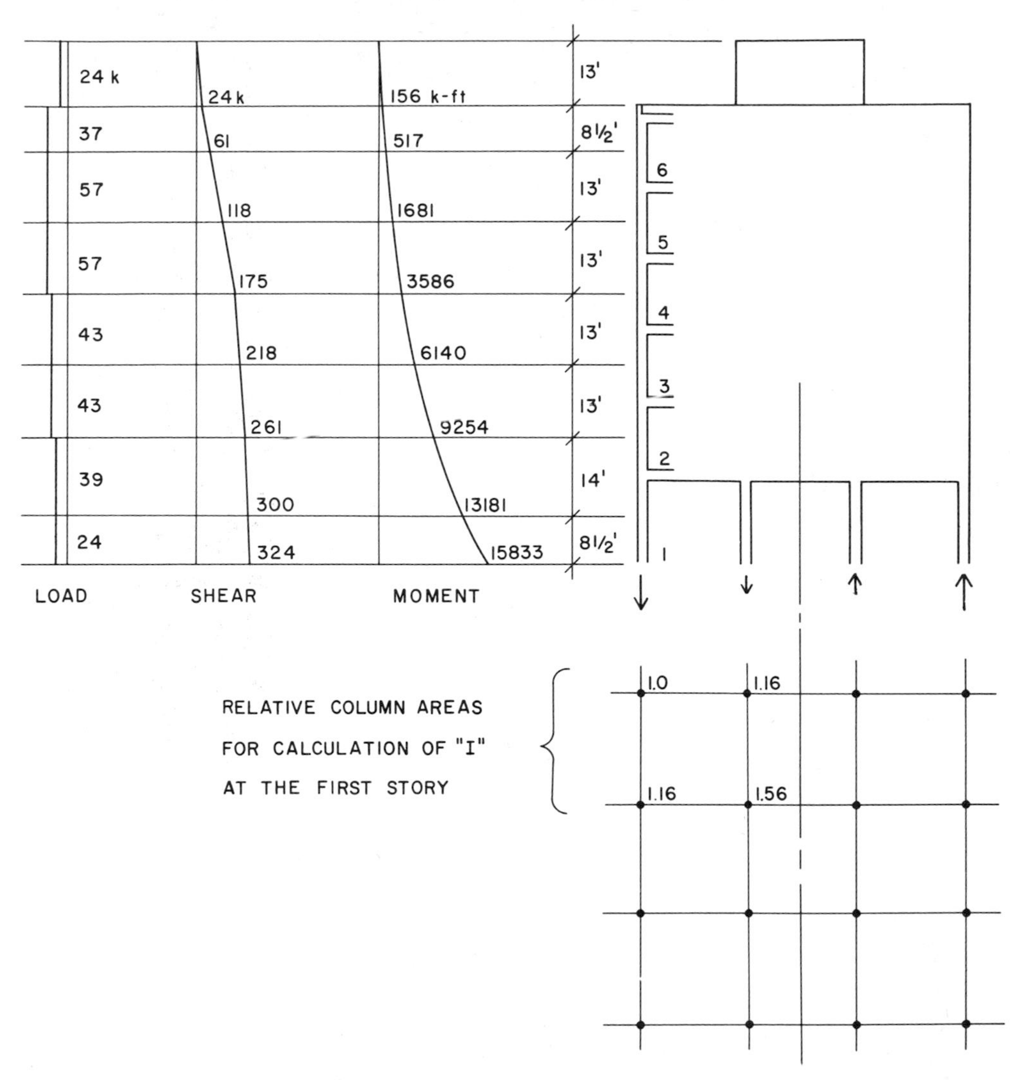

FIGURE 19.15. Analysis for axial column loads due to wind.

TABLE 19.8. Calculation of I for Axial Loads due to Wind

Column	Relative A	Total A	Distance to Axis	$I = A(D)^2$
A	1.56	6.24	18	2,022
B	1.16	4.64	18	1,503
C	1.16	4.64	54	13,500
D	1.00	4.00	54	11,664
Total I				28,719

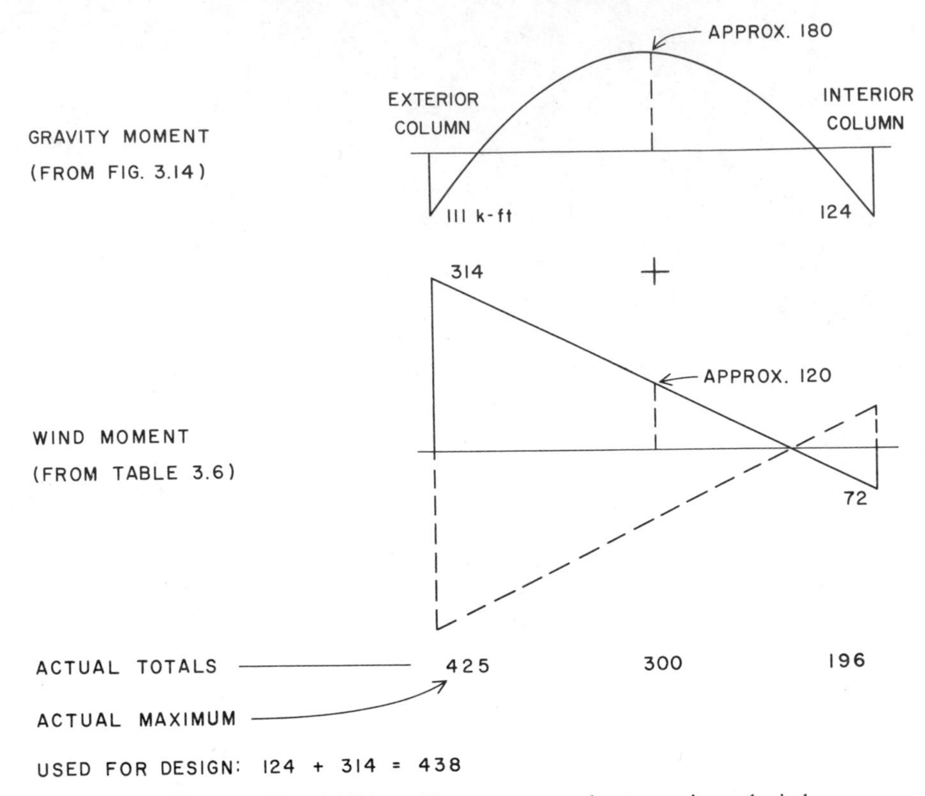

FIGURE 19.16. Addition of beam moments due to gravity and wind.

For the beam designs we will use the maximum gravity moment plus the maximum wind moment. This may be slightly conservative in some cases, since these two may not occur at the same place. Figure 19.16 illustrates this problem, using the east–west exterior bent at the second floor. The gravity moments are taken from the analysis in Figure 19.8 and the wind moments from Table 19.7. Since the error is small, we will simply use the tabulated maximums and add them for our approximate design.

The north–south interior girder:
At the second floor:

$$M_g = 446 \text{ k-ft (from Table 19.5)}$$

or

$$M_g + M_w$$
$$= (\tfrac{3}{4})(446 + 329)$$
$$= 581 \text{ k-ft [788 kN-m]}$$

Because the combination is higher, we use it to pick W 24 × 117 (from Appendix A, Table A.1).

At the third floor:

$$M = 449 \text{ k-ft}$$

or

$$M = (\tfrac{3}{4})(449 + 275)$$
$$= 543 \text{ k-ft [736 kN-m]}$$

Use W 24 × 117.

At the fourth floor:

$$M = 449 \text{ k-ft}$$

or

$$M = (\tfrac{3}{4})(449 + 234)$$
$$= 512 \text{ k-ft [694 kN-m]}$$

Use W 24 × 104.

At the fifth floor:

$$M = 451 \text{ k-ft}$$

or

$$M = (\tfrac{3}{4})(451 + 182)$$
$$= 475 \text{ k-ft [644 kN-m]}$$

Use W 24 × 104.

At the sixth floor:

$M = 451$ k-ft (wind not critical)

Use W 24 × 104.

At the roof:

$M = 385$ k-ft (wind not critical) [522 kN-m]

Use W 24 × 84.

The north–south exterior girder:

At the second floor:

$$M = 240 \text{ k-ft}$$

or

$$M = (\tfrac{3}{4})(240 + 111)$$
$$= 263 \text{ k-ft [357 kN-m]}$$

Use W 24 × 62.
Wind is not critical from the third floor through the sixth floor.

$$M = 240 \text{ k-ft [325 kN-m]}$$

Use W 24 × 62.

At the roof:

$$M = 225 \text{ k-ft [305 kN-m]}$$

Use W 24 × 55.

The east–west interior beam:
At the second floor:

$$M = 113 \text{ k-ft}$$

or

$$M = (\tfrac{3}{4})(113 + 347)$$
$$= 345 \text{ k-ft [468 kN-m]}$$

Use W 24 × 76.

At the third floor:

$$M = 114 \text{ k-ft}$$

or

$$M = (\tfrac{3}{4})(114 + 264)$$
$$= 284 \text{ k-ft [385 kN-m]}$$

Use W 24 × 68.

At the fourth floor:

$$M = 114 \text{ k-ft}$$

or

$$M = (\tfrac{3}{4})(114 + 213)$$
$$= 245 \text{ k-ft [332 kN-m]}$$

Use W 24 × 62.

At the fifth floor:

$$M = 115 \text{ k-ft}$$

or

$$M = (\tfrac{3}{4})(115 + 155)$$
$$= 202 \text{ k-ft [274 kN-m]}$$

Use W 24 × 55.

At the sixth floor:

$$M = 115 \text{ k-ft}$$

or

$$M = (\tfrac{3}{4})(115 + 95)$$
$$= 158 \text{ k-ft [214 kN-m]}$$

Use W 21 × 44.

At the roof:

$$M = 92 \text{ k-ft}$$

or

$$M = (\tfrac{3}{4})(92 + 33)$$
$$= 94 \text{ k-ft } [127 \text{ kN-m}]$$

Use W 18 × 35.

The east–west exterior beam:
At the second floor:

$$M = 124 \text{ k-ft}$$

or

$$M = (\tfrac{3}{4})(124 + 314)$$
$$= 329 \text{ k-ft } [446 \text{ kN-m}]$$

Use W 24 × 76.

At the third floor:

$$M = (\tfrac{3}{4})(124 + 264)$$
$$= 291 \text{ k-ft } [395 \text{ kN-m}]$$

Use W 24 × 68.

At the fourth floor:

$$M = (\tfrac{3}{4})(124 + 229)$$
$$= 265 \text{ k-ft } [359 \text{ kN-m}]$$

Use W 24 × 68.

At the fifth floor:

$$M = (\tfrac{3}{4})(124 + 182)$$
$$= 230 \text{ k-ft } [312 \text{ kN-m}]$$

Use W 24 × 62.

At the sixth floor:

$$M = (\tfrac{3}{4})(124 + 111)$$
$$= 176 \text{ k-ft } [239 \text{ kN-m}]$$

Use W 21 × 50.

At the roof:

$$M = 72 \text{ k-ft}$$

or

$$M = (\tfrac{3}{4})(72 + 38)$$
$$= 83 \text{ k-ft } [113 \text{ kN-m}]$$

Use W 18 × 35.

The second estimate of beams and girders is summarized in Table 19.9. We now proceed to consider a second design of the columns, using the more critical of three loading conditions as follows:

Case 1: Axial gravity load plus gravity moment on both axes.

Case 2: Axial gravity load plus gravity and wind moment on the x axis and gravity moment only on the y axis.

Case 3: Axial gravity load plus gravity moment only on the x axis and gravity plus wind moment on the y axis.

Assuming wind to be critical for the lower columns, we consider the critical column to be the first-story column. We then proceed for column A as follows.

TABLE 19.9. Second Approximation of the Bent Beams and Girders

Level	North–South Interior	North–South Exterior	East–West Interior	East–West Exterior
R	W 24 × 84	W 24 × 55	W 28 × 35	W 18 × 35
6	W 24 × 104	W 24 × 62	W 24 × 44	W 21 × 50
5	W 24 × 104	W 24 × 62	W 24 × 55	W 24 × 62
4	W 24 × 104	W 24 × 62	W 24 × 62	W 24 × 68
3	W 24 × 117	W 24 × 62	W 24 × 68	W 24 × 68
2	W 24 × 117	W 24 × 62	W 24 × 76	W 24 × 76

Case 1: $P = 816$ k, $M_x = 71$ k-ft, $M_y = 16$ k-ft. Assume $B_x = 0.180$, $B_y = 0.470$.

$$P + P'_x + P'_y = 816 + (0.180)(71 \times 12) + (0.470)(16 \times 12)$$
$$= 816 + 154 + 90$$
$$= 1060 \text{ k } [4715 \text{ kN}]$$

Case 2: Use $M_x = 71 + 255 = 326$ k-ft.

$$P + P'_x + P'_y$$
$$= 816 + (0.18)(326 \times 12) + 90$$
$$= 816 + 704 + 90 = 1610 \text{ k } [7161 \text{ kN}]$$

Case 3: Use $M_y = 16 + 110 = 126$ k-ft.

$$P + P'_x + P'_y$$
$$= 816 + 154 + (0.47)(126 \times 12)$$
$$= 816 + 154 + 711 = 1681 \text{ k } [7477 \text{ kN}]$$

Case 3 is critical; for design use $(\frac{3}{4})(1681) = 1261$ k [5609 kN]. Assuming a K of 1.5 and using Appendix A, Table A.2, we select the section as follows:

$$KL = 1.5(15) = 22.5, \text{ say } 22 \text{ ft } [6.7 \text{ m}]$$

Use: W 14 × 257.

With the splice above the second floor, the critical consideration for the second tier of the column is the second-story axial load and column moment due to wind and the gravity moment from the third-floor framing.

Case 1: $P = 682$ k, $M_x = 67$ k-ft, $M_y = 16$ k-ft. Assume $B_x = 0.185$, $B_y = 0.480$.

$$P + P'_x + P'_y = 682 + (0.185)(67 \times 12) + (0.48)(16 \times 12)$$
$$= 682 + 149 + 92$$
$$= 923 \text{ k } [4106 \text{ kN}]$$

Case 2: Use $M_x = 67 + 169 = 236$ k-ft

$$P + P'_x + P'_y$$
$$= 682 + (0.185)(236 \times 12) + 92$$
$$= 682 + 524 + 92 = 1298 \text{ k } [5774 \text{ kN}]$$

Case 3: Use $M_y = 16 + 72 = 88$ k-ft.

$$P + P'_x + P'_y$$
$$= 682 + 149 + (0.48)(88 \times 12)$$
$$= 682 + 149 + 507 = 1338 \text{ k } [5951 \text{ kN}]$$

Case 3 is critical; for design use $(\frac{3}{4})(1338) = 1004$ k [4466 kN]. Assuming KL = 1.5(13) = 19.5, say 19 ft [5.8 m], use W 14 × 193.

With the second splice above the fourth floor, the critical loads for the top tier are the axial load and wind moment at the fourth story plus the gravity moments from the fifth-floor framing.

Case 1: $P = 416$ k, $M_x = 63$ k-ft, $M_y = 15$ k-ft. Assume $B_x = 0.185$, $B_y = 0.50$.

$$P + P'_x + P'_y = 416 + (0.185)(63 \times 12) + (0.5)(15 \times 12)$$
$$= 416 + 140 + 90$$
$$= 646 \text{ k } [2873 \text{ kN}]$$

Case 2: Use $M_x = 63 + 109 = 172$ k-ft.

$$P + P'_x + P'_y$$
$$= 416 + (0.185)(172 \times 12) + 90$$
$$= 416 + 382 + 90 = 888 \text{ k } [3950 \text{ kN}]$$

Case 3: Use $M_y = 16 + 42 = 58$ k-ft.

$$P + P'_x + P'_y$$
$$= 416 + 140 + (0.5)(58 \times 12)$$
$$= 416 + 140 + 348 = 904 \text{ k } [4021 \text{ kN}]$$

Case 3 is critical; for design use $(\frac{3}{4})(904) = 678$ k [3106 kN]. Assuming KL = 19 ft, use W 14 × 145.

From these calculations, and similar ones for the other columns, we summarize a new set of column sizes as shown in Table 19.10.

Having obtained a reasonable approximation of column and beam sizes, the next step would be to perform a more rigorous analysis for the gravity and wind loads on the frames. Assuming that

TABLE 19.10. Column Sizes—Second Approximation

		Column A		Column B		Column C		Column D	
Level	Assumed Critical KL	Design Load[a] (kips)	Choice	Design Load (kips)	Choice	Design Load (kips)	Choice	Design Load (kips)	Choice
R 6 5	19 ft	678	W 14 × 145	661	W 14 × 132	485	W 14 × 99	831	W 14 × 176
4 3	19 ft	1004	W 14 × 193	886	W 14 × 176	701	W 14 × 145	810	W 14 × 176
2 1 B	22 ft	1261	W 14 × 257	1024	W 14 × 211	865	W 14 × 176	890	W 14 × 193

[a] Design load = $P + P'_x + P'_y$.

the approximate analysis and design has been carefully performed, this should not result in any startling changes in member sizes. Because an "exact" analysis and design of the bents is beyond the scope of this book, we will proceed with the design using the approximate answers so far obtained. In any event, these sizes are reasonably adequate for use in cost estimates, development of architectural details, and preliminary design of structural connections.

As was mentioned previously, the complete structural design and detailing of this building would be a task involving an amount of work several times the total size of this book. We will, however, briefly discuss some of the additional considerations to be made in the design of the steel frame.

Connections for the steel frame could be all of a relatively routine form. Most members not involved in the bents would use the "standard" framed connections—probably double angles either bolted or welded. The beam-to-column connections in the bents must be of a moment-resistive form. The variety of possibilities for these is considerable; the commonest types are well illustrated in the AISC Manual. Actually, it is usually desirable to involve the steel fabricator in the final development of these details, since the variables of field conditions and fabricating shop facilities are important considerations. In theory, just about any of the ordinary types of connection could be used for this structure.

The column splices may also be done with a variety of techniques, as shown in the steel manual. The situation is simplified here by the fact that all the columns are of the same nominal size. Because of the rigid frame action, the splices must develop some significant moment resistance in both directions as well as transmit the axial compression.

19.6. ADDITIONAL DESIGN CONSIDERATIONS

The column base connection consists essentially of a rectangular steel bearing plate that functions to transmit the highly concentrated bearing stress from the column into a much lower, distributed bearing stress on the top of the concrete foundation. This involves considerable bending in the plate if it cantilevers significantly from the face of the column. The three basic determinants for the plate size and thickness are therefore the size of the column, the value of the total axial compression load, and the allowable bearing on the concrete.

Approximate sizes for the plates can be obtained from the tables in the steel manual. Individual plate sizes are tabulated for each of the wide flange column sizes. However, the axial load used assumes a full development of maximum compressive stress in the column. Because our column sizes are considerably increased by the bending moments, this means that the size of plate shown for the columns is larger than necessary. The approximate plate size should therefore be selected by using the tabulated "maximum load" rather than the column size. This will result in a plate of reasonably correct total bearing area size, but may result in a conservative thickness, because the bending may be less if the column size is larger.

The steel frame and metal deck must be protected to obtain the fire ratings required by the code. The means for this, as assumed in the determination of dead loads and shown later in the construction drawings, are:

Top surface of the metal deck and beams is protected by the concrete fill. At openings and at the spandrel this is sometimes carried around the exposed face of the beam to the level of the bottom flange.

Bottom of the metal deck and bottom and sides of the beams are protected by sprayed-on fireproofing.

Depending on the architectural details, the columns may be protected by cast concrete jackets, by sprayed-on fireproofing, or by lath and plaster encasement.

The exterior curtain wall design involves a number of structural detailing considerations. The wall itself must be designed for the gravity and wind forces and the attachments to the structural frame must transmit these forces to the frame. The variety of possibilities is virtually endless, with the variables of the materials and details of the wall, the location of the wall relative to the columns and spandrel beams, the incorporation of HVAC elements and the facilitation of interior partitioning. The construction details shown present a fairly common solution, involving an exterior wall of metal and glass and the facilitation of a modular interior partitioning system.

Another detail consideration is that of the support necessary for the various items that must be hung from the underside of the roof and floor structure. This includes the ceiling systems, HVAC ducts and components, lights, electrical conduits, and so on. Support will usually be provided by brackets from the beams or by wires from the deck. With the use of a modular ceiling system most of this support will be achieved by hanging a basic framing system that accommodates elements of the modular ceiling. Lights and HVAC registers will usually be a part of the ceiling system and will be supported by this frame. Large ducts and wiring for the lights may be hung separately within the ceiling void space. Individual support is usually provided for any heavy fans, reheat units, and similar features that are incorporated in this space. The needs for this equipment and the means for supporting it must be considered in establishing the floor-to-floor dimension and the details of the floor framing.

As shown in Figure 18.3, we have provided a total of 48 in. for the distance from the bottom of the ceiling to the top of the floor fill. Some preliminary sizing and coordination of the structure, the HVAC and lighting systems, and the architectural details would have to be done to establish this dimension. A major consideration is often the size of the large supply ducts for the air handling system that must pass under the beams. If cooling is achieved with the air system, the ducts will be insulated and of larger size, which further increases the clearance required.

A final consideration for the steel frame is that of the tolerable vertical deflection of the floor and roof systems. This is a complex issue involving considerable judgment and not much factual criteria. Some of the specific considerations are:

The Bounciness of the Floors. This is essentially a matter of the stiffness and fundamental period of vibration of the deck. Use of static deflection limits generally recommended will usually assure reasonable lack of bounce.

Transfer of Bearing to the Curtain Wall and Partitions. The deformations of the frame caused by live gravity loads and wind must be considered in developing the joints between the structure and the nonstructural walls. Flexible gaskets, sliding connections, and so on, must be used to permit the movements caused by these loads as well as those due to thermal expansion and contraction.

Dead Load Deflection of the Floor Structure. As shown in Figure 19.17, there is a cumulative deflection at the center of a column bay because of the deflection of the girders plus the deflection of the beams. Within the normally permitted deflections, this could add up to a considerable amount for the 36-ft spans. If permitted to occur, the result would be that the top of the metal deck would be several inches below that at the columns. Because the top of the concrete fill must be as flat as possible, this could produce a much thicker fill in the center of the bays. The usual means for compensating for this is to specify a camber for the beams approximately equal to the calculated dead load deflection.

Where heavy permanent walls and openings for stairs, elevators, and duct shafts occur, special framing must be provided, as shown in the framing plans in Figure 19.1 and in the construction drawings. Light partitions may be supported directly by the deck, but heavy masonry walls should be placed over framing members.

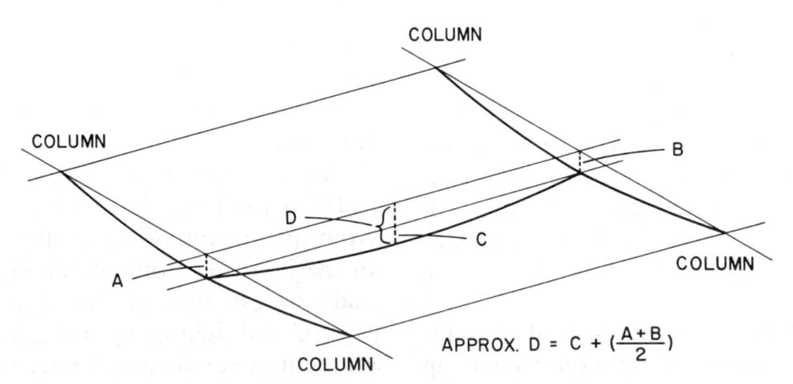

FIGURE 19.17. Deflection of the floor.

19.7. DESIGN OF THE FIRST FLOOR, BASEMENT, AND FOUNDATIONS

The basement plan, shown in Figure 18.1, shows that part of the first floor will be framed as a spanning structure over the basement area and part will be placed over unexcavated (actually backfilled) earth. Several options are possible for each of these conditions. We will assume that the portion over the basement will consist of the same basic system as the typical floor: steel beams with metal deck and concrete fill. The beams would frame directly to the steel columns or would be seated on the concrete basement walls.

Because of the considerable amount of backfill that would be required, it is probably not advisable to use a simple slab on fill for the remainder of the floor, since considerable settlement would occur. The other options are to use a framed floor over a crawl space or a concrete framed system poured directly on fill. The plan and some details for the latter option are shown in the construction drawings. The system consists of a two-way concrete slab on edge supports with a beam system providing support on 18-ft centers in each direction. The beams are supported by the steel columns, the basement walls, and a series of small piers in the center of the 36-ft spans.

Because the building plans indicate that the first floor will also be used for rental office space, it is assumed that the distribution of power and communication in the floor is required. In the floor area over the basement this would be accomplished as for the typical floor, using the metal deck voids and the concrete fill. In the remainder of the floor it would be necessary to deal with this as for a concrete framed floor, with the distribution done in a lightweight concrete fill on top of the structural concrete. Details of this system are shown in Figure 19.2 as utilized for the concrete floor system in the next example.

The partial basement also produces two different conditions for the steel columns and the footings. Those in the area of the basement must be dropped below the level of the basement floor. Those in the remainder of the plan could theoretically be quite higher; only a short distance below the first floor structure. However, for a number of reasons they would probably be fairly deep. Some of the considerations are:

- To obtain the relatively high allowable soil bearing and to assure equalized settlement of the foundations.
- To avoid influence of the pressure of the higher footings on the deeper adjacent ones.
- To allow for tying or strutting of the isolated footings, as would be required if seismic loads were critical.

Because of the elevator pit and the driveway on the west side of the basement, some of the footings would be dropped an additional distance. The result is that the bottoms of the footings for this building will occur at several different elevations. The foundation plan shown in Figure 20.1 indicates this condition.

The typical footings are as shown in Table 19.11. Sizes are given for the three basic column load conditions and for the first-floor intermediate piers. The design loads include the previously tabulated column loads plus the loads from the first-floor and the basement walls. The footing sizes and reinforcing are derived from the tables in *Simplified Design of Building Foundations* (Ref. 11), using f'_c of 3 ksi and maximum soil pressure of 5 ksf.

TABLE 19.11. Typical Footings[a]

Footing for	Load on Footing (kips)	Footing Dimensions		Reinforcing Each Way
		Width	Thickness	
Column A	1100	16 ft, 0 in.	43 in.	20 No. 11
Column B	600	11 ft, 6 in.	34 in.	13 No. 10
Column C	400	9 ft, 6 in.	28 in.	12 No. 9
First-floor pier	100	5 ft, 0 in.	16 in.	7 No. 6

[a] Maximum soil pressure of 5 ksf, $f'_c = 3$ ksi, Grade 40 reinforcing.

The basement walls and the grade walls between columns in the unexcavated area will be placed on wall footings. However, they will also tend to span between columns because of their high relative stiffness as deep beams. It is probably advisable to consider them as spanning and to provide adequate reinforcing for this action as well as some additional size for the column footings.

CHAPTER TWENTY

Construction Drawings for the Steel Structure

The illustrations in this chapter show some of the construction details for the building with the steel frame. Since we did not discuss the design of the roof and penthouse, we have shown the structural plans for a typical floor, the first floor, and the basement and foundations only.

There are obviously many details that need coordination between the structure and the various architectural elements. We have shown some possibilities for some of the architectural elements of the ceilings, walls, floors, and the exterior curtain wall in order to illustrate the relations that need consideration. Because our principal concern is for the structure, however, we have not shown the complete building construction details.

20.1. CONSTRUCTION ILLUSTRATED IN OTHER CHAPTERS

Some of the illustrations shown in other chapters show materials applicable to the construction of this structure. The following is a description of some of these drawings.

1. The building structure must be explained at a variety of scales. With limits of a particular size of drawing (overall total dimension) the precision and detailed completeness of a construction section will be less the larger the scope of the section. It is necessary, however, to explain both the overall form of the structure and the precise details of its assembly. Drawings such as those in Figures 18.2 and 18.3 help to give an overall picture of the structure. From that level of scale it may be necessary to graduate up progressively until the required information can be shown. For small elements such as those used for window frames, it may be necessary to go to half size or even full size drawings.

2. The exterior wall details developed later in this chapter demonstrate the use of a wall surface developed with metal panels. An alternative surface development with precast concrete elements is given in Figure 22.9. Although the concrete elements are shown attached to a concrete structure, they could as well be used with the steel structure.

3. An alternative to the steel floor deck used here is shown in Figure 23.1, which consists of a solid concrete slab formed and poured directly on top of the steel framing. A second alternative,

showing the use of precast concrete deck units, is given in Figure 23.2.

20.2. STRUCTURAL PLANS

Figures 20.1, 20.2, and 20.3 show partial structural plans for the basement, first floor, and typical upper level floor structures. The locations of the details presented in the next section are given on these plans.

Basement and Foundation Plan

The partial plan in Figure 20.1 shows most of the typical situations for the basement and foundation construction. The foundations consist of bearing footings for the walls and the concentrated pier and column loads. Particular characteristics of the soil strata, the groundwater condition, site grading, and so on, may influence the detailing of these subgrade elements. We have assumed a rela-

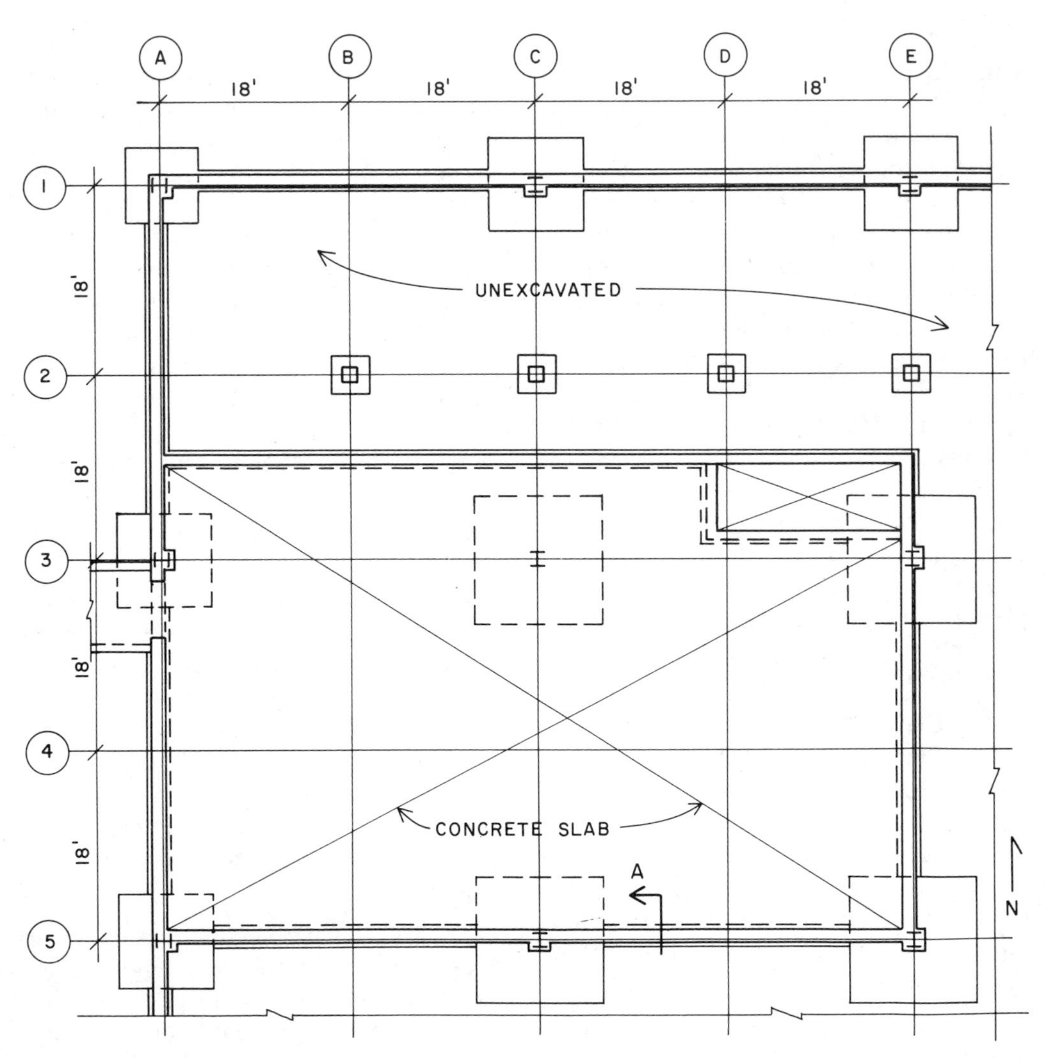

FIGURE 20.1. Basement and foundation plan.

tively high soil capacity of 8000 psf and have developed the construction details in response to various structural and building planning considerations. Some of these considerations are as follows:

1. *Depth of Column Footings.* Where the basement occurs the column footings must be dropped a distance sufficient to permit the steel base plate and anchor bolts to fit above the footing without protruding above the floor slab. Although the basement slab itself could be used to provide part of the cover, we have shown the footing dropped to allow for a separate encasement of the base plate and the anchor bolts. By having the floor slab float freely above this encasement, it is possible for the granular subbase and a waterproofing membrane to be placed continuously under the floor slab. If a serious water condition exists, more details would be required to prevent water from entering the basement.

Where there is no basement, the column footings could be raised. The height of these footings would depend on:

The level at which the desirable soil bearings can be found.

The need for frost protection (depth of the footing below the adjacent finished grade).

Depth of adjacent excavation or construction: other column footings, walls, elevator pit, utility trenches, tunnels, and so on.

2. *Depth of Wall Footings.* There are two basic types of walls: the basement walls and walls that serve as grade walls at the edges of the ground floor. The basement walls must be carried below the level of the basement floor slab as a minimum. Since they will rest directly on the column footings at some points, the minimum location of the bottom of the walls would probably be at the tops of the column footings, as shown in Detail A (Figure 20.4). If the wall footing is placed at this depth, it would be considerably higher than the adjacent deep column footings. Its effectiveness would therefore be limited in the vicinity of the column footings. To allow for this the wall footing can be stepped down to the level of the bottom of the column footing or the walls can be designed to span across this short distance.

Where there is no basement the walls would probably be reduced in height. They may be designed for bearing on their own footings, or may be designed as beams spanning between the column footings. In either case, a wall footing would probably be used as a construction platform for the wall forms.

3. *Use of Walls as Ties.* In addition to their functions as walls or grade beams, the concrete walls serve as ties, holding the whole subgrade construction together as a structural entity. This is considered as a critical requirement when seismic forces are the major lateral load condition. In this case there would probably be some additional tie walls or struts provided between the interior column footings to supplement the basement and grade walls shown in the plan. Where wind is the major lateral load consideration, the basement floor slab and the first-floor concrete construction as shown would probably be sufficient.

First Floor Framing Plan (Fig. 20.2)

The first-floor area over the basement is shown with a structure similar to that for the typical upper level floors. At the unexcavated portions the first-floor structure consists of a concrete slab and beam system poured directly over fill. The typical beam is formed as shown in Detail C (Fig. 20.5). This network of beams on 18-ft centers is supported by the basement

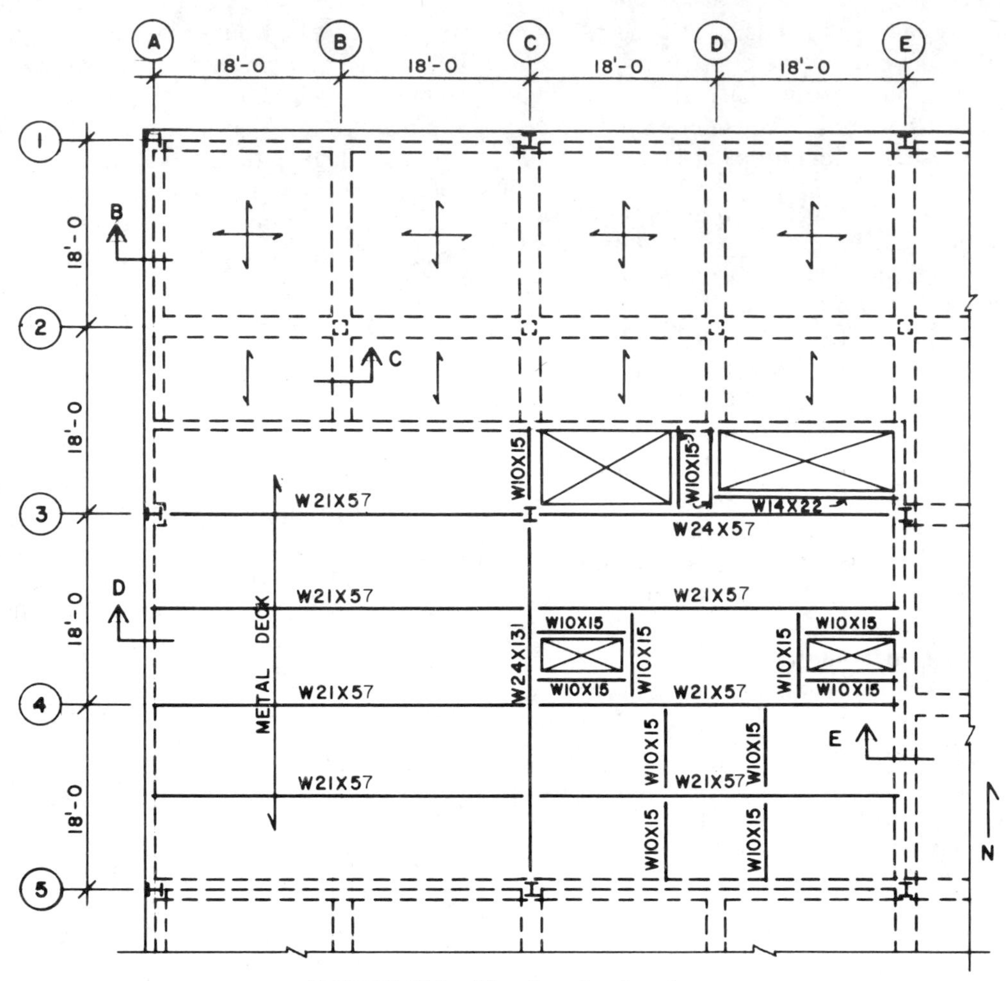

FIGURE 20.2. First-floor framing plan.

and grade walls and by a series of intermediate piers. The typical slab is a two-way slab on edge supports with a few one-way slabs at the north side of the basement.

These two systems interface at the interior basement walls, as shown in Detail E (Fig. 20.6). The concrete fill is continuous over both systems. A special detail required is the support for the steel beams at the concrete walls, as shown in Detail D and E (Figs. 20.6). This consists of a pocket in the wall with some erection bolts and a bearing plate for the end of the beam.

Typical Floor Framing Plan—Upper Levels (Fig. 20.3)

The typical floor consists of the steel deck placed over the network of beams that provide supports at 9-ft centers as well as at the building edges and at the edges of large openings. Some of the considerations in the detailing of this system are as follows:

1. *Attachment of the Steel Deck.* The deck units are normally welded to the steel beams in the valleys of the deck corrugations. These connections must

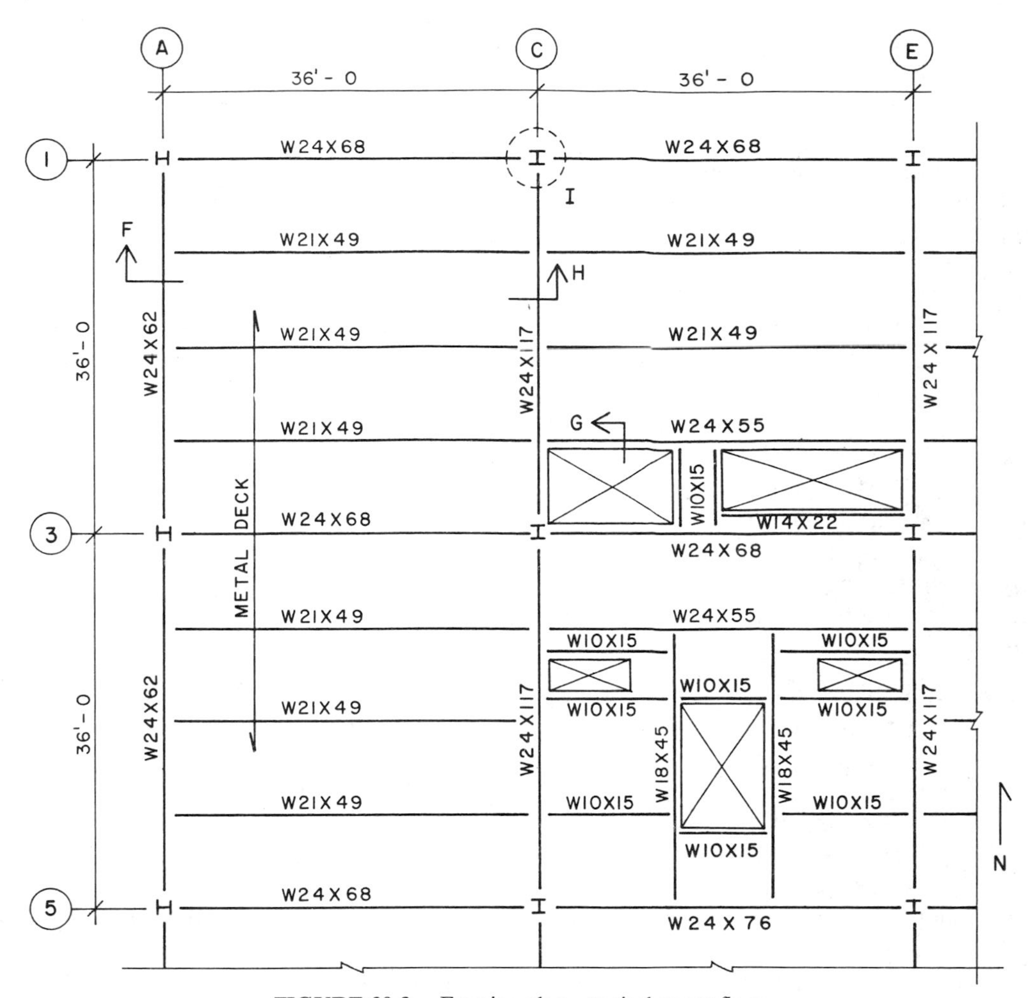

FIGURE 20.3. Framing plan—typical upper floor.

transfer the lateral loads from the floor, acting as a diaphragm, to the steel beams, which in turn transfer the loads to the column-beam bents. Details for this attachment and load ratings for the connections and deck are usually provided by the deck manufacturers.

2. *Support for the Hung Ceiling, Lights, Ducts, and So On.* The usual method of support for the ceiling is by wires that are installed through holes in the steel deck. Wire sizes and spacing and details for installation will depend on the type of ceiling and the specific type of deck used. Lighting units, ceiling registers, and small ducts may be supported as part of the ceiling construction, using the same wire hangers. Heavy ducts and equipment elements of the HVAC system will usually be supported by brackets or hangers attached directly to the beams.

3. *Fireproofing of the Floor Construction.* As discussed previously, the system used here consists of utilizing the concrete fill on top of the deck to protect the upper surface of the deck as well as the faces of beams at the building edge

and the edges of openings. The underside of the deck and the remaining exposed surfaces of beams are protected by sprayed-on fireproofing.

4. *Support for Interior Walls.* Planning of this building envisions two basic types of interior walls—permanent and demountable. Permanent walls are limited to those in the core and would consist of masonry or metal-framed plastered partitions. The weight of these plus their permanency would dictate that the floor structure provide both vertical and lateral support as part of the permanent structural system. Construction and finish of these walls would be influenced by desired architectural details, as well as by considerations of structural design, required code fire rating, acoustic separation, and similar factors.

Demountable walls may consist of a variety of constructions. Some of them may consist of masonry or plastered partitions, although the design of the floor deck should consider this if such is the case. It is expected, however, that most walls will consist of some relatively light construction, including possibly the use of some patented, modular system of relocatable units. The choice of the ceiling system and its detailing would need to consider the necessity for providing attachment and lateral support for whatever walls are anticipated.

The framing plans shown are abbreviated for clarity. Complete construction drawings would include sufficient information to establish the exact location of all beams, to establish the elevations of beam tops, to indicate required camber of beams, and to identify the type of connection for each beam. Some of this information may also be provided in details or schedules and be referred to by notes or symbols on the plans.

20.3. CONSTRUCTION DETAILS

The locations for the details described in this section are indicated on the plans in the preceding section.

Detail A (Fig. 20.4)

This section shows the relations between the basement floor, the basement wall, and the steel column and its footing. The basement floor is a paving slab over compacted fill. Necessity for a moisture

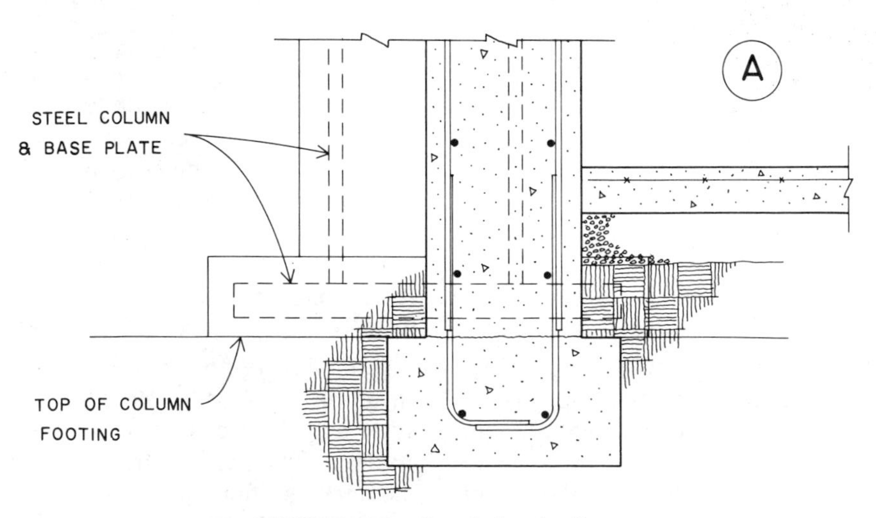

FIGURE 20.4. Foundation detail.

barrier under the slab and moisture penetration-resistant treatment at the slab-to-wall intersection would depend on the groundwater conditions at the site.

The basement wall at this point is a retaining wall, spanning vertically from the basement floor to the first-floor construction. Location of the wall footing was discussed previously in regard to the foundation plan. As shown in the detail, and also discussed previously, the top of the column footing is shown dropped so that the base plate and anchor bolts can be encased below the floor slab.

Detail B (Fig. 20.5)

This shows the general condition at the building edge adjacent to the unexcavated lower level. To accommodate the rental areas on the first floor the first-story wall is assumed to be similar in detail to that for the typical floor, as shown in Detail F (Fig. 20.6).

The first floor structural slab, although poured over fill, is a spanning slab and is supported vertically by the grade wall. In cold climates there should bc some insulation for the floor at the building edge and a thermal break between the floor slab and the exterior wall.

Detail C (Fig. 20.5)

This shows the typical beam for the unexcavated portion of the first floor. If the fill material is reasonably cohesive, the lower stem of this beam may be exca-

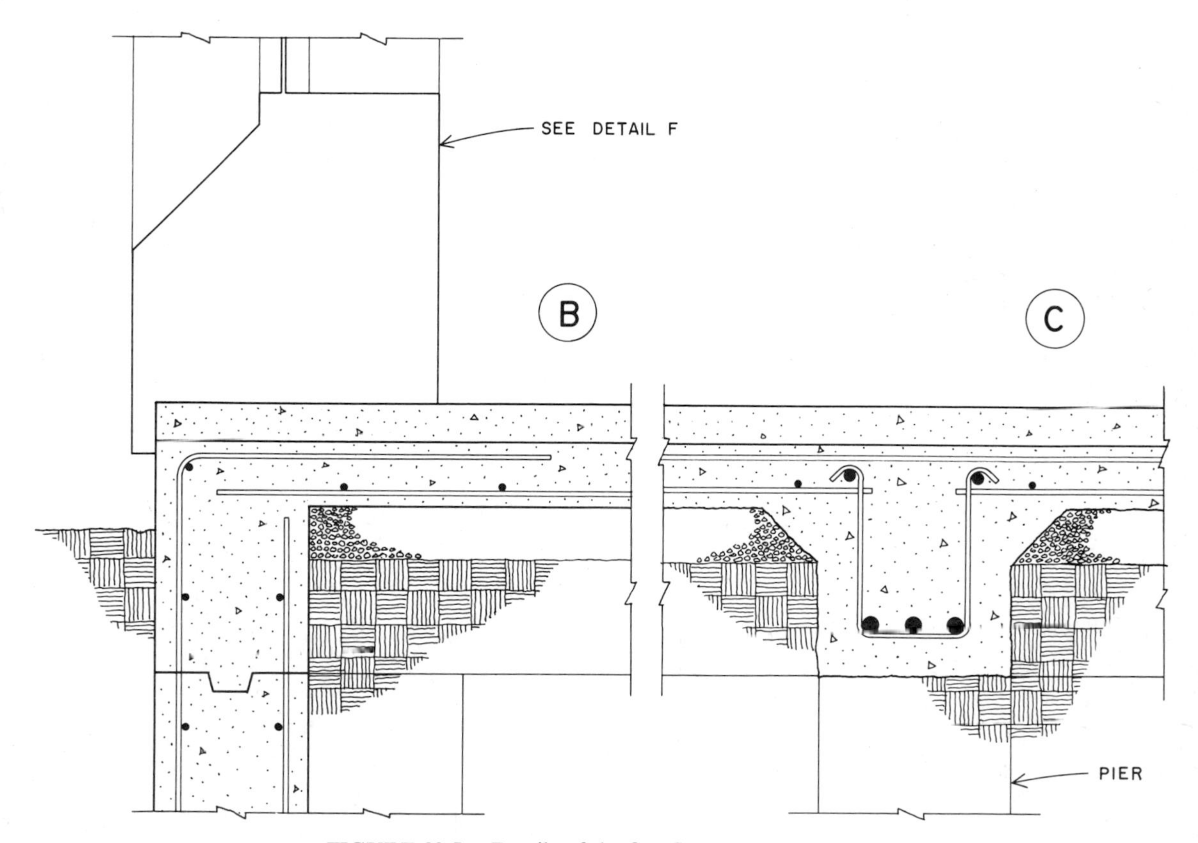

FIGURE 20.5. Details of the first-floor structure on fill.

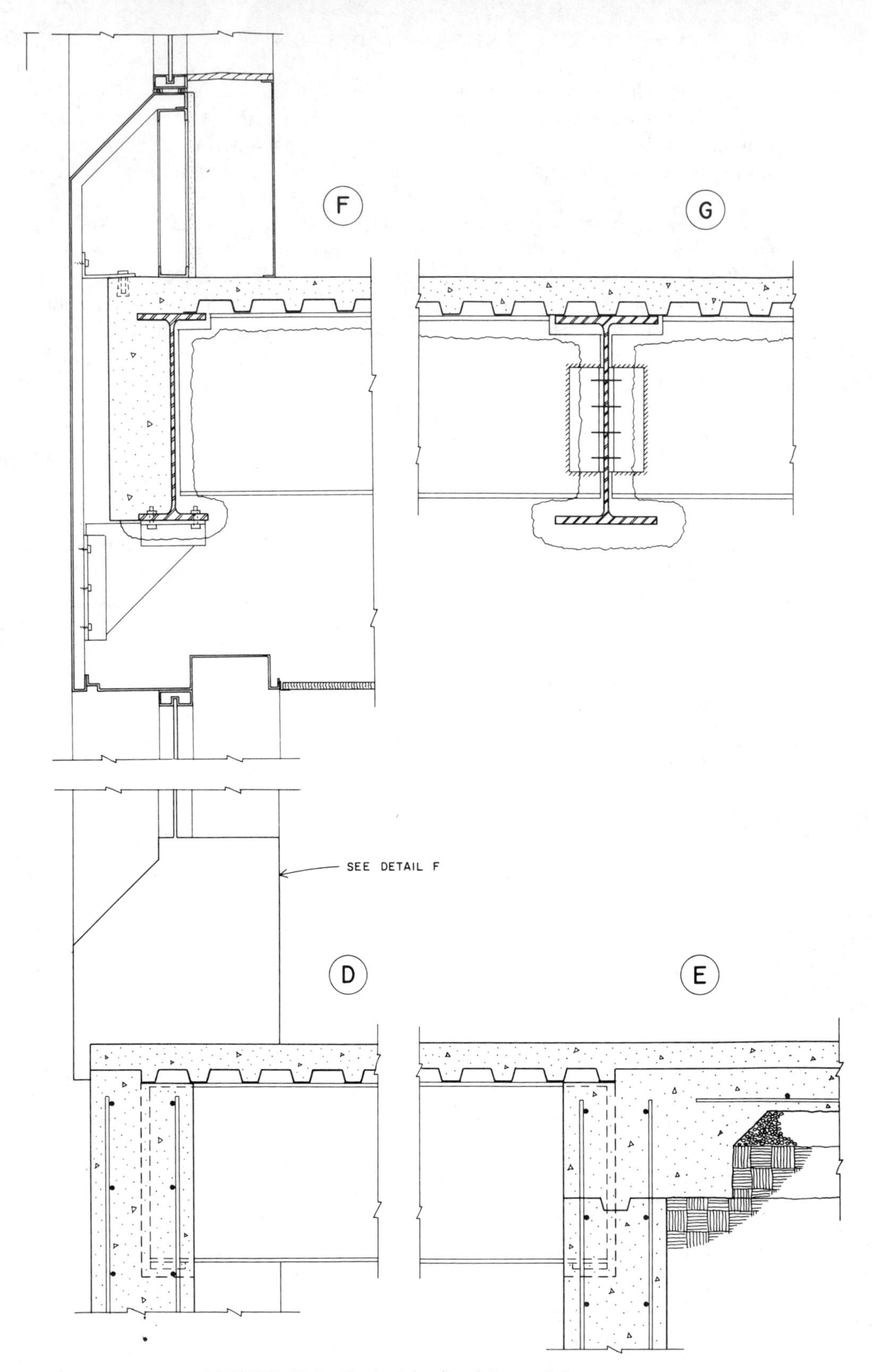

FIGURE 20.6. Typical details of the steel floor structures.

vated by trenching as shown. The reinforcing in the beam requires the usual 3-in. cover as for footings.

The piers for these beams (midway between columns) would probably be poured with column forms before the fill is placed, with the pour stopped at the level of the bottom of the beams and the pier vertical reinforcing extending into the beam.

Detail D (Fig. 20.6)

This shows the edge condition adjacent to the steel framed floor. The first-story wall is assumed to be similar to that for Detail B (Fig. 20.5). The steel beams are shown as supported by the wall, with a pocket, bearing plate, and erection anchor bolts.

Some consideration should be given to the transfer of horizontal force between the top of the wall and the floor construction if the basement wall is a retaining wall. At the section cut this is not the case because of the truck dock area.

Detail E (Fig. 20.6)

This shows the intersection between the two types of floor construction at the top of the basement wall. Because the concrete fill is continuous over both structures, the tops of the steel deck and the structural concrete slab are at the same level.

The key at the top of the wall pour should be adequate to provide for the lateral force due to the retained fill. The top of the wall pour is dropped to the level of the bottom of the concrete beams to allow the bottom reinforcing in the beams to extend over the supporting wall.

Detail F (Fig. 20.6)

This demonstrates the typical spandrel condition at the upper floors. The metal framed window wall is shown centered on the column line with the finished face of the spandrel brought out flush with the finished face of the column. Although not shown in detail, it is assumed that the building skin at the spandrels and columns consists of insulated units with an exterior metal facing. These units are shown supported by brackets from the floor and the spandrel beam. The window wall units rest on a short steel stud wall with a wide sill brought out to the finished face of the column. The space under these sills may house HVAC

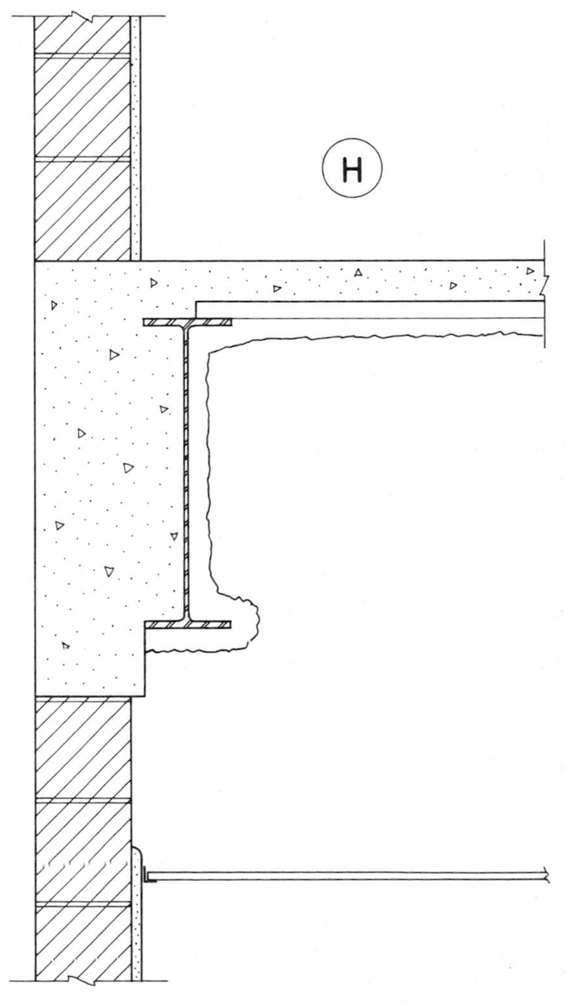

FIGURE 20.7. Framing at the core walls.

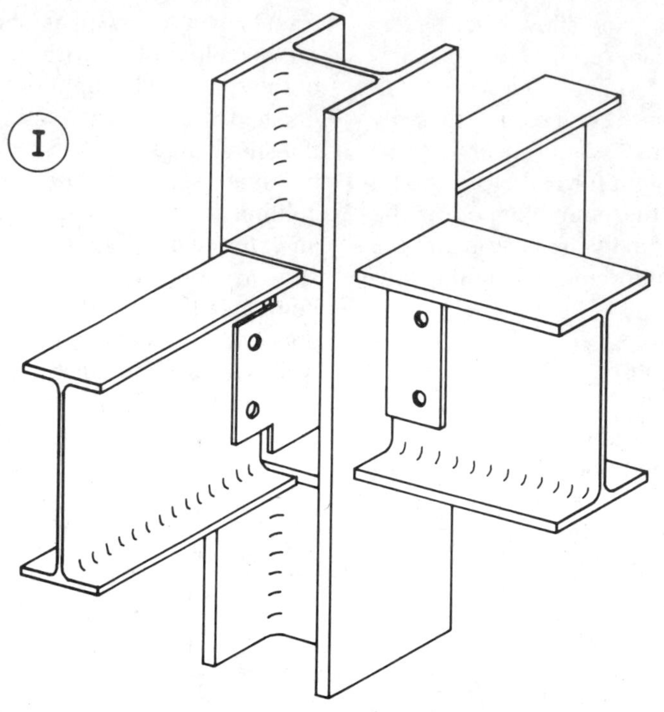

FIGURE 20.8. Detail of the column–beam connections.

units, if such a system is used. Lateral support must also be provided for the top of the window units. One way to achieve this would be to add some additional elements to the bracket that is attached to the bottom of the spandrel beam.

These details are of major concern in the architectural design and are subject to considerable variation without significant change in the basic structural system for the building.

Detail G (Fig. 20.6)

This illustrates a typical beam-to-girder connection using standard double angle connectors. As shown, the angles are typically welded to the beams in the shop and field connected to the girders with bolts.

Detail H (Fig. 20.7)

This gives one possibility for the floor edge condition at the large openings for the stair, elevators, and duct shafts. Although the section has concrete block for the wall, other materials may be used if a thinner wall is desired. One variation would be to stop the concrete closer to the steel beam and to run the wall past the face of the concrete.

Detail I (Fig. 20.8)

This shows one possibility for the beam-to-column and girder-to-column connections to achieve the column-beam bents. For erection and shear transfer, vertical plates would be added to connect the beam webs to the columns. The moment connections are achieved by butt welding the beam and girder flanges to the face of the column or to the stiffener plates that are coped and welded to the column web and inside face of the flanges.

Because shop fabrication and field erection practices vary somewhat locally, as well as between different steel fabricators, the final details and specifications for these connections should be worked out cooperatively between the building designer and the steel fabricator.

CHAPTER TWENTY-ONE

Design of the Concrete Structure

Figure 21.1 shows a framing plan for the typical floor using a reinforced concrete slab and beam system and reinforced concrete columns. The basic system consists of a series of beams on 12-ft centers supporting a multiple span, one-way slab. The orientation of the beams was elected to preserve the continuity of the majority of the beams, the only interrupted beams being at the opening for the elevators. As with the steel system, the beams and girders on the column lines form a series of rigid bents for resistance to lateral loads.

The beam spacing is related to the slab thickness, which must usually be a minimum of $4\frac{1}{2}$ in. for a required two-hour fire rating. Because the girders will be quite large, not much gain is to be made by forcing the beams to a minimum depth. Other options for the beam spacing would be 9 ft or 18 ft. For various reasons, the 12-ft spacing seems to be the best.

As with the steel structure, the exterior curtain wall could be a skin system completely covering the structure. To show some different construction detailing situations, however, we will use an exposed structure at the building exterior surface. On the interior of the building the structure will be mostly covered by the floor fill and finish flooring, by the suspended ceiling, and possibly by plaster on the interior column surfaces.

A framing system for the roof would most likely be similar to that used for the floor. A similar system would also be used for the portions of the first floor occurring over the basement. The basement, foundations, and the first floor over the unexcavated areas would be essentially the same as for the steel structure. The structure for the penthouse would most likely be a light steel frame—the same as for the steel structure.

A minimum concrete design strength with f'_c of 4 ksi [27.6 MPa] will be used. This is a matter of some variation regionally, being affected by economics as well as practical considerations of available materials and construction practices. It is advisable to use a slightly higher than average value for the concrete quality if the surfaces are to be exposed to the weather. To keep the sizes of the lower story columns down, it is possible to use a higher strength of the concrete for those columns alone.

Grade 40 reinforcing will be used for the beams and slabs. Grade 60 bars will

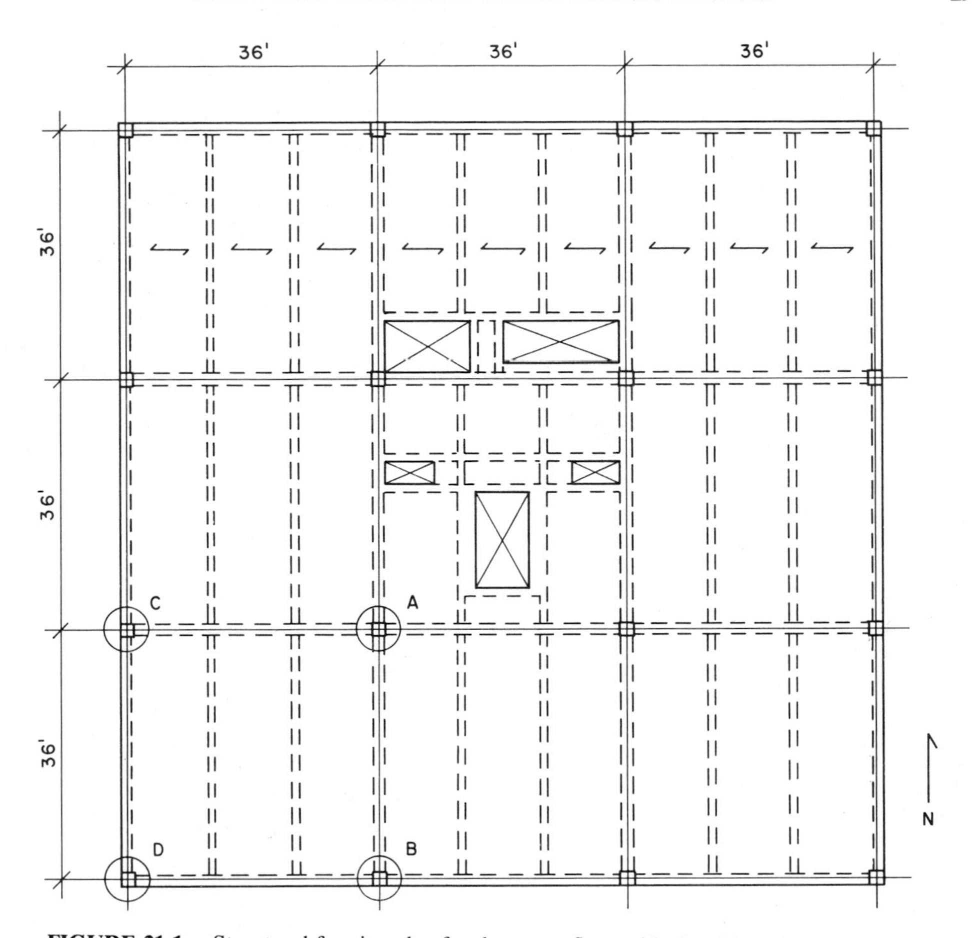

FIGURE 21.1. Structural framing plan for the upper floor with the slab and beam system.

be used for the columns, except in the lower stories where a higher strength will be used.

21.1. THE SLAB AND BEAM FLOOR SYSTEM

Because the majority of the beams are multiple span and/or part of the indeterminate column bents, exact analysis would most likely be done with computer programs. For a reasonably approximate design we will use the coefficients given in Chapter 8 of the ACI Code (Ref. 9). For simplification we will also use the working stress method for most of the design, which will produce somewhat conservative results in most cases when compared to those that would be obtained using the strength method.

Design of the Floor Slab

Live loads: 50 psf for the office areas
100 psf for lobbies, corridors, stairs
2000-lb concentrated load per UBC 2304(c)

Dead loads: Slab (estimate 5 in.) = 63 psf
3-in. lightweight fill

= 30

Ceiling, lights, ducts

= 15

Partitions [UBC 2304(d)]

= 20

Total dead load:

= 128 psf + beam stems [6.13 kPa]

The total design load for the slab is thus 178 psf [8.52 kPa] for the typical office areas. With a beam width of 15 in. [381 mm] the clear span for the interior spans will be 10 ft 9 in. Because of the position of the exposed spandrels, the end spans will be slightly larger. We will assume them to be 11 ft 3 in. In referring to Figure 21.3, the maximum moment in the slab will be $wL^2/10$ in the end span. We may thus check the slab for this condition as

$$\text{maximum } M = (\tfrac{1}{10})(178)(11.25)^2$$
$$= 2253 \text{ ft-lb } [3.06 \text{ kN-m}]$$

Using $f'_c = 4$ ksi and $f_s = 20$ ksi for a 1-ft strip,

$$\text{required } bd^2 = \frac{M}{K} = \frac{2253(12)}{324}$$
$$= 83.44 \text{ in.}^3$$

With b for the 1-ft strip equal to 12 in.,

$$\text{required } d = \sqrt{\frac{83.44}{12}} = 2.64 \text{ in. } [67 \text{ mm}]$$

As shown in Figure 21.2, with a 1-in. cover and maximum bar size of No. 5, the actual d will be 3.94 in. This means that the concrete stress in flexure will not be critical and all sections will be in the classification of underreinforced. This makes the true j values all higher than that for the balanced section.

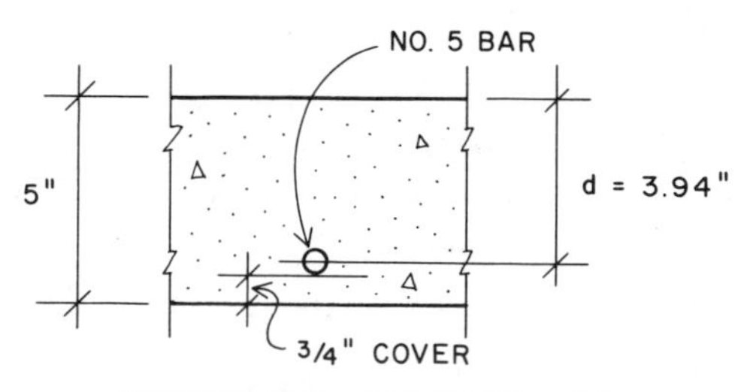

FIGURE 21.2. Detail of the slab.

Except for very short, highly loaded spans, shear stress will not be critical for a one-way spanning slab. Limiting deflection may be checked against the requirements in Chapter 9 of the ACI Code. These call for a minimum thickness of $\frac{1}{24}$ of the span with one end discontinuous and $\frac{1}{28}$ of the span with both ends continuous. Thus

$$\frac{11.25(12)}{24} = 5.625 \text{ in. for the end spans}$$

$$\frac{10.75(12)}{28} = 4.607 \text{ in. for the interior spans}$$

Since the large spandrel beams will actually provide considerable restraint, it seems reasonable to use the 5-in. slab.

Using an approximate value for j, the required steel areas at the various critical sections are determined and the bars selected as shown in Figure 21.3. The steel areas may be derived directly from the moment coefficients as

$$M = CwL^2 = C(178)(11.25)^2$$
$$= C(22{,}528) \text{ for the end span}$$

$$A_s = \frac{M}{f_s jd} = \frac{C(22{,}528)(12)}{(20{,}000)(0.88)(3.94)}$$
$$= C(3.90) \text{ for the end span and } C(3.56) \text{ for interior spans}$$

To save steel tonnage, an alternating system of long and short bars may be used, as shown in the bottom part in Figure 21.3.

Design of the Typical Beam

The beams not on the column lines are mostly three-span members carrying uniform load from a 12-ft-wide strip of floor. We will design these for the span of 36 ft

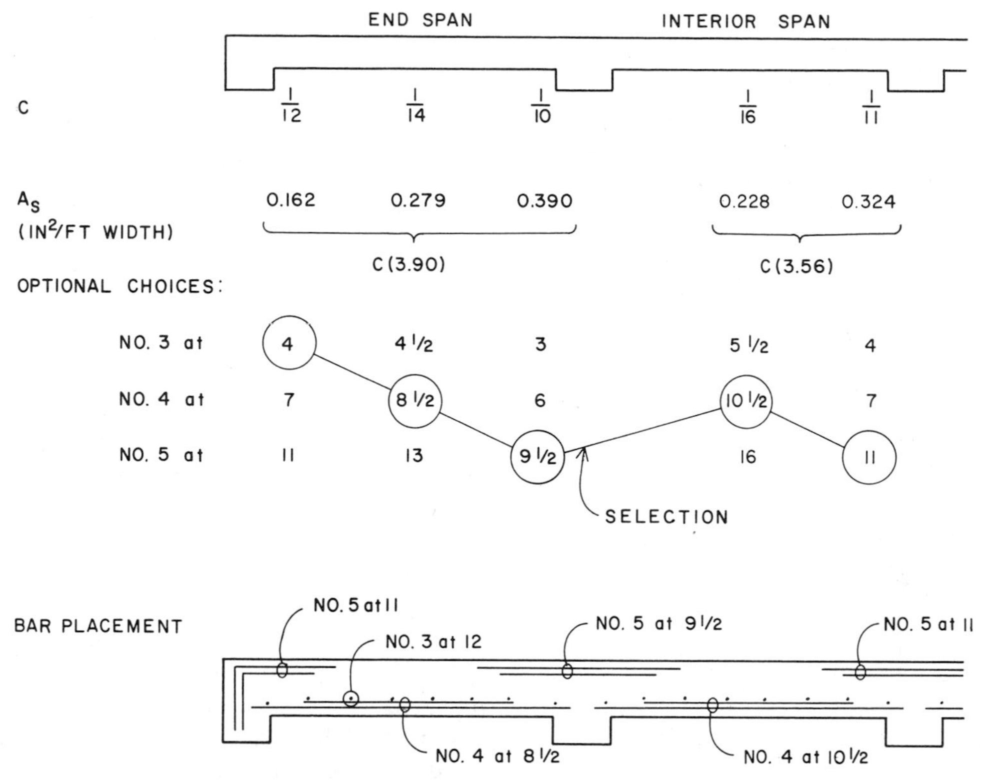

FIGURE 21.3. Design of the continuous slab.

using the moment coefficients from the ACI Code. Without deflection calculations the minimum depth is limited to $L/18.5$, or approximately 24 in. For a preliminary calculation we will assume a 15-in.-wide by 24-in.-high section. The design load will then be:

Dead load: Slab at 128 psf × 12
= 1536 lb/ft
Beam stem (19 by 15)
= 297
Total dead load
= 1833 lb/ft
Live load: 33 psf × 12
= 396 (reduced for area)
Total design load
= 2229 lb/ft [32.5 kN/m]

The maximum moment at the first interior support will be:

$$M = (\tfrac{1}{10})(wL^2) = (\tfrac{1}{10})(2.229)(36)^2$$

$$= 289 \text{ k-ft } [392 \text{ kN-m}]$$

For a rough guide the section chosen should be capable of about two thirds of this moment without compressive reinforcing. We thus compare the critical moment just determined with the balanced moment capacity of the section:

$$M = Kbd^2 = \frac{(0.324)(15)(20.5)^2}{12}$$

$$= 170 \text{ k-ft } [231 \text{ kN-m}]$$

Since this indicates that the balanced moment capacity of the trial section is only 59% of the critical required moment, we will reselect the section. Note that the actual value for d, as shown in Figure 3.39, is taken as approximately 3.5 in. less than the overall height for negative moment. This permits the top reinforcing in the girders to be placed above that in the intersecting beams, giving the heavier loaded girders the advantage. For positive moment, however, this problem does not exist, since the beam and girder will have different depths. Thus the d for positive moment will be approximated by deducting 2.5 in. from the height.

Based on the preliminary calculations, we select the beam size for an approximate balanced moment capacity of 200 k-ft [271 kN-m]. Then, if

$$M = 200 \text{ k-ft} = Kbd^2$$

$$\text{required } bd^2 = \frac{M}{K} = \frac{200(12)}{0.324} = 7407$$

And if $b = 12$ in.,

$$d = \sqrt{\frac{7407}{12}} = \sqrt{617} = 24.8 \text{ in. [630 mm]}$$

if $b = 15$ in.,

$$d = \sqrt{\frac{7407}{15}} = \sqrt{494} = 22.2 \text{ in. [564 mm]}$$

For a second trial we select a 15 × 26-in. overall size with a d of approximately 22.5 in. for negative moment. This beam should now be checked for shear stress as follows. From the ACI Code the maximum shear is taken as $1.15(wL/2)$. Using a clear span of 34.5 ft, the maximum end shear will be

$$V = 1.15\left(\frac{wL}{2}\right) = 1.15(2260)\left(\frac{34.5}{2}\right)$$

$$= 44{,}833 \text{ lb [199 kN]}$$

For design the critical shear may be taken as that at a distance of d from the support. We thus deduct (22.5/12)(2260), or 4238 lb from the shear at the end of the span for a design shear of 40,595 lb. The maximum design shear stress is thus

$$v = \frac{V}{bd} = \frac{40{,}595}{(15)(22.5)} = 120 \text{ psi [827 kPa]}$$

With the f'_c of 4000 psi the allowable shear stress on the concrete is 70 psi. Deducting this from the maximum shear stress leaves 50 psi to be carried by the shear reinforcing. With No. 3 U stirrups and an allowable stress of 20 ksi, the required spacing at the end of the beam is

$$s = \frac{(A_v)(f_v)}{(v')(b)} = \frac{2(0.11)(20{,}000)}{(50)(15)}$$

$$= 5.87 \text{ in. [149 mm]}$$

This is quite reasonable for a beam of this size, so the section is not critical for shear. We now proceed to a preliminary selection of the beam reinforcing using the ACI Code coefficients for the critical moments. The required steel areas will be determined from the moments as follows: For positive moment the section is a T, and we will use an approximate jd equal to d minus $t/2$. Thus

$$jd = d - \frac{t}{2} = 23.5 - 2.5 = 21.0$$

$$A_s = \frac{M}{f_s jd} = \frac{M(12)}{20(21)} = 0.0286M \quad (M \text{ in k-ft})$$

For negative moment, as discussed previously, $d = h - 3.5$. Thus, using the balanced section value for j, we have

$$d = 26 - 3.5 = 22.5 \text{ in.}$$

$$A_s = \frac{M}{f_s jd} = \frac{M(12)}{20(0.86)(22.5)} = 0.0310M$$

This may be used for moment up to 200 k-ft (the balanced capacity) with an additional reinforcing for the double reinforced section for moment in excess of 200 k-ft. This addition will be

$$A_s = \frac{M}{f_s(d - d')} = \frac{M(12)}{20(22.5 - 2.5)}$$
$$= 0.30M$$

Actually the A_s values could be derived directly from the ACI Code coefficients. The author, however, prefers to see the values of the moments calculated for a better sense of the design forces.

It is obvious that a design using SI units would not be done with a design strip width of 12 in. We have therefore not provided direct unit conversions for this work.

Figure 21.4 shows the calculations for determining the critical moments and steel areas for the typical beam. Before final selection of the reinforcing the bond

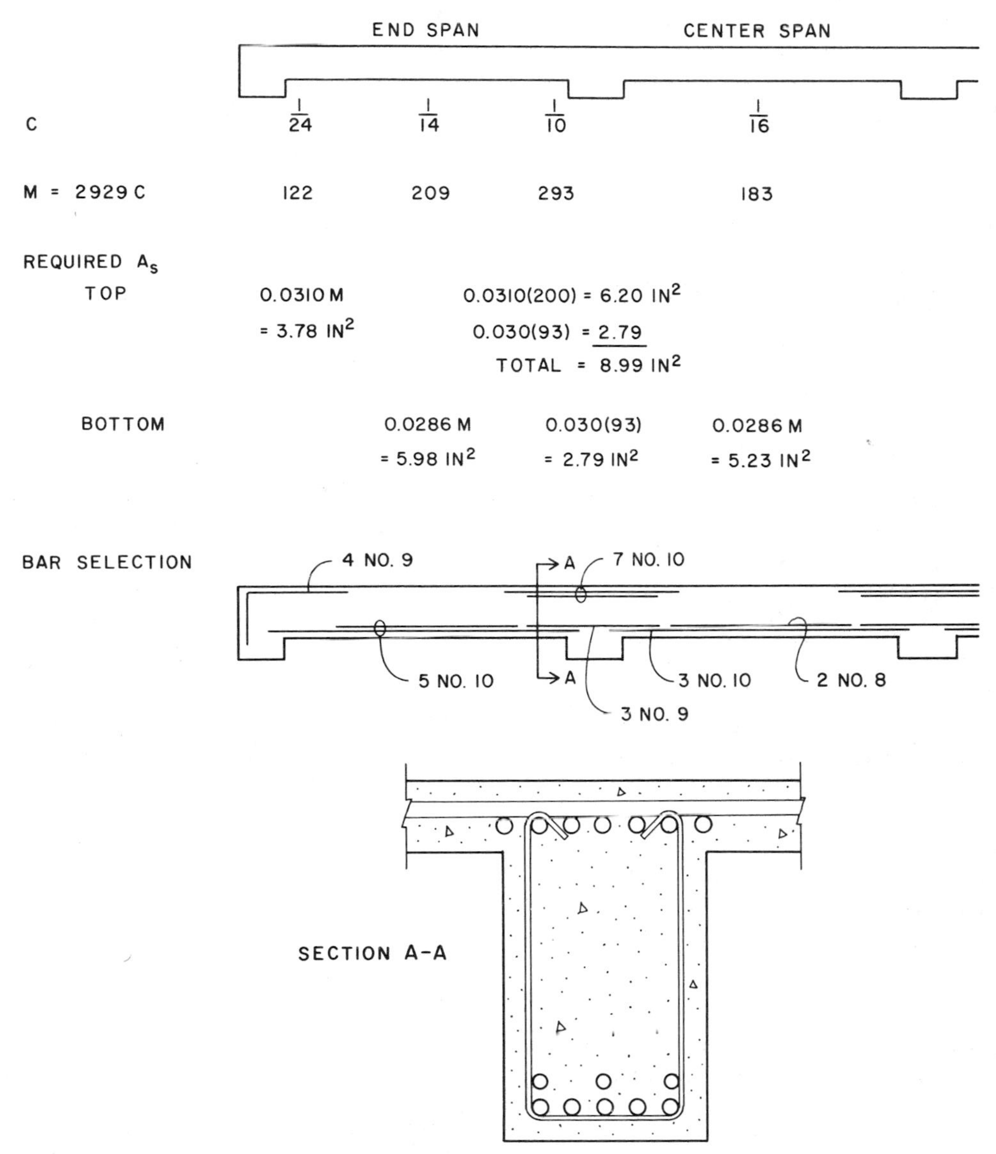

FIGURE 21.4. Design of the beam.

stresses must be checked and the allowable stress for the compressive reinforcing verified. The placement of the reinforcing must also be selected with the details of the girder and its reinforcing.

With regard to the beam itself, some of the placement considerations are shown in Figure 21.4. To keep the reinforcing in one layer at the bottom of the beam in the exterior span, the beam width must be increased. An alternative to this would be to increase the depth, if headroom is not critical. The depth increase would slightly reduce the steel area required and would add less additional concrete to the section.

In the top of the beam it is possible to place some of the reinforcing outside the stirrups as shown, as long as the minimum cover is maintained. For the negative moment at the interior support the added bars shown would have to be placed in a second layer, which slightly reduces the $(d - d')$ moment arm that was assumed in the calculations.

The reinforcing for the compression side of the double reinforced section may be provided by a separate set of bars as shown. If so, they would have to be placed in a second layer. An alternative would be to bend the bars from the interior span to fit over those in the end span. Doing this at both ends, however, would result in rather long bars, which may not be desirable for handling.

21.2. BEAMS AND GIRDERS IN THE VERTICAL BENTS

Design of the Column Bent Beam

The interior beam on the column line will have the same combination of wind and gravity moments as was illustrated in the design of the steel structure. We will use a process of designing the beams for the gravity moments first and then investigate the need for any changes required for the added wind moments. As will be seen, this works reasonably well for this building, but may not be as useful for buildings of greater height or shorter beam spans.

For the gravity load design the only difference for the column-line beam will be an increase of the end moment at the outside column to $(\frac{1}{16})wL^2$. This is an increase of 50% over that used for the end moment at the spandrel girder for the typical beam.

Some general design considerations for the column-line beams are:

1. The beam reinforcing must pass through the columns. The size and spacing of the column reinforcing must be carefully coordinated with those for the beam to allow for this.
2. Some of the top reinforcing should be made continuous to provide for reversal moment and to add torsional strength to the bents.
3. Some of the bottom reinforcing in the end span should be extended into the exterior column and bent into the outside face of the column. This is to provide for the reversal wind moment.
4. Full loop stirrups (similar to column ties) should be provided through the length of the beam to increase its torsional strength.

With the exposed structure, as discussed previously, the design of the spandrel beams and girders must be coordinated with the general architectural design of the exterior walls. Figure 21.5 shows one possibility for the spandrel. Although it would be poured with a construction joint as shown, the section could be considered as a single structural unit with doweling of the vertical reinforcing and a series of horizontal shear keys at the pour joint. This very deep section, with an approximate d of 70 in., would result in a much higher stiffness for the exterior column–beam bents, which means that they would take a higher percentage of the total wind force on the building. This effect will be considered later in the wind design.

With the section as shown in Figure 21.5, the wall load on the spandrel will be

Spandrel: (14 by 75 approximately)	= 1094 lb/ft	15.96 kN/m
Window wall: 6 ft at 15 psf	= 90	1.31
Total wall load	= 1184 lb/ft	17.27 kN/m

Added to this will be approximately one half of the dead and live loads for the typical beam. Because of the slightly smaller area of floor supported by the spandrel beam, the live load reduction will be less. The design loads are thus

Total dead load: $1184 + \frac{1}{2}(1536)$	= 1952 lb/ft	28.49 kN/m
Life load: 42 psf $\times$ 6	= 252 lb/ft	3.68
Total design load	= 2204 lb/ft	32.17 kN/m

Because this total is approximately the same as that for the typical beam, the design gravity moments will be similar. The spandrel has such a large d, however, that the calculated steel areas will be quite small. Because of its depth, its exposure, and the high wind moments in the exterior bents, the spandrels should be reinforced with continuous top and bottom reinforcing, with other minimal reinforcing as for a wall, and with vertical loop ties throughout their length.

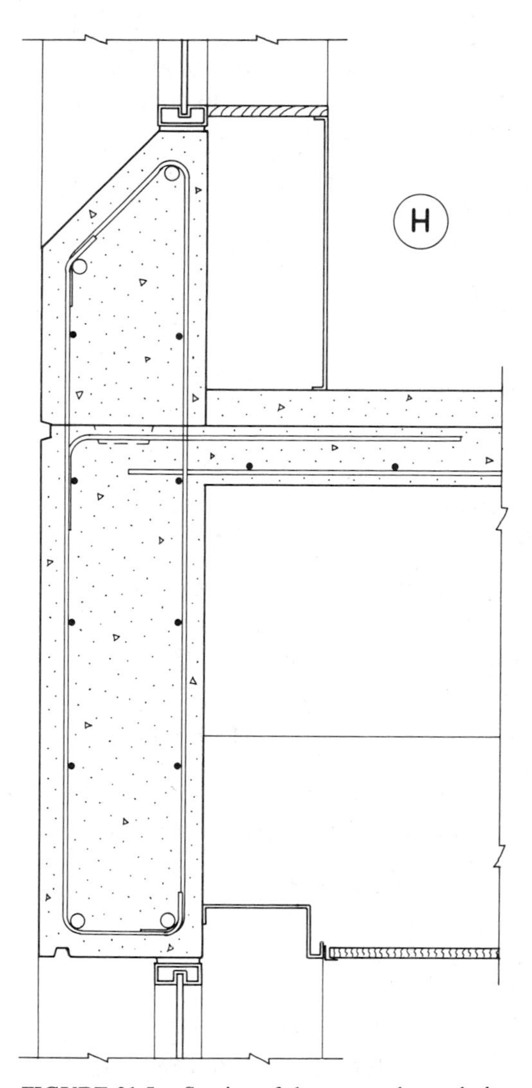

FIGURE 21.5. Section of the exposed spandrel.

Design of the Girders

As in the steel structure, the girders carry a combination of uniform and concentrated loads. The proportion of the uniform to the concentrated load is slightly higher here because of the heavier girder and the two-way action of the slab which pulls some of the floor load to the girder at the ends of the beams. For an approximate design we will determine the total load on the girder and consider it to be carried as uniform load, using the ACI Code coefficients for design moments. The load on the interior girder is as follows:

Beams: Dead load = 2 × 34.5 × 1860 lb/ft
= 128,340 lb
Live load = 2 × 34.5 × 12 × 30 psf
= 24,840 lb

Assuming an 18 × 30 girder,
Stem dead load = (18 × 25)(150/144)(34)
= 15,938 lb
Floor dead load = 1.5 × 34 × 128 psf
= 6,528 lb
Floor live load = 1.5 × 34 × 30
= 1,530 lb
Total load on girder
= 177,176 lb [788 kN]

On the spandrel girder the load will be approximately one half of that due to the beams plus the spandrel dead load as calculated for the spandrel beam. Thus

One half of beam load
= 76,590 lb
Spandrel + wall = 40,256
Total load = 116,846 [520 kN]

For the spandrel girder the maximum moment and area of steel required are

$$M = (\tfrac{1}{10})WL = (\tfrac{1}{10})(117)(34)$$

$$= 397.8 \text{ k-ft } [539 \text{ kN-m}]$$

$$A_s = \frac{M}{f_s jd} = \frac{397.8(12)}{20(0.86)(70)}$$

$$= 3.96 \text{ in.}^2 \; [2555 \text{ mm}^2]$$

The selection of the reinforcing should be delayed until the wind analysis is made.

For the interior girder, assuming an 18 × 30 in. section with a d of 27.5 in.:

$$M = (\tfrac{1}{10})(177)(34)$$

$$= 601.8 \text{ k-ft } [816 \text{ kN-m}]$$

$$A_s = \frac{601.8(12)}{20(0.86)(27.5)}$$

$$= 15.3 \text{ in.}^2 \; [9872 \text{ mm}^2]$$

This is a lot of reinforcing; furthermore, the balanced moment capacity of the section is only about 60% of the critical moment. At this point it would have to be established whether the limiting depth has been reached for the girder. Referring to Figure 18.3, it may be seen that a total of 48 in. has been allowed from the finished floor to the bottom surface of the ceiling. Subtracting for the floor fill and ceiling construction, this leaves approximately 42 in. With a 30-in.-deep girder there would be a maximum clearance of 12 in. below the girder

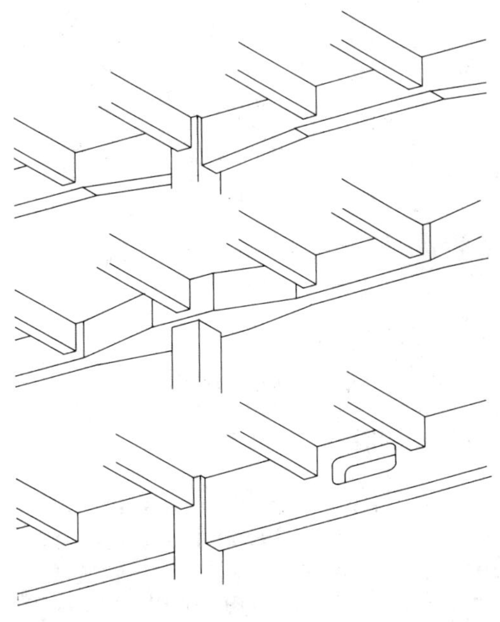

FIGURE 21.6. Options for increasing the girder strength.

for a duct. If the duct layout can be arranged so as to avoid having the largest ducts pass under the girders, this may be adequate. However, any reduction of this clearance by increase in the girder depth seems unfeasible.

The sketches in Figure 21.6 show three possible solutions to this problem. The first two consist of increasing the negative moment capacity of the girder by either widening it or increasing its depth at the ends. The other possibility shown is that of using the entire depth available and providing holes through the web of the girder. A rough rule of thumb for these holes is that they should not exceed one third of the beam depth in height.

On the basis of these considerations we assume a 40-in. overall height girder with holes provided as required for ducts up to 12-in. deep. The weight of the girder will be slightly more, so we will use a total design load of 184 kips for the calculations. The determination of the critical moments and required steel areas for the girders is summarized in Figures 21.7 and 21.8. Selection of the final rein-

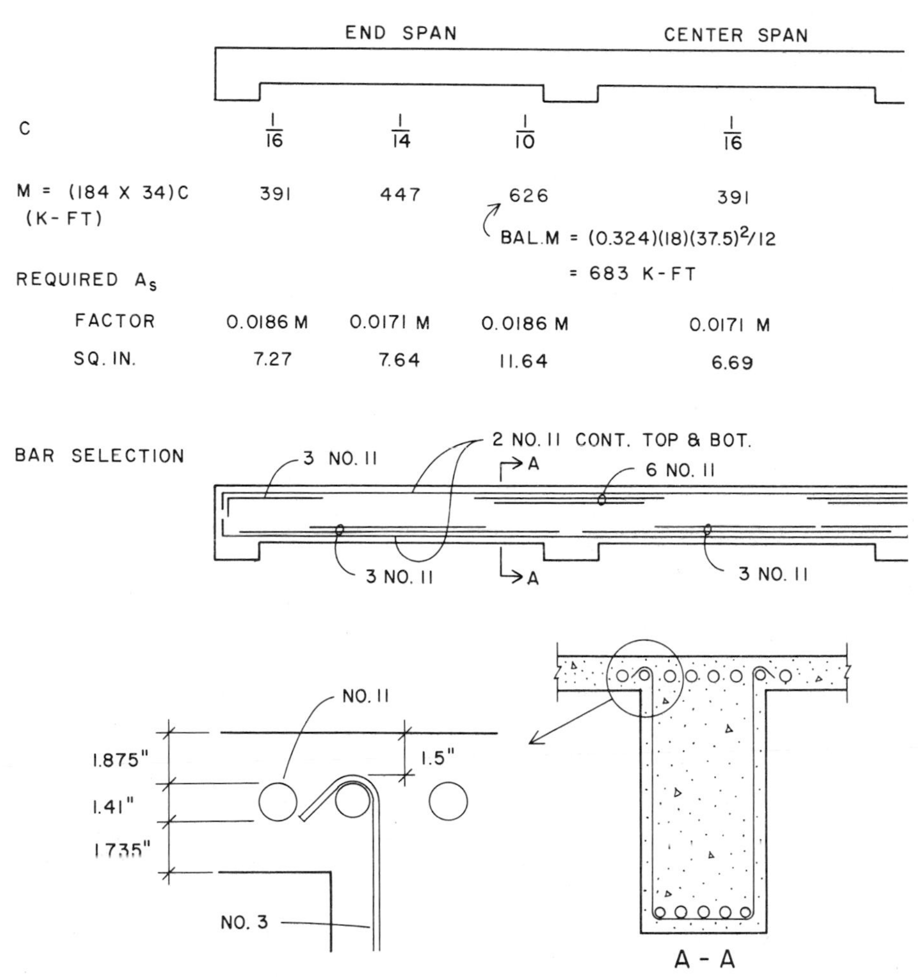

FIGURE 21.7. Design of the interior girder.

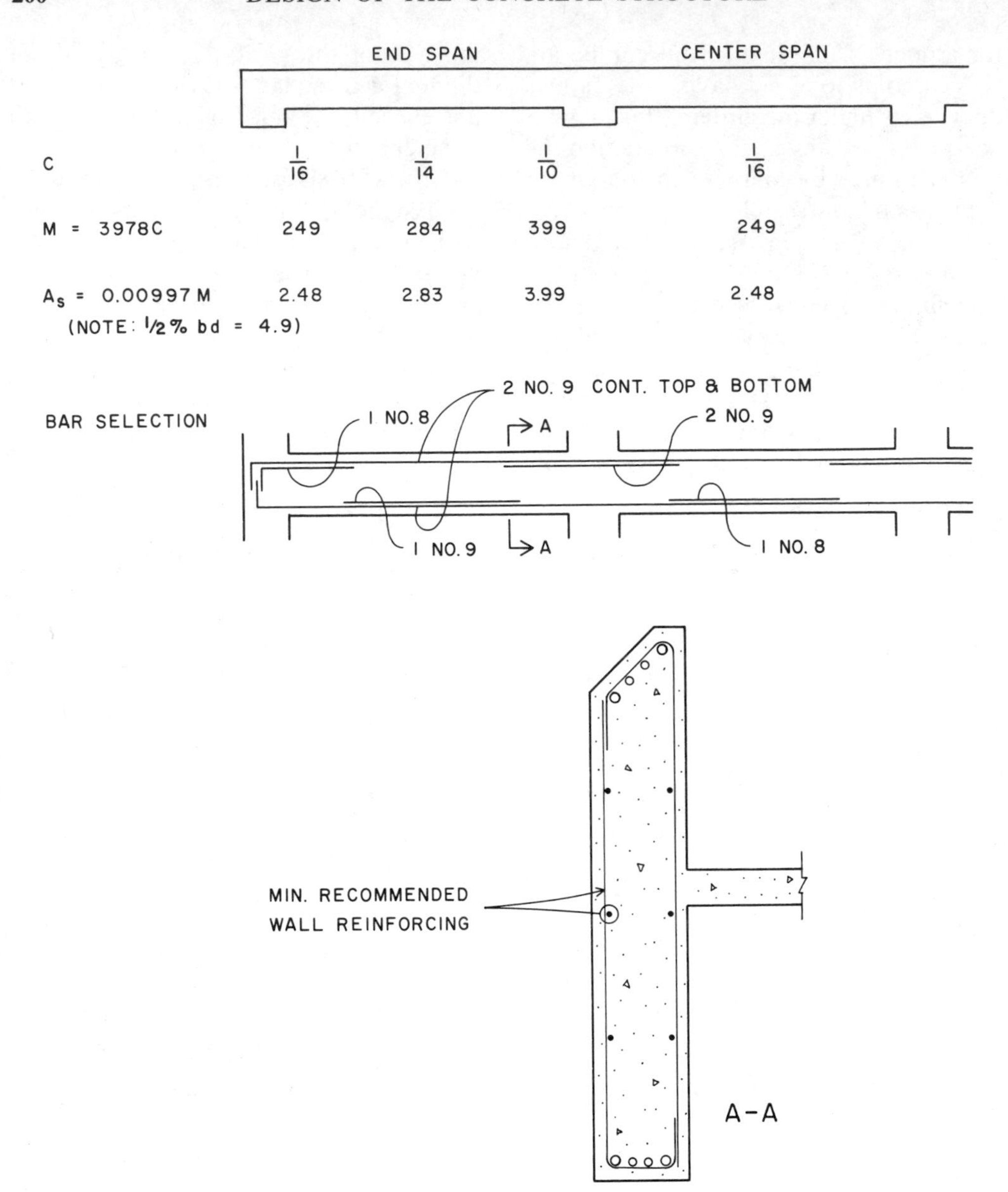

FIGURE 21.8. Design of the spandrel girder.

forcing should be delayed until the wind analysis is made.

21.3. THE COLUMNS

As with the steel structure, the columns must be designed for the combination of gravity axial compression, rigid frame moments due to gravity, and the bending and axial force due to wind. Because of the higher gravity dead loads with the concrete structure, it is less likely that wind will be a critical factor in the bent design. It is even more logical, therefore, to use the procedure illustrated in the design of the steel structure: that of designing first for gravity alone to obtain approximate sizes of members.

A tabulation of the column axial loads for the concrete structure could be done using the format illustrated for the steel structure in Table 19.3. Since the basic building plan is the same, the live loads for the columns would be the same for both structures. Thus the only new tabulation necessary is for the dead loads. To save effort, we will approximate these by comparing the unit loads for the two structures.

Referring to the earlier calculations for the slab design, the unit dead load for the typical floor is 128 psf, which includes the slab, the fill, the partition load, the ceiling, and the suspended equipment. Added to this will be the weight of the beam and girder stems. For the typical 36-ft square bay these will be

3 beams/bay at 15 in. by 21 in. × 34 ft = 33,469 lb

1 girder/bay at 18 in. by 35 in. × 34 ft = 22,313 lb

total stem weight/bay = 55,782 lb

or

$$\frac{55{,}782}{(36)^2} = 43 \text{ psf average}$$

Adding this to the other dead load, the total dead load is 171 psf for the typical floor. Since the column weights and the exterior wall loads will also be higher, we will approximate the dead loads for the concrete structure by using twice the tabulated loads for the steel structure. Table 21.1 gives the column loads determined on this basis using the design live loads plus twice the dead loads for each column as determined in Table 19.3.

TABLE 21.1. Gravity Loads for Concrete Columns (kips)

	Column A	Columns B and C	Column D
Level	DL + %LL = Design Load	DL + %LL = Design Load	DL + %LL = Design Load
P, R			
	70 + 10 = 80		
R			
	230 + 90 = 320	90 + 13 = 103	52 + 7 = 59
6			
	442 + 62 = 504	224 + 18 = 242	134 + 12 = 146
5			
	656 + 88 = 744	360 + 32 = 392	216 + 16 = 232
4			
	870 + 114 = 984	496 + 45 = 541	300 + 23 = 323
3			
	1084 + 140 = 1224	632 + 58 = 690	384 + 30 = 414
2			
	1300 + 166 = 1466	768 + 71 = 839	474 + 37 = 511
1			
	1640 + 192 = 1832	922 + 85 = 1007	576 + 44 = 620
B			

Variation of the column strength from top to bottom of the building is a different matter for the concrete structure. It was reasonable to accomplish this variation in the steel structure with no change in the finish size of the columns. This is less reasonable here, especially for the exterior columns which are exposed architecturally. For the interior columns the variation in size may cause some variation in the plan dimensions of the core layout, although there are possibilities for accommodating this. For the exterior columns the method of variation, as well as the actual dimensions, must be coordinated with the detailing of the exterior walls and the spandrel-to-column relationships.

Figure 21.9 shows some of the relationships and the scheme that will be used for effecting dimensional changes in the exterior columns. For reasons of simplification of the window detailing, the external face of the columns will be kept at a constant width of 24 in., with changes occurring in the other direction. The exterior face may project beyond the spandrel and the interior face may be brought in from the window sill edge as shown. The simplified spandrel section shows the basis for these limits. Because of its two exposed faces, the corner column will be maintained at a constant size of 24 in. square.

Because the spandrel and column have the same requirements for cover of the reinforcing, the bars closest to the exposed face would normally be the same distance from the edge if the spandrel and column are flush. This problem is eliminated if the column face is a few inches outside the spandrel. However, if the flush face condition is desired, the outside bars in the spandrel must be bent to pass the column verticals, as shown in the sketch in Figure 21.9.

As has been mentioned previously, there is a general problem of coordina-

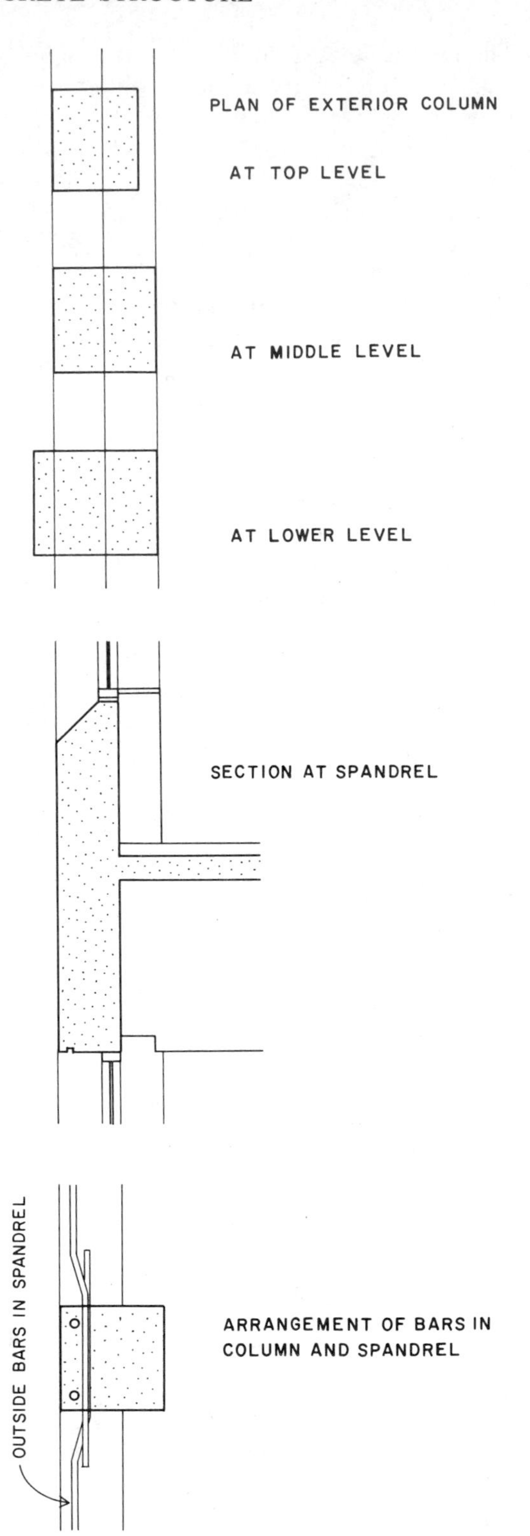

FIGURE 21.9. Details of the exterior columns.

tion of the column reinforcing with that of the intersecting beams and girders. At every column there is a three-way intersection of reinforcing and the column verticals must be placed so that the beam and girder bars can pass through. With a high percentage of reinforcing in a column, and especially with spiral columns, this can be a severe problem. Carefully drawn layouts or mockups should be done to avoid nasty calls from the construction site.

Figure 21.10 shows a first trial design for the interior columns using only the axial gravity loads previously tabulated. Three sizes have been used, varying from 20-in. square at the top to 28-in. square at the bottom. Some of the considerations made in the reinforcing selection are:

1. The number of bars used in each column has been coordinated to avoid adding splice bars; that is, there are no columns with more reinforcing than in the column below.

2. The number of bars and the tie layouts have been selected to avoid having complex tie layouts with ties cluttering up the center space, which makes

COLUMN A - FIRST TRY - AXIAL LOAD ONLY

	DESIGN LOAD	SIZE (INCHES)	REINFORCING	f'_c	ACTUAL CAPACITY (e = 3")	LAYOUT
R–6	320	20 X 20	4 NO. 9	4	317	
6–5	504	24 X 24	8 NO. 11	4	554	
5–4	744	24 X 24	16 NO. 11	4	694	
4–3	984	30 X 30	16 NO. 11	4	1157	
3–2	1224	30 X 30	20 NO. 11	5	1230	
2–1	1466	34 X 34	20 NO. 11	5	1490	
1–B	1832	36 X 36	20 NO. 14	5	1796	

FIGURE 21.10. Design of column A for gravity load only.

concrete placing difficult. Although the use of spiral columns, especially in the lower stories, could reduce the column size, we have avoided them principally because they make placing of the beam and girder reinforcing more difficult.

3. Grade 60 bars have been used for all columns. However, the design strength of the concrete (f'_c) has been increased in the lower story columns from 4 to 5 ksi. This is principally in the interest of reducing the column sizes required.

We must, of course, consider the combined effects of axial compression and bending on all columns. Figure 21.11 shows the general relation of these effects in the form of the typical interaction graph. Three general cases are described in the illustration. Case 1 occurs when the moment is minor compared to the axial load. This is the case for most interior columns, especially at the lower stories. In this case the moment does not cause significant change from the design for axial load only.

Case 3 occurs when the moment is major compared to the axial load. This is likely to occur with exterior columns at the upper stories and possibly with wind moments in the lower stories. In this case it is reasonable to design essentially for the moment only, with a little bonus for the axial load.

Case 2 is the general case, when both the axial load and moment are significant. There is no simple way to approach this design other than by trial and error. Use of handbooks or canned computer programs is to be recommended if more than a few sections must be designed.

Some approximation of axial loads and moments must be obtained before any judgment can be made about the situation for a particular column. We will therefore develop approximate gravity moments and approximate wind moments for the bents before attempting to design the exterior columns.

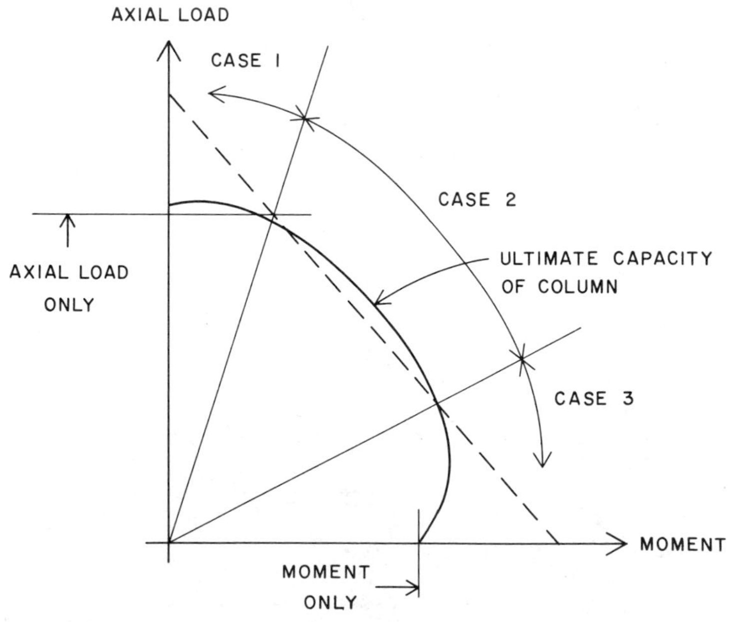

FIGURE 21.11. Column interaction: axial load plus bending.

TABLE 21.2. Approximation of Gravity Moments on Exterior Columns

	Column B		Column C		Column D	
	N–S	E–W	N–S	E–W	N–S	E–W
Moment at roof $M = (\frac{1}{16})wL^2$ (k-ft)	183	×	×	391	183	249
Moment at floor $M = (\frac{1}{32})wL^2$ (k-ft)	92	×	×	196	92	125

For a first approximation of gravity moments in the bents we may use the end moments from the beam and girder designs, as determined from the ACI Code coefficients. At the roof this results in a moment of $(\frac{1}{16})wL^2$ on the sixth-story column. At all other levels the moment will be $(\frac{1}{32})wL^2$ because two columns resist the beam end. For an approximate design we have assumed the total roof design load to be equal to that for the typical floor. Table 21.2 tabulates these moments for the four typical columns.

The two most critical bending conditions occur at the roof level at columns C and D. A quick check should be made to assure that these two conditions will be possible with the size limits established in the preliminary design. For column C the condition is that of a major moment from the interior girder combined with a small axial load and a small moment from the spandrel beam. This column can be designed quite literally as a 24-in.-square doubly reinforced beam for the large moment of 391 k-ft.

With the values of 4000 psi for f'_c and 60 ksi for f_y, the balanced moment capacity for the section will be

$$M = Kbd^2 = \frac{(0.295)(24)(21)^2}{12} = 260 \text{ k-ft}$$

For this moment the area of steel required in tension will be

$$A_s = \frac{M}{f_s jd} = \frac{(260)(12)}{(24)(0.85)(21)} = 7.28 \text{ in.}^2$$

Since this moment is only about 62% of the required moment, we must rely on the compressive reinforcing with some additional tension reinforcing to develop the additional resistance. The additional area of steel required is

$$A_s = \frac{M}{f_s(d - d')} = \frac{(131)(12)}{(24)(18)} = 3.64 \text{ in.}^2$$

The total area of tension reinforcing required is thus 10.92 in.^{2}. If provided by No. 11 bars, the number required is

$$N = \frac{10.92}{1.56} = 7$$

If placed in a single layer in the outside face of the column, this would require a column 28-in. wide. If it is necessary to keep the column width of 24 in., we must place reinforcing in two layers in the outside face or increase the dimension perpendicular to the plane of the outside wall. Choosing the second alternative, we will increase this dimension to 30 in., which makes the column face project 6 in. beyond the spandrel. The balanced moment capacity of the column now increases to

$$M = Kbd^2 = \frac{(0.295)(24)(27)^2}{12} = 430 \text{ k-ft}$$

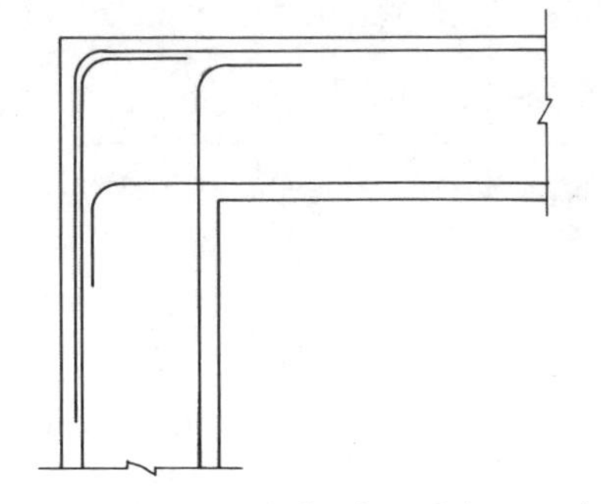

FIGURE 21.12. Reinforcing of the exterior columns at the roof.

This means that in theory the compressive reinforcing is not required and the area of tension reinforcing may be simply calculated as follows:

$$A_s = \frac{M}{f_s jd} = \frac{(391)(12)}{(24)(0.85)(27)} = 8.52 \text{ in.}^2$$

This area can be provided by six No. 11 bars, with a total of 9.36 in.^{2} Actually the small axial compressive force slightly increases the moment capacity, as shown in the interaction graph in Figure 21.11. The design for moment alone is therefore conservative.

COLUMN C – FIRST TRY – GRAVITY LOAD ONLY

Level	AXIAL LOAD (KIPS)	MAJOR MOMENT (K-FT)	e	CASE FOR M/N (SEE FIG. 21.11)	SIZE (INCHES)	REINFORCING
R–6	103	391	46"	3	24X30	6 NO. 11
6–5	242	196	9.7"	2	"	6 NO. 11
5–4	392	"	6.0"	2	"	10 NO. 11
4–3	541	"	4.4"	2	"	10 NO. 11
3–2	690	"	3.4"	1	"	10 NO. 11
2–1	839	"	2.8"	1	"	10 NO. 11
1–B	1007	"	2.3"	1	"	10 NO. 11

LAYOUT

f' = 4 ksi (above level 3)

f' = 5 ksi (below level 3)

FIGURE 21.13. Design of column C for gravity load only.

As shown in Figure 21.12, part of the tension reinforcing in the outside face of the column may be provided by bending down the top bars from the girder. Thus the column is actually shown in the column design table as having only four No. 11 bars, one in each corner. The two outside corner bars plus the four bars bent down from the girder provide the necessary six bars for the tension reinforcing in the column.

At the sixth floor level the moment in the column drops to half that at the roof. Considering this moment only, the tension reinforcing requirement drops to three No. 11 bars in the outside face. The minimum reinforcing used from this point down is therefore three No. 11 bars in the outside column face.

Figure 21.13 presents a summary of the design for column C. The size of 24 by 30 in. is maintained throughout the height of the column, since it is adequate for the axial load at the bottom as well as for the moment at the top.

Column B has less moment because the beam end moments are smaller. However, for the purpose of balance in the architectural details, it would probably also be made a constant size of 24 by 30 in. throughout its height.

Column D, the corner column, sustains considerable bending in both directions. As with columns B and C, the large gravity bending moments at the roof level are major design considerations. As has been previously discussed, the shape and size of this column are a matter of architectural detailing and construction considerations as well as structural behavior. The sketches in Figure 21.14 show some of the possibilities for this column.

In the upper sketch the exterior width of 24 in. is maintained on both column faces, matching the width of the intermediate columns on the building elevation. This limits the column size to 24 by 24 in. and would require considerable reinforcing for the moments at the roof.

The second sketch shows the outside faces pulled out 6 in. beyond the spandrel, as was done for the intermediate columns. This creates a 30-in.-square column, which is fine for the moments at the roof but a bit large for the axial loads at lower levels. It also means that the column face in the building elevation is wider than the intermediate columns.

The third sketch is a compromise, with a 6-in. square nick taken out of the

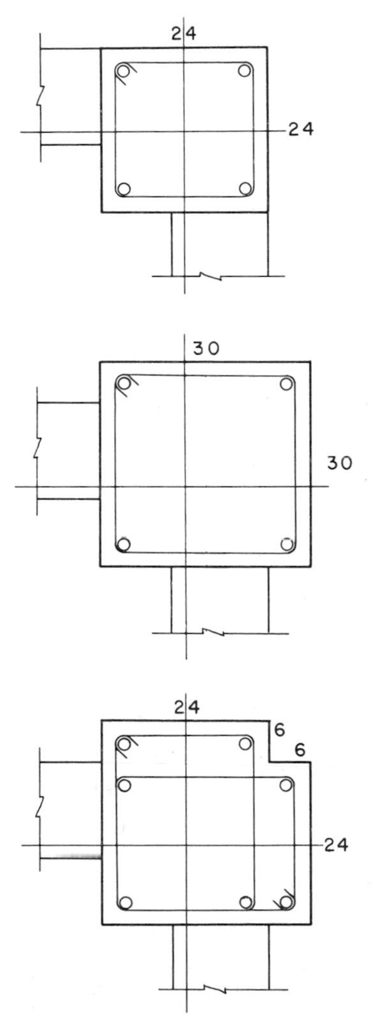

FIGURE 21.14. Options for the corner columns.

COLUMN D – FIRST TRY – GRAVITY LOAD ONLY

30" SQUARE – FULL HEIGHT

	AXIAL LOAD (KIPS)	MOMENT X-AXIS/Y-AXIS (K-FT)	e (INCHES)	CASE FOR M/N (SEE FIG. 21.11)	REINFORCING
R–6	59	322 / 183	65 / 37	3	8 NO. 10
6–5	146	161 / 92	13 / 7.6	3	8 NO. 10
5–4	232	"	8.3 / 4.8	2	8 NO. 10
4–3	323	"	6 / 3.4	2	8 NO. 10
3–2	414	"	4.7 / 2.7	2	8 NO. 10
2–1	511	"	3.8 / 2.2	2	8 NO. 11
1–B	620	"	3.1 / 1.8	2	8 NO. 11

LAYOUT

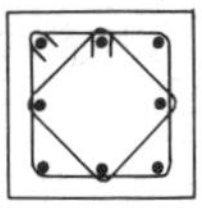

f'_c = 4 ksi (above level 3)

f'_c = 5 ksi (below level 3)

FIGURE 21.15. Design of column D for gravity load only.

corner of the 30-in. square. This leaves the 30-in. depth for bending in both directions and presents a face width to match the intermediate columns on the elevation. Reinforcing placement and tie layouts are a little more complicated for this option, but can be handled. For axial load design the section is still essentially a 30-in. square with a loss of only about 4% of its area in one corner.

As with the intermediate exterior columns, the moment at the roof level can be partly developed by bending down the top end bars from the spandrel beams and girders. The location of these bars is shown in the sketches in Figure 21.14.

Figure 21.15 presents a summary of the design of column D, based on the use of a 30-in. square section for the full height of the column. As it turns out, the reinforcing required at the top story for the gravity bending at the sixth floor is approximately the same as that required for the total load combination at the basement. It is probably practical, therefore, to use the same reinforcing for the full height to simplify bar placement, dowelling, and tie layouts.

21.4. DESIGN FOR WIND

The wind load on the building was previously determined for the design of the steel structure, as shown in Figure 19.10. As for the steel structure, an approximate wind design may be done by assuming the total wind shear at each story to be distributed to the columns in proportion to their individual stiffnesses. Because of the larger dead loads in the concrete structure, it develops that there are only a few considerations that need to be made to alter the design for the gravity loads to have adequate resistance of the bents to wind. To demonstrate this, and avoid the work of a complete wind analysis, we will determine the maximum wind shears and moments at the first story and the second-floor level.

Figure 21.16 illustrates the method for determination of the shear distribution to the first-story columns. The column heights used are based on the interior girder depth of 40 in. and the spandrel girder depth of 7 ft. Because of the deep spandrels, the exterior bents will resist a higher proportion of the total story shear.

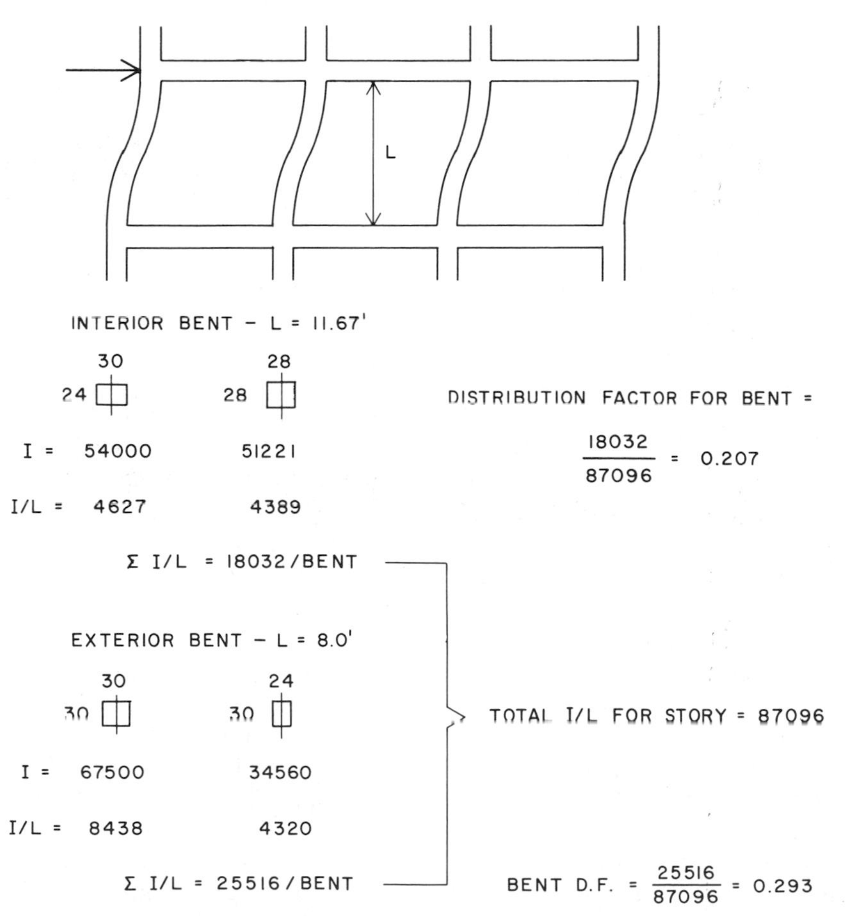

FIGURE 21.16. Shear stiffness of the first-story bents.

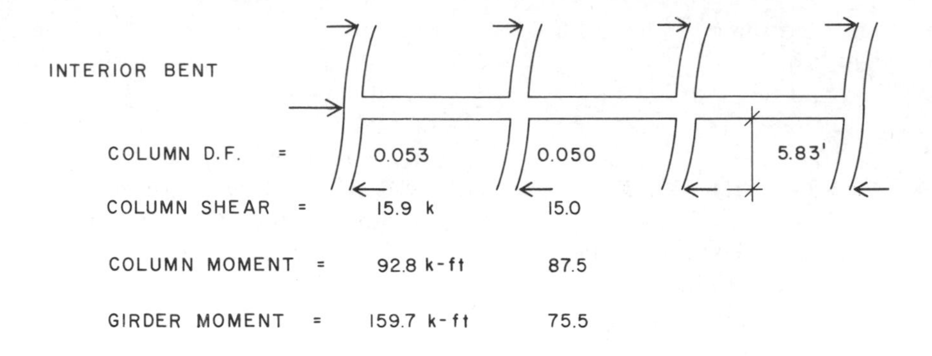

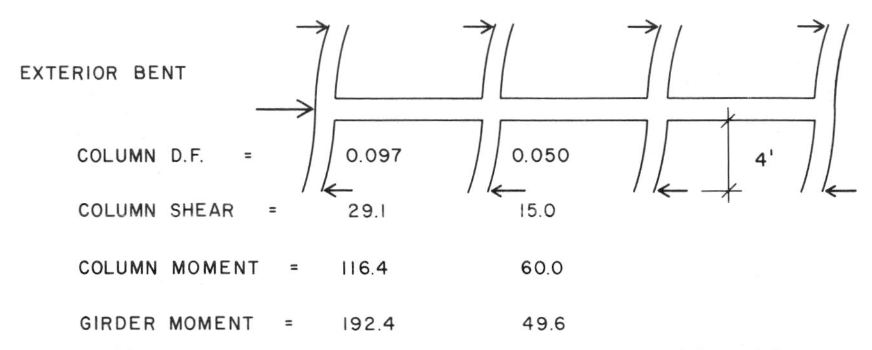

FIGURE 21.17. Wind analysis: first-story columns and second-floor girders.

Figure 21.17 illustrates the basis for determination of the column shears and moments and the girder moments. Comparison of these column moments with those approximated for the gravity loads, as given in Table 21.2, should indicate that the wind loads will not be a factor in the column designs.

A similar comparison of the girder moments with those used in the girder designs in Figures 21.7 and 21.8 will indicate that the only consideration necessary is a slight increase in the end moment at the corner column for the spandrel girder.

None of the wind moments calculated will result in reversals of the sign of the end moments in the girders. There will, however, be some redistribution of the moments throughout the length of the spans that will cause some shifts from the moment variations assumed in the development of typical details for bar cutoffs and extensions. For this reason, as well as to add a general increased toughness to the bents, some continuous top and bottom bars should be used in all the column-line beams and girders. These bars will be made continuous through the interior columns and will be hooked into the exterior columns. For the same reasons the column reinforcing in the top story will be bent into the girders. The sketches in Figure 21.18 show some of these details.

The effects of wind moments on the column line beams in the north–south direction will be somewhat more critical, because the gravity loads are less on these members. The same continuous bars and hooked end details should be used in these bents. If headroom per-

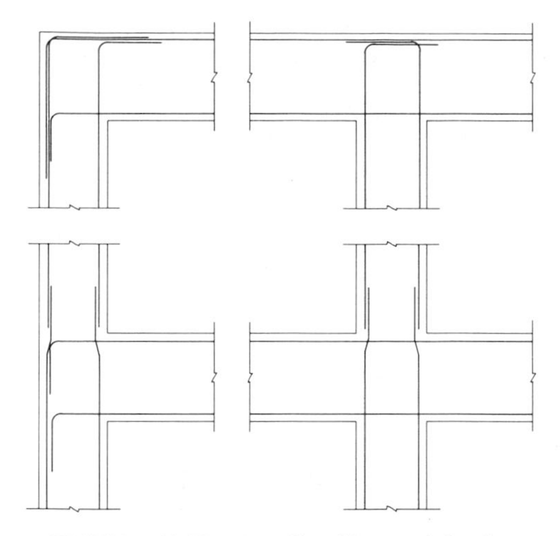

FIGURE 21.18. Details of bent reinforcing.

mits, it would be advisable to increase the depth of the column-line beams in the lower levels. If that is done, the same reinforcing could probably be used in all the beams, with the required additional strength being gained by the depth increase.

21.5. THE FIRST FLOOR, BASEMENT, AND FOUNDATIONS

The lower level construction for the concrete structure will be similar to that for the steel structure, with the exception of the first-floor area over the basement and

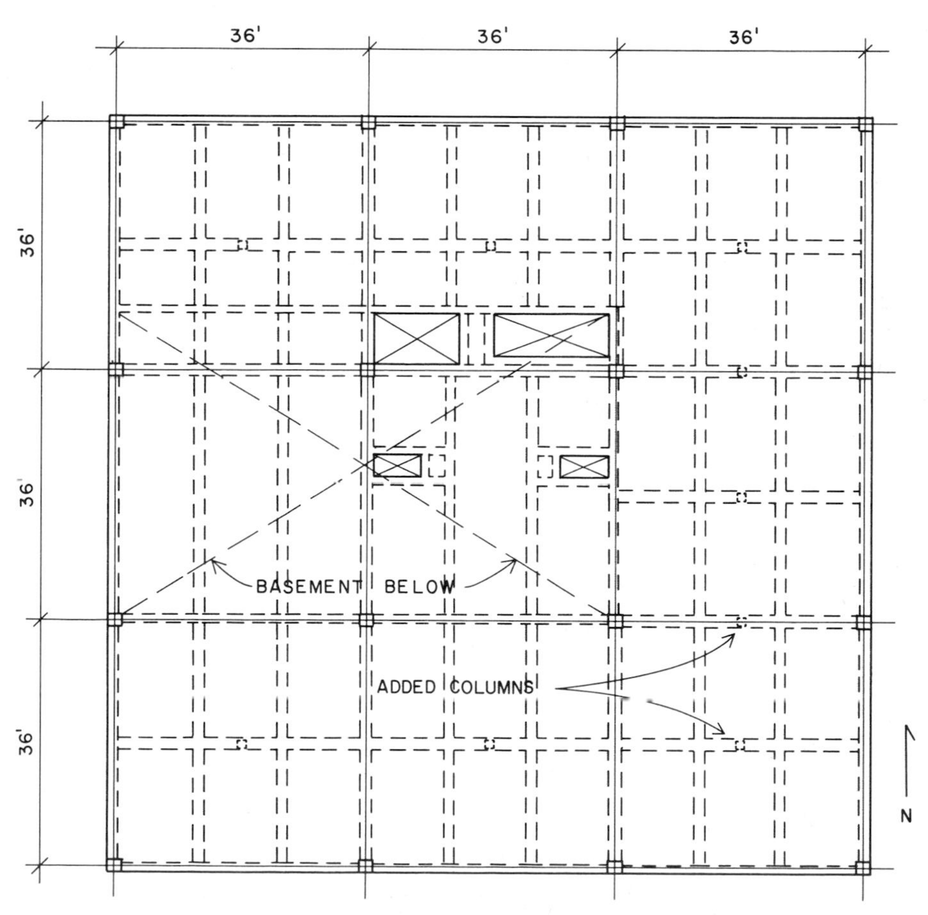

FIGURE 21.19. Framing plan: first floor.

the elimination of the steel columns and base plates at the basement. A slightly different layout may be used for the floor structure over the unexcavated area since it is possible to develop continuity with the framed system over the basement in this structure. The plan in Figure 21.19 shows a layout for this type of system.

One problem in the use of a continuous structure over the two first-floor areas is that the span length changes abruptly at the edge of the basement area. This causes a condition of moment reversal and high shear in the first span of the short beams. Although it is possible to design for this condition, there are some alternatives worth considering.

One alternative consists of using a construction joint between the two areas, effectively interrupting the structural continuity between them. Being thus made independent of the other system, the structure over the unexcavated area could be designed as for the steel structure, as shown in Figure 20.2, or could use a variety of layouts.

If it is possible to introduce some additional columns in the basement area, another alternative would be that shown in the plan in Figure 21.20. The shortened spans in the basement area are now in balance with those over the unexcavated area, which eliminates the problem discussed for the first system. The beams and girders at the basement area would be considerably smaller than those required for the longer spans, although this cost savings would be offset by the need for additional columns and footings.

The three alternatives discussed are

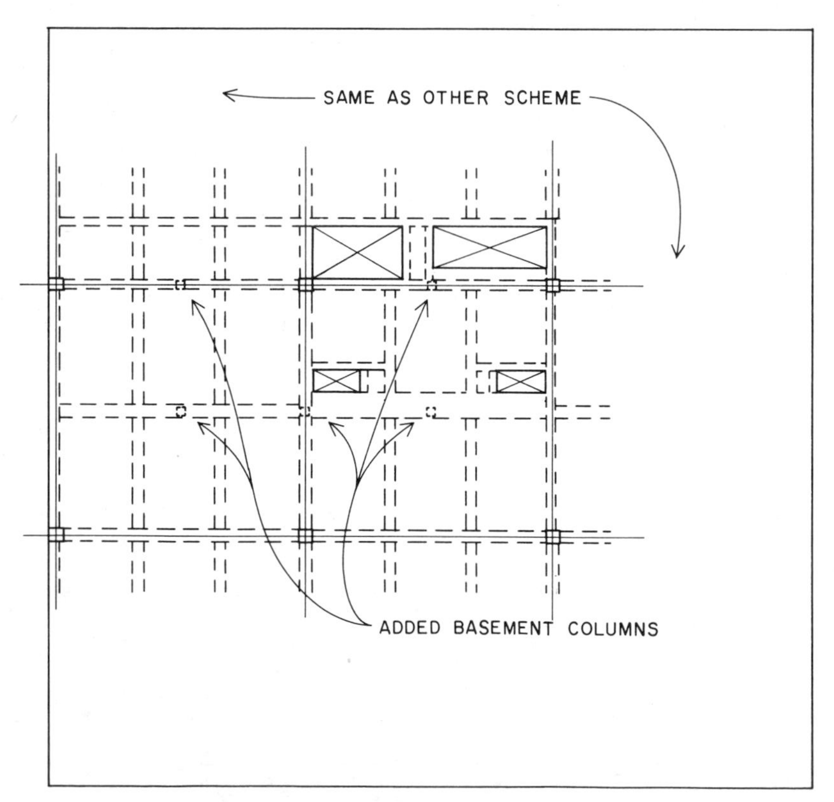

FIGURE 21.20. Alternate framing plan: first floor.

reasonably competitive. Assuming that the additional basement columns are not a problem, the author prefers the third solution and has shown it in the construction drawings.

The basement walls, basement floor slab, grade walls, and footings would be essentially similar to those for the steel structure. The continuity of horizontal reinforcing in the walls is somewhat simplified, because the steel columns are not encased in the concrete sections. The tops of the column footings may be slightly higher, since the steel base plates and anchor bolts need not be accommodated.

The column footings will be somewhat larger because of the greater dead loads, making the feasibility of spread footings marginal. If the allowable soil pressure is any less than the 8000 psf assumed, the use of piles or piers would probably be advisable.

One additional consideration in the footing design is the need for a footing depth sufficient to provide for the required dowelling extension of the column bars in the bottom story. Since this length is a function of both the bar size and the f_y value, the choice of these in the column design must be coordinated with the footing design.

CHAPTER TWENTY-TWO

Construction Drawings for the Concrete Structure

The illustrations that follow show some of the details for the typical floor and lower level construction for the concrete structure. As with the other examples, the plans and details are essentially illustrative only and do not include all the details and notation that would be required for complete construction documentation.

22.1. CONSTRUCTION ILLUSTRATED IN OTHER CHAPTERS

Some of the illustrations in other chapters show materials that are applicable to the construction of this structure. A description of some of these drawings follows.

1. The details in Figures 17.1 and 17.2 for Building Three would be equally applicable for this structure if the general wall configuration is acceptable. It is also possible to encase the concrete structure totally and to use a metal exterior skin as shown in Figure 20.6 for the steel structure for this building. The construction as shown in Figure 21.5 was the basis for the design of the spandrel framing. Another possibility is presented later in this chapter—using skin wall units of precast concrete. All of this is just the tip of the iceberg; variations are almost endless, although we have shown the two basic systems—the exposed structure and the encased structure.

2. The ground-floor construction for the concrete structure would be essentially the same as for the steel structure. Figure 20.2 shows the plan of the system for the steel structure with a framed floor poured on grade for the portions of the plan with no basement. It may also be possible to use a simple slab on grade floor for these areas, as shown for Building Three in Figure 17.2.

22.2. STRUCTURAL PLANS

Partial Basement and Foundation Plan (Fig. 22.1)

This plan is essentially similar to that for the steel structure as shown in Figure 20.1. Note the added columns and footings in the basement area to accommodate the first-floor system using the scheme shown in Figure 21.20.

Some of the footing details and wall details would be different because of the omission of the steel columns, the column base plates and anchor bolts, and the seats for the first-floor steel beams. The location of the exterior walls is slightly different because of the different treatment of the skin of the building; see Figures 20.5 and 22.5 for comparison.

Partial Framing Plan—Ground Floor (Fig. 22.2)

This plan shows the use of a continuous system for the two areas with a basic column module of 18 ft. The location of the north wall of the basement causes some disruption of this system, although the only significant considerations are the change in direction of the slab spans and the special reinforcing for the beams that are supported by this wall. The latter condition is shown in Figure 22.7.

Partial Framing Plan—Typical Floor (Fig. 22.3)

This shows the typical upper level floor system as designed in the calculations. The core framing at the large openings

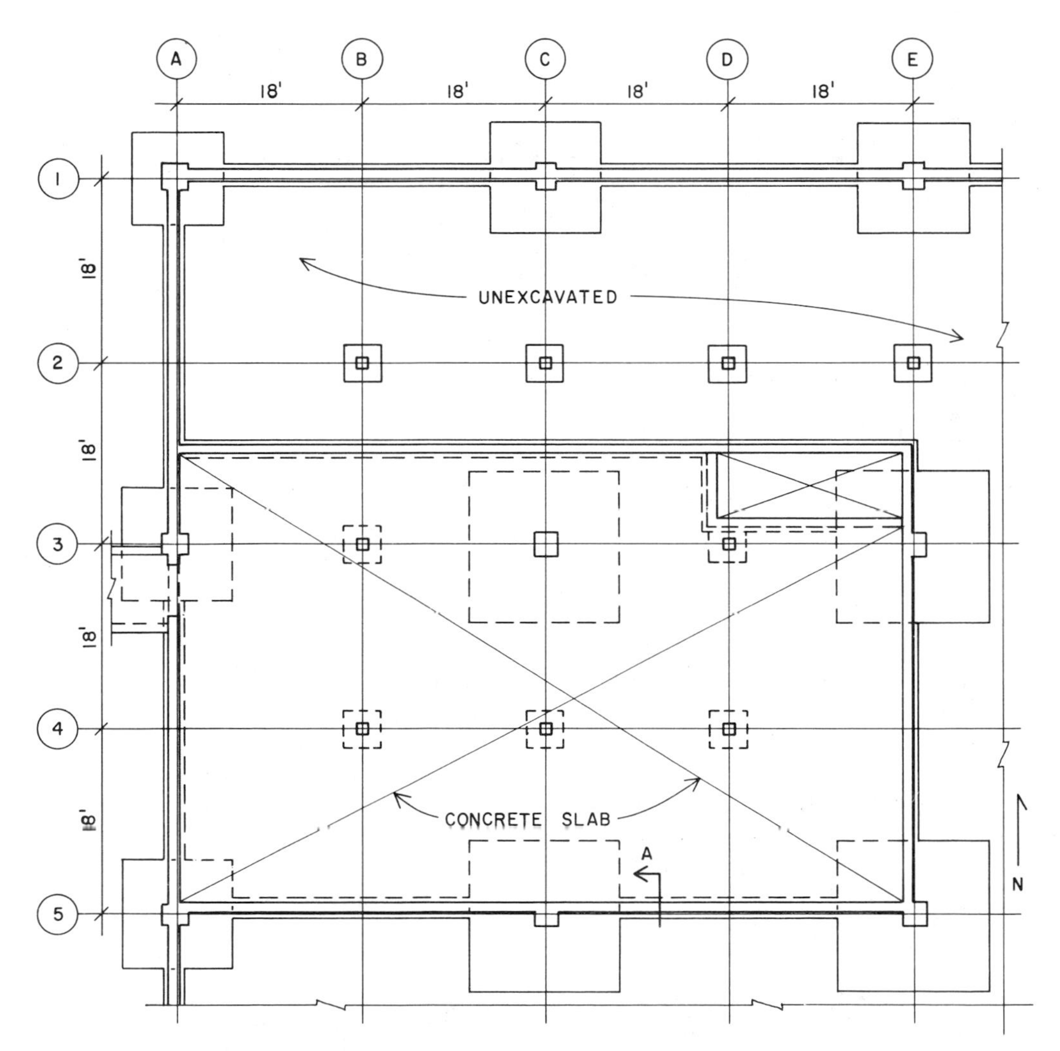

FIGURE 22.1. Basement and foundation plan.

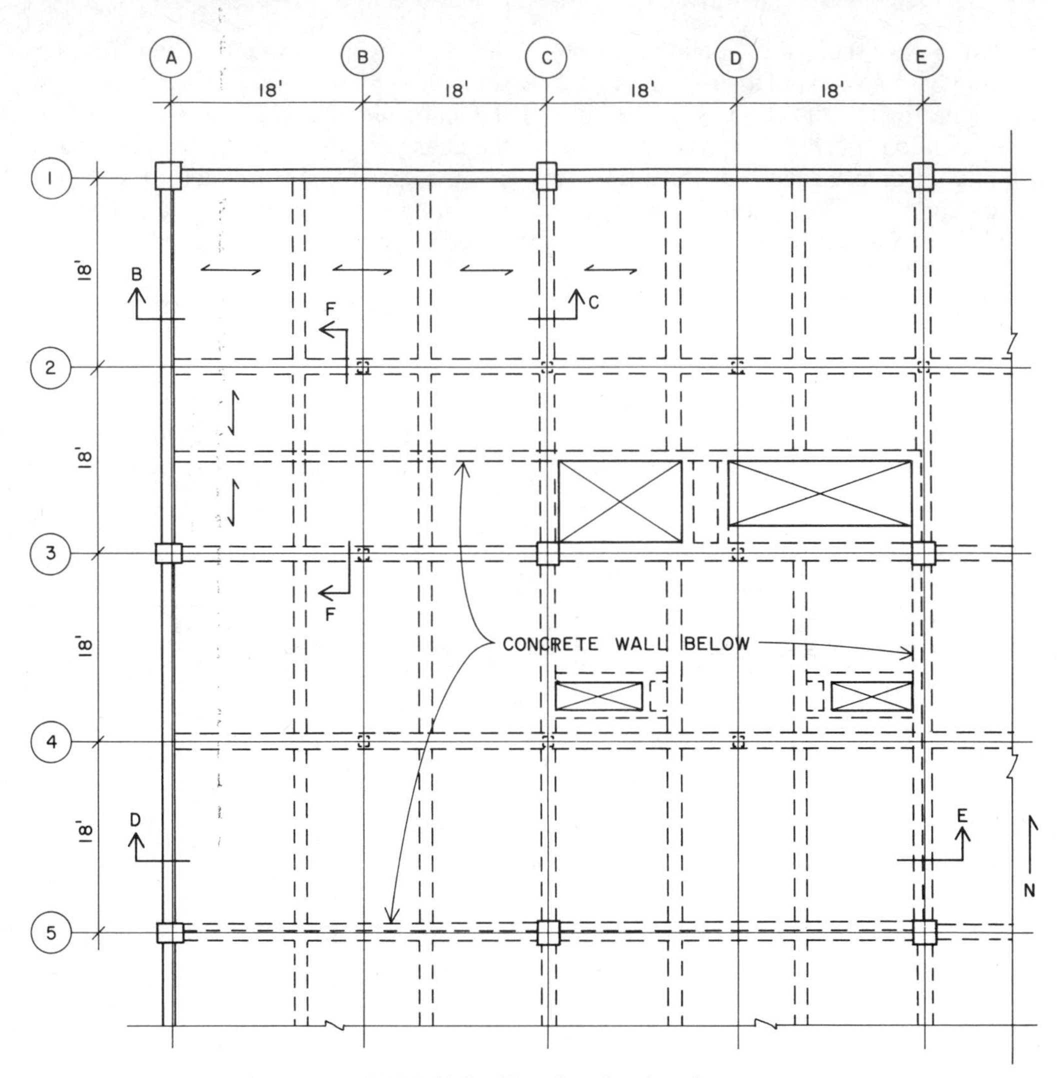

FIGURE 22.2. First-floor framing plan.

has been developed to cause the least disruption of the basic system layout.

22.3. CONSTRUCTION DETAILS

The locations for the various details that follow are indicated by section marks on the structural plans.

Detail A (Fig. 22.4)

This shows the typical basement floor slab and wall construction. The detail is essentially the same as that for the steel structure, except that the tops of the column footings are raised because of the absence of the steel base plates and anchor bolts. The need for some water sealing of the slab-to-wall joint and a moisture barrier under the slab would depend on specific site conditions.

Detail B (Fig. 20.5)

This section is also essentially the same as in the steel structure, with the possible exception of the wall portion above

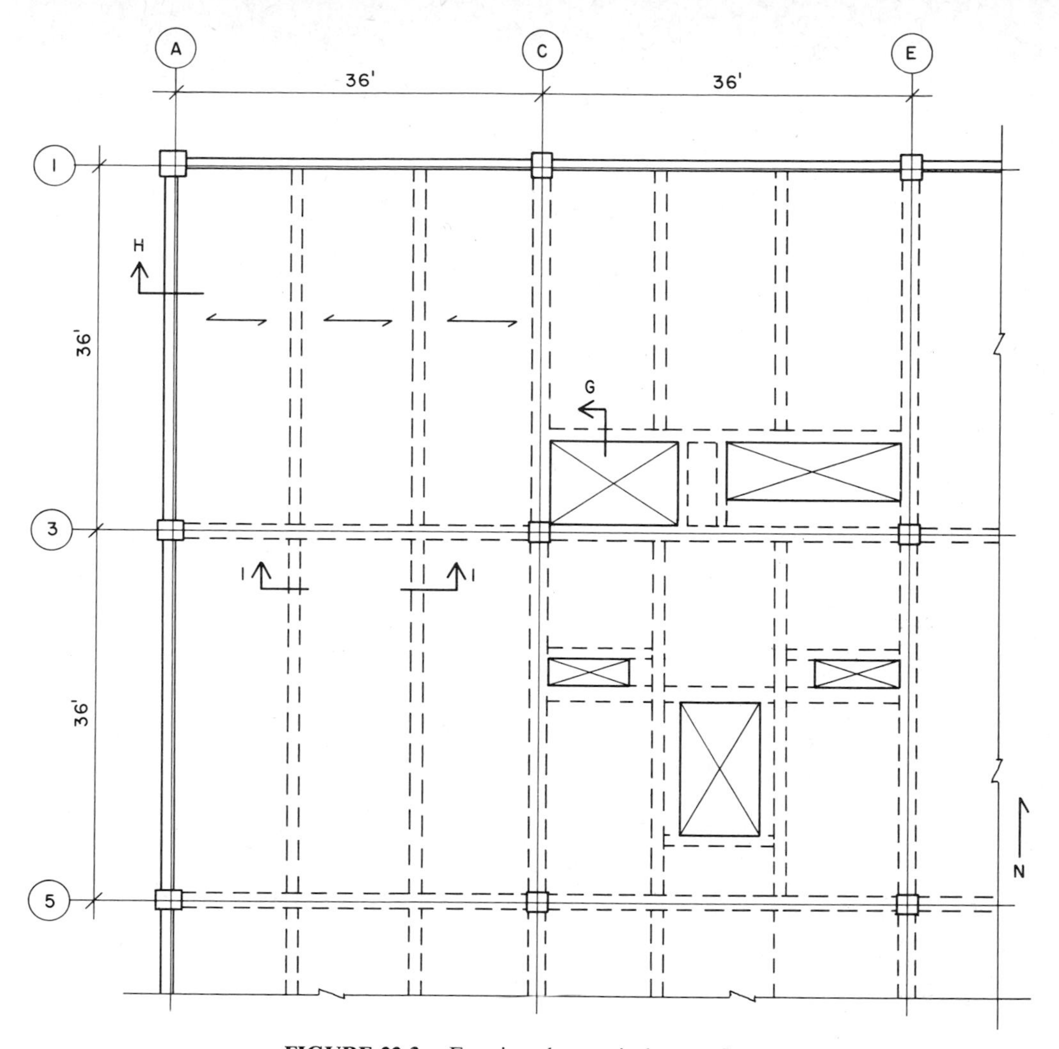

FIGURE 22.3. Framing plan: typical upper floor.

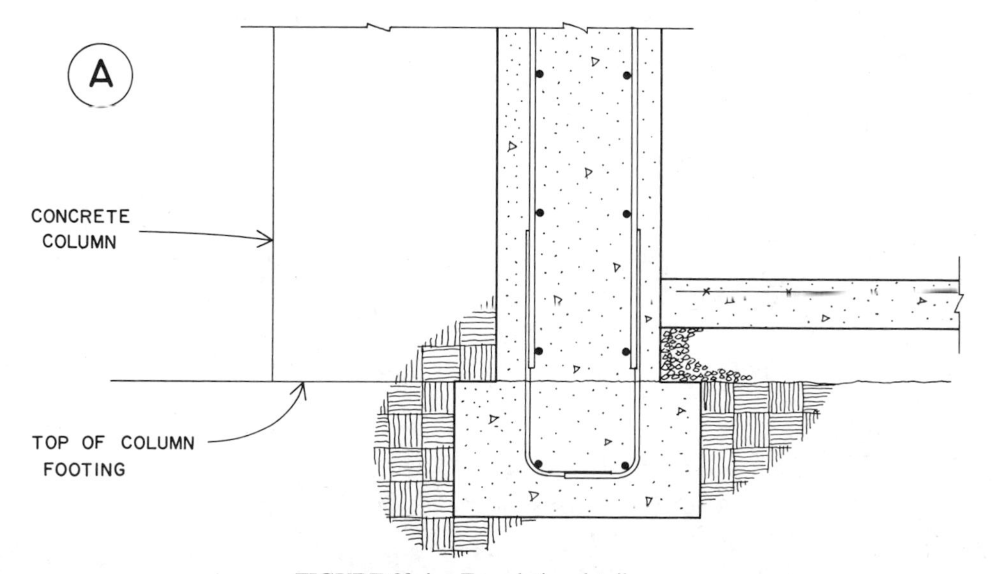

FIGURE 22.4. Foundation detail.

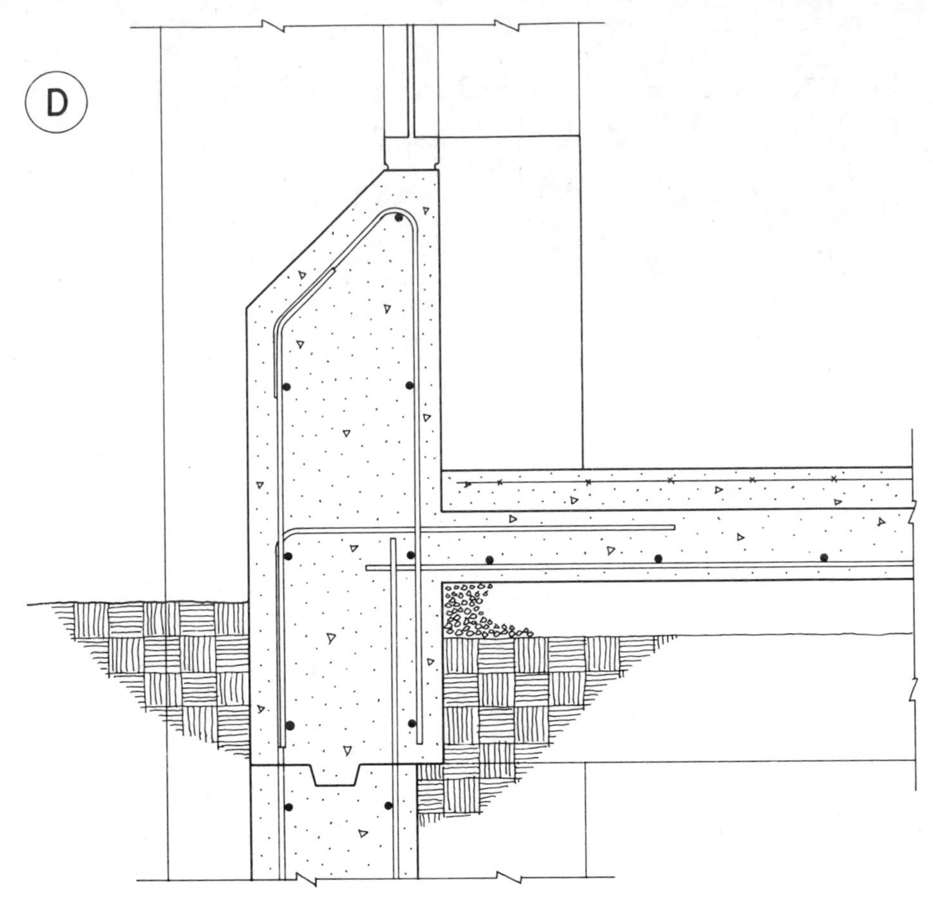

FIGURE 22.5. Wall detail at first floor.

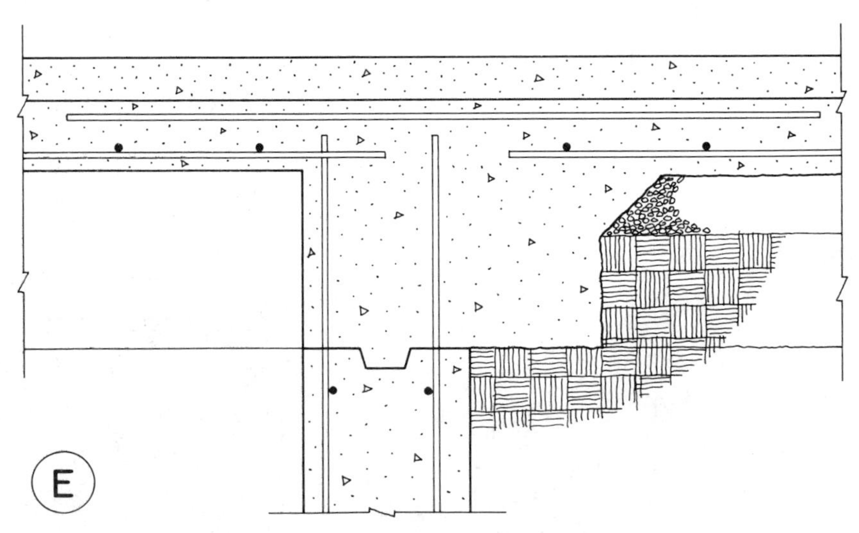

FIGURE 22.6. Detail at interior edge of basement.

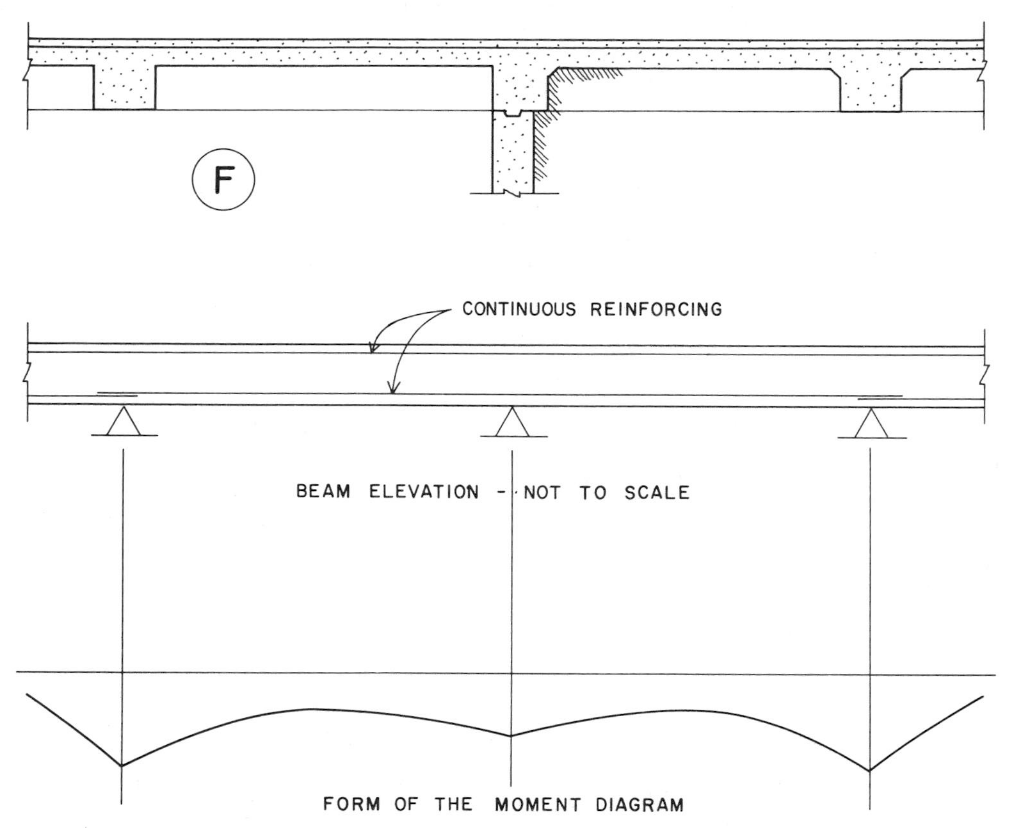

FIGURE 22.7. Effect of continuity in the first-floor structure.

the floor level. If a metal curtain wall is used, the detail is the same as that in Figure 20.5. If the exposed concrete wall is used, the detail is similar to that in Figure 22.5. If a precast concrete skin is used, the detail is some modification of that shown in Figure 22.9 for the typical floor.

Detail C (Fig. 20.5)

With the slab and beam system poured on fill, the typical details are the same as shown for the steel structure. The precise form and detail of these members would depend on the nature of the fill material and on the need for moisture protection and/or the thermal insulation.

The piers for this system would be formed and poured before the backfill is placed. An alternate to poured concrete piers would be to use piers of concrete masonry units with the cores filled with concrete.

Detail D (Fig. 22.5)

This shows the typical exterior wall condition at the basement area. A pour joint for the walls and columns is made at the level of the bottom of the concrete beam system for the framed floor. The remainder of the walls and columns, up to the top of the floor slab, would be poured with the beam and slab system.

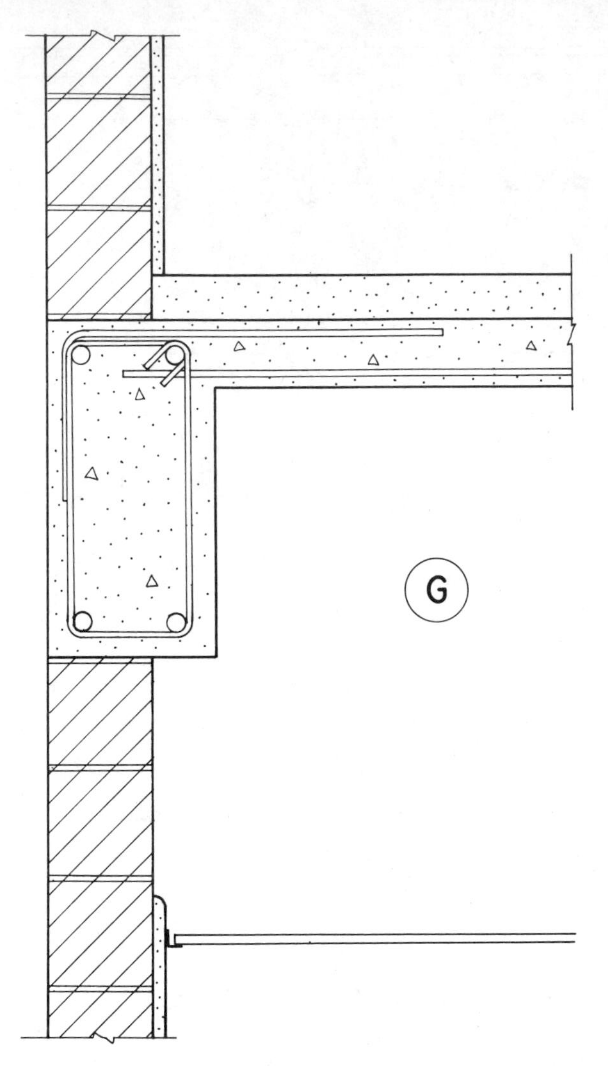

FIGURE 22.8. Framing at the core walls.

Detail E (Fig. 22.6)

This shows the transition between the two floor systems and indicates that they are poured monolithically, with the pour joint in the wall at the bottom of the beams. For continuity of the beam reinforcing in the bottom of the beams the detailed position of the bottom of the beams is slightly lower in the system on fill, because additional cover is required for the concrete that is deposited on the soil.

Detail F (Fig. 22.7)

This shows the condition that occurs where the basement wall causes a disruption of the regular 18-ft span for the floor system. This results in a high shear and a condition of virtually continuous negative moment for these short spans.

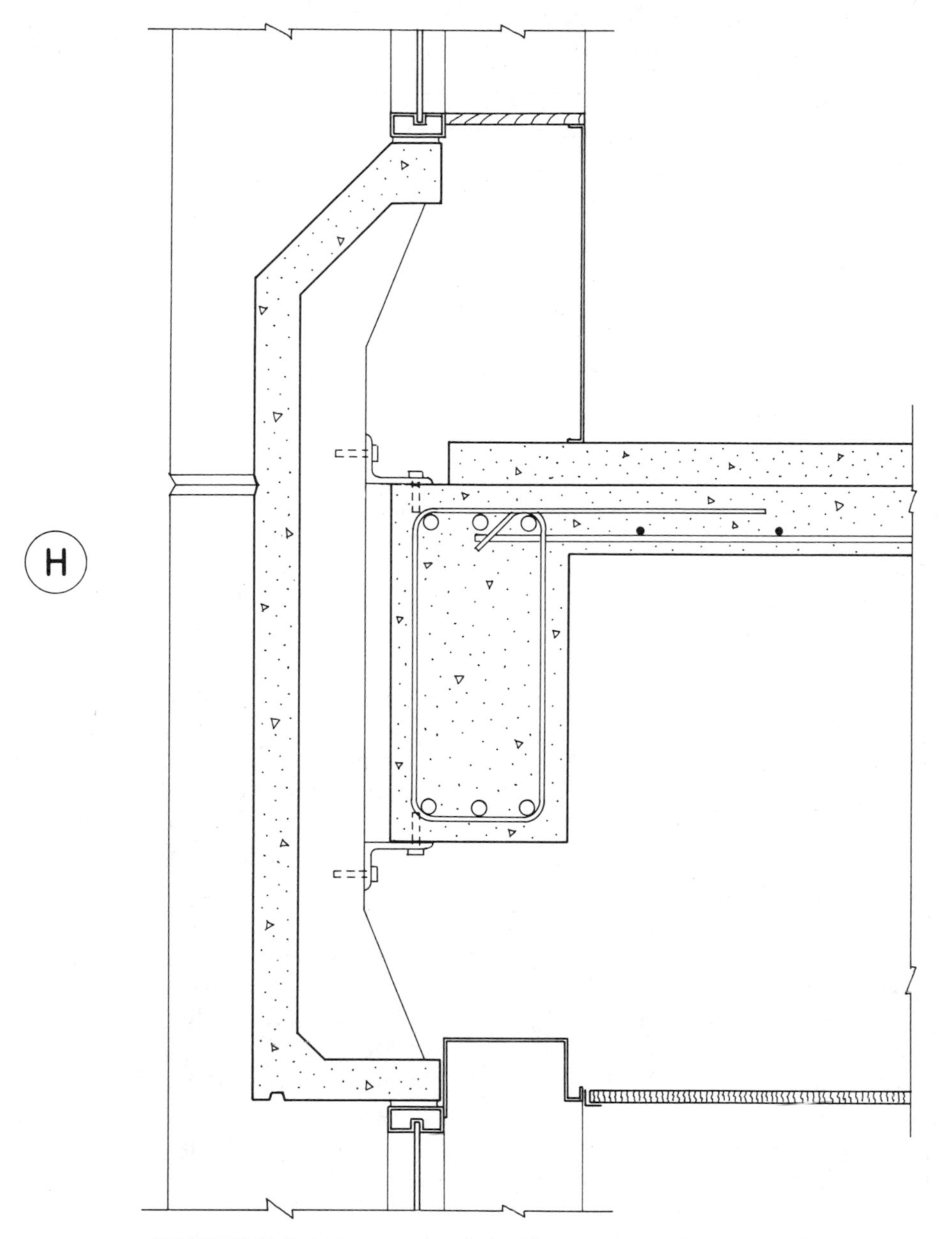

FIGURE 22.9. Alternate spandrel with precast concrete cover.

They would probably be reinforced as shown by extending some of the bottom bars and by making the negative moment top bars continuous. The high shear stress would probably require a continuous series of closely spaced stirrups.

Detail G (Fig. 22.8)

This section is similar in some details to that for the steel structure as shown in Figure 20.7. From the point of view of the wall construction, the concrete beam

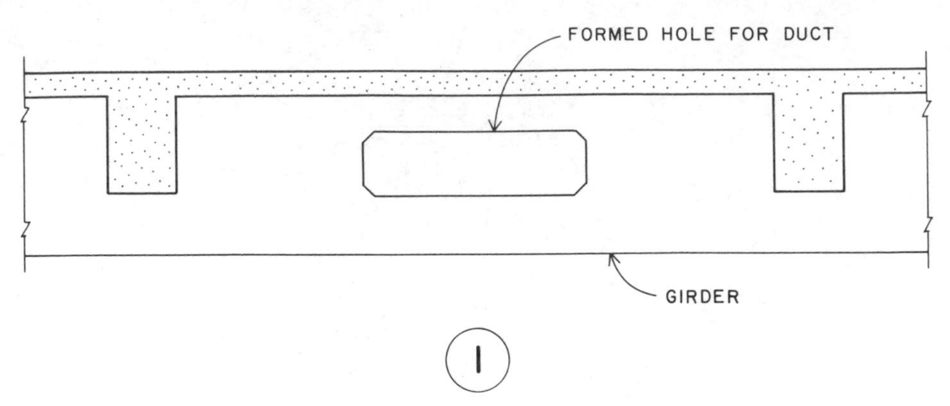

FIGURE 22.10. Detail for opening through the deep girder.

is similar to the concrete encasement for the steel beam. As with the steel structure, it would also be possible to pull the face of the beam back and run the wall past it, if the exposed concrete is not desired.

Detail H (Fig. 22.9)

This is actually an alternate to the section as shown in Figure 21.5. It indicates the use of a precast concrete facing member that is supported by the spandrel beam in a manner similar to that shown for the metal skin unit with the steel structure, as in Figure 20.6. One advantage of the construction in Figure 22.9 over the solid poured spandrel would be a reduction in dead load and volume of concrete.

It would also be possible to use the metal curtain wall similar to that for the steel structure. Attachment would be similar to that for the precast members, using preset anchors or inserts in the concrete structure.

Detail I (Fig. 22.10)

This shows the use of an opening through the web of the deep interior girder to accommodate air handling ducts, as was discussed in the design calculations. The usual rule of thumb is to keep these openings within the middle third of the span and within the middle third of the depth. This allows for continuity of the major tension and compression forces due to moment and avoids the highest shear condition near the end of the span. In our floor system layout, the location of the beams must also be considered; the middle third is conveniently open for this condition.

It is desirable to provide a radius or chamfer at the corners of the openings to help relieve the concentration of stresses at these points.

CHAPTER TWENTY-THREE

Alternate Construction

The design examples for Building Four have illustrated two options for the structure—one in steel and the other in poured-in-place concrete. In theory there are a considerable number of options involving variations in materials, system type, form, layout, and specific details of construction. In real design situations the truly feasible choices are often limited by particular dictates of economics, speed of construction, local code requirements and construction practices, and desired architectural details.

If the building plan and the 36-ft square column bay system are adhered to, these factors alone will establish the priority of some solutions over others. Add to these the particular floor live loads, the desire to include wiring in the floors, a demountable partitioning system, and a modular ceiling system, and the need for a particular fire rating for the floor assembly and the choices are narrowed further.

The discussion that follows presents a few other possibilities for the construction.

23.1. ALTERNATE FLOOR SYSTEMS

Poured Concrete Slab on Steel Beams

An alternative to the metal deck used with the steel beams would be a poured-in-place concrete slab. Using welded attachments on the tops of the steel beams would permit the development of composite action of the concrete slab and steel beams, which could result in a reduction in size of the typical beams. This composite action does not aid in the rigid frame action, which was a major factor in design of the column line beams and girders.

One advantage of this system would be the elimination of the sprayed-on fireproofing on the underside of the deck, since the concrete slab could develop the necessary fire resistance by itself. The applied fireproofing would therefore be used only on the steel beams and columns.

Dead load of this structure would be higher because the metal deck and fill would be replaced by a structural con-

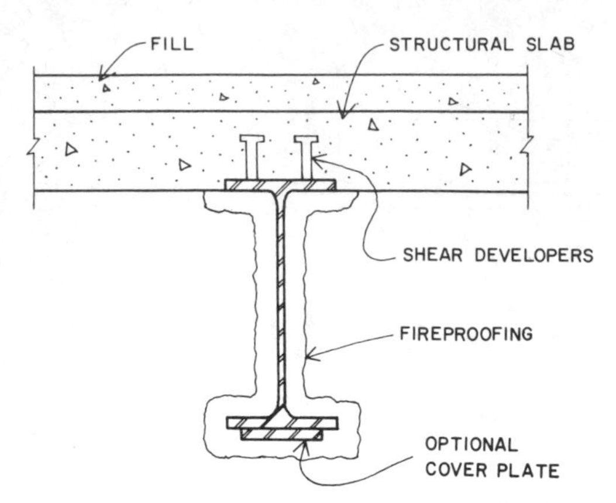

FIGURE 23.1. Steel frame with poured-in-place concrete slab.

crete slab and separate fill. The total dead load of this system would be somewhere between that for the two design examples since the steel beams weigh less than the concrete beam stems.

The layout of the framing system would be essentially the same as with the metal deck, although the spacing of the beams could be increased because minimum slab thickness required for fire rating would probably permit longer spans. Although this is a structural possibility, the wider spaced, heavier loaded beams would have to be deeper, which may make the change questionable.

A detail for this type of system is shown in Figure 23.1. Design of the slab is essentially the same as was illustrated in the concrete structure example. Design of the composite steel beams is well illustrated in textbooks and handbooks.

Precast Concrete Deck on Steel Beams

Another alternative with the steel frame is to use precast concrete deck units. A concrete fill would also be used with this system, serving the purposes of leveling and bonding of the units as well as the previous ones of incorporation of wiring. Although the precast units are usually voided, the weight reduction would be only slightly below that of the solid concrete slab. While it is possible to use some of the void spaces in the precast units for incorporation of wiring or plumbing and even for air distribution, this requires very careful coordination during the design and detailing processes and is not often done.

These precast units can easily achieve longer spans than those used in the design examples. However, the heavier, deeper beams produced by wider spacing might cause a problem. Effective use of this type of construction would probably require a general revision of the framing layout, with something different from the square bay system.

As with the poured slab, it may be possible to eliminate the sprayed-on fireproofing, depending on the fire rating required. Some details for this type of construction are shown in Figure 23.2. Design of the precast units would be done using the load tables and suggested details and specifications provided by the manufacturer.

One-Way Concrete Joist System

Figure 23.3 shows a layout and some details for a system using a one-way concrete joist system supported by girders in one direction. While the joist system itself is quite efficient, the loads on the girders are high, making the system less feasible for the square bay layout. A rectangular bay system, with short span girders and long span joists would improve the system. If the deep spandrels and the maximum depth interior girders, as illustrated in the design example, are used the system would be workable, although not optimal.

One potential problem results from the fact that this system is not given a very high fire rating by building codes, making it possible only with some fire

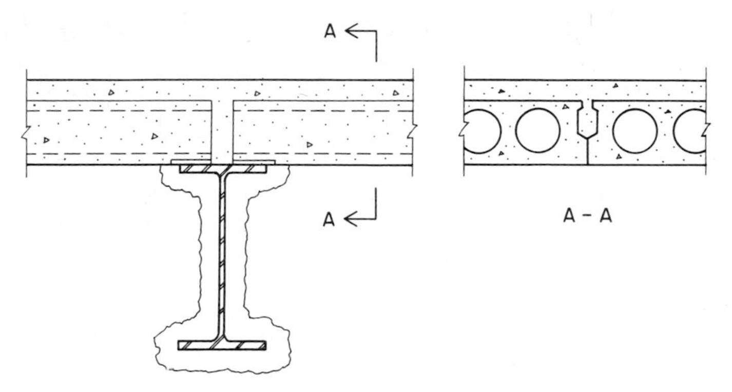

FIGURE 23.2. Steel frame with precast concrete deck units.

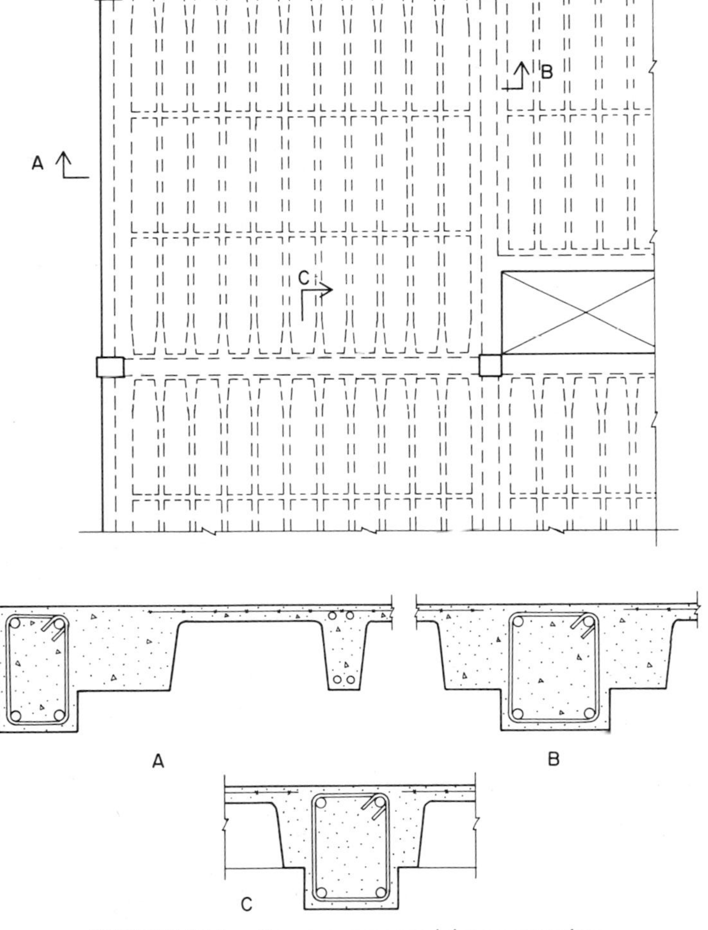

FIGURE 23.3. One-way concrete joist construction.

protection. This may require the use of a fully plastered ceiling or of some other assembly capable of the higher rating.

The large girders not withstanding, this system would probably be slightly lower in weight than the beam and slab system in the design example. The joists and slab may theoretically be made quite thin, although fire rating requirements may not permit their optimal reduction.

For purposes of developing the rigid frame action in the direction parallel to the joists, as well as providing for the framing of the large openings, it may be necessary to provide column line beams, as shown in the framing plan in Figure 23.3.

Design of the joist system itself can be done from various handbook tabulations, if the span, load, and concrete strength can be matched to the handbook examples. Even where the match-up is not exact, the handbooks can usually be used to establish an approximate design, thus eliminating several early approximation stages of the design.

Two-Way Concrete Joist (Waffle) System

The square column bays make it reasonable to consider the use of a two-way spanning concrete system. The size of the span and the relatively low live load make a waffle system, rather than a solid slab system, most feasible. A principal drawback, as with the one-way joists, is the limited fire rating of the system.

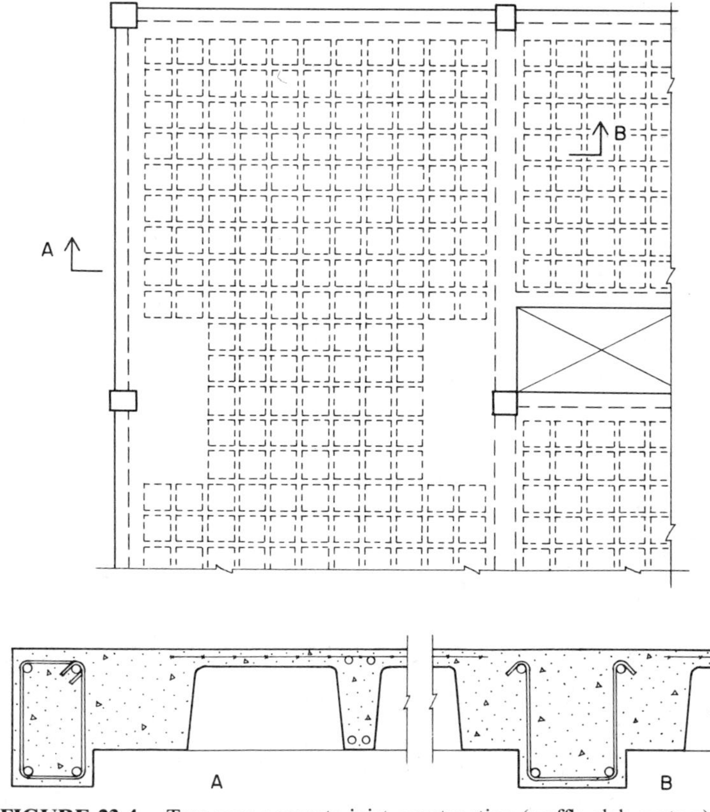

FIGURE 23.4. Two-way concrete joist construction (waffle slab system).

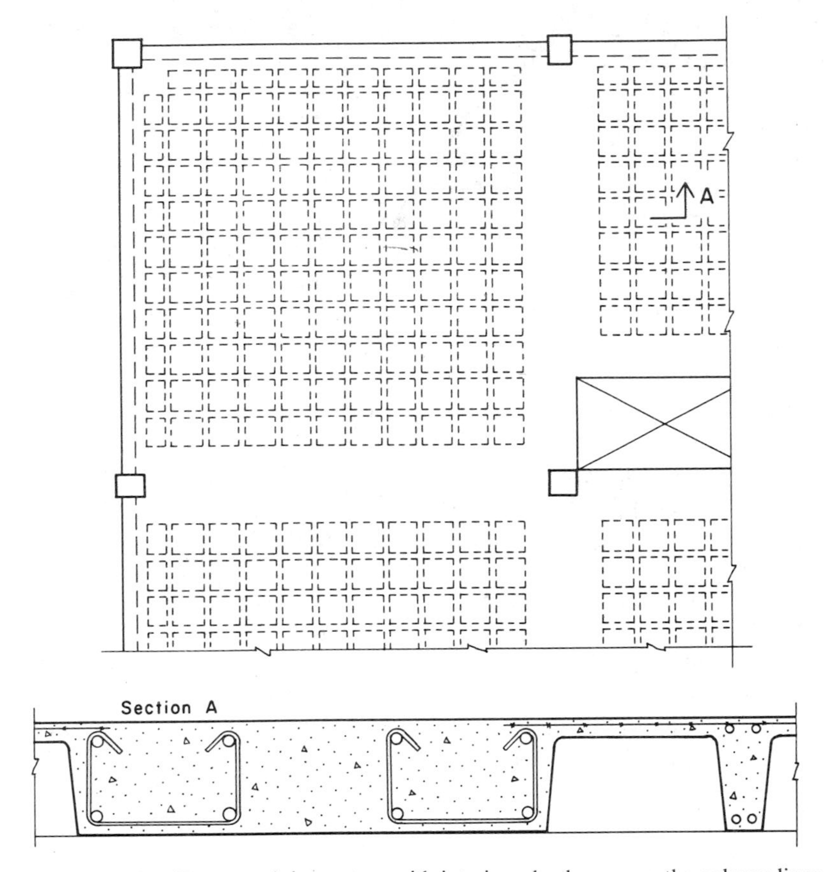

FIGURE 23.5. Two-way joist system with interior edge beams on the column lines.

Figure 23.4 shows a layout and some details for a system that is analogous to the flat slab system with drop panels. The large openings for stairs and elevators would require a transition to a beam system in part of the plan, with the joists in this area functioning essentially in one-way span action.

Figure 23.5 shows a layout and some details for a system that is analogous to the two-way slab on edge supports. The edge supports in this case consist of the spandrels and the wide column-line beams created by the unvoided waffle spaces.

Only a complete design of both systems could demonstrate the relative efficiency and cost of the two alternatives. Both offer the potential advantage of reducing the depth of the floor construction somewhat, because except for the core framing, the structure does not have the deep beams and girders on the interior that were part of the two design examples.

Design of the typical waffle joists in these two systems may also be done from handbook tables.

23.2. ALTERNATE LATERAL BRACING SYSTEMS

For both the steel and concrete structures we have used a full frame system with moment-resisting joints for this building. Alternatives for the multistory building were discussed and illustrated for Building Three in Section 13.3. The range of possibilities presented there is also applicable to this building, although shear walls would be less feasible in zones of high seismic risk.

23.3. ALTERNATES FOR THE EXTERIOR WALLS

As we have mentioned frequently, numerous variations are possible for the form and materials of the exterior walls. In recent times the heightened concern for heat loss and gain and the use of natural light have had strong influence on the construction of the solid portions of exterior walls and on the form and details of windows. Adding these concerns to the usual architect's concern for the appearance of the building's exterior surface creates a situation in which the structural designer is considerably restrained and thus must work closely with others in the development of the structure.

When the exterior wall is essentially not structural, the development of the structure must generally be done with the idea of keeping it from interfering with the various nonstructural functions of the wall. The structure must support the wall, but otherwise generally just keep out of the way. The wall must acknowledge the basic requirements of the structure (sizes required, alignments, fire protection, etc.), but is otherwise free to flow around it.

When the structure is exposed, or the wall is a structural bearing or shear wall, the wall–function and structural–function relationships become more intertwined. For a successful building design, this situation calls for a very close cooperation between the architect and the structural designer.

References

1. Harry Parker, *Simplified Engineering for Architects and Builders,* 6th ed., Wiley, New York, 1984.
2. Harry Parker, *Simplified Mechanics and Strength of Materials,* 3rd ed., Wiley, New York, 1977.
3. Harry Parker, *Simplified Design of Structural Steel,* 5th ed., Wiley, New York, 1983.
4. Harry Parker, *Simplified Design of Reinforced Concrete,* 5th ed., Wiley, New York, 1984.
5. Harry Parker, *Simplified Design of Structural Wood,* 3rd ed., Wiley, New York, 1979.
6. Harry Parker, *Simplified Design of Building Trusses for Architects and Builders,* 3rd ed., Wiley, New York, 1982.
7. *Uniform Building Code,* 1985 ed., International Conference of Building Officials, 5360 South Workman Mill Road, Whittier, CA 90601. (Called simply the UBC.)
8. Charles G. Ramsey and Harold R. Sleeper, *Architectural Graphic Standards,* 7th ed., Wiley, New York, 1981.
9. *Building Code Requirements for Reinforced Concrete,* (ACI 318-83), American Concrete Institute, Box 19150, Redford Station, Detroit, MI 48219, 1983. (Called the ACI Code.)
10. *Manual of Steel Construction,* 8th ed., American Institute of Steel Construction, 400 N. Michigan Ave., Chicago, IL 60611, 1980. (Called simply the AISC Manual.)
11. James Ambrose, *Simplified Design of Building Foundations,* Wiley, New York, 1981.
12. James Ambrose and Dimitry Vergun, *Simplified Building Design for Wind and Earthquake Forces,* Wiley, New York, 1980.
13. *Steel Deck Institute Design Manual for Composite Decks, Form Decks, and Roof Decks,* Steel Deck Institute, PO Box 3812, St. Louis, MO 63122.
14. *Masonry Design Manual,* 3rd ed., Masonry Institute of America, 2550 Beverly Boulevard, Los Angeles, CA 90057, 1979.
15. *Standard Specifications, Load Tables, and Weight Tables for Steel Joists and Joist Girders,* Steel Joist Institute, Suite A, 1205 48th Ave., North, Myrtle Beach, SC 29577, 1984.
16. *CRSI Handbook,* Concrete Reinforcing Steel Institute, Schaumburg, IL, 1982.

APPENDIX A

Design Data

This appendix contains some of the materials that have been used in the design examples. These materials are provided for the convenience of readers to whom the references may not be available. If they are available, however, it is recommended that information be taken directly from the references, which are frequently revised and contain additional explanations and examples for their use. Following is a list of the materials contained in this appendix with the reference sources from which they have been reprinted or adapted with permission of the publishers.

From the AISC Manual (Ref. 10):

Table A.1 Section Modulus and Moment of Resistance for Selected Rolled Structural Shapes

Table A.2 Allowable Column Loads for Selected W Shapes

Table A.3 Allowable Column Loads for Standard Steel Pipe of A36 Steel

Table A.4 Load–Span Values for Steel Beams

From *Masonry Design Manual* (Ref. 14):

Table A.5 Rigidity Coefficients for Cantilevered Masonry Walls

Table A.6 Rigidity Coefficients for Fixed Masonry Walls

Figure A.1 Flexural coefficients *K*-chart for reinforced masonry.

From *Simplified Engineering for Architects and Builders*, 6th ed. (Ref. 1):

Figure A.2 Deflection of steel beams with bending stress of 24 ksi [165 MPa]

Table A.7 Balanced Section Properties for Rectangular Concrete Sections with Tension Reinforcing Only

Figure A.3 k factors for rectangular concrete beams with tension reinforcing—as a function of n and p.

Table A.8 Weights of Building Construction

S_x	Shape	F_y = 36 ksi		
		L_c	L_u	M_R
In.^{3}		Ft.	Ft.	Kip-ft.
1110	**W 36x300**	**17.6**	**35.3**	**2220**
1030	**W 36x280**	**17.5**	**33.1**	**2060**
953	**W 36x260**	**17.5**	**30.5**	**1910**
895	**W 36x245**	**17.4**	**28.6**	**1790**
837	**W 36x230**	**17.4**	**26.8**	**1670**
829	W 33x241	16.7	30.1	1660
757	**W 33x221**	**16.7**	**27.6**	**1510**
719	**W 36x210**	**12.9**	**20.9**	**1440**
684	**W 33x201**	**16.6**	**24.9**	**1370**
664	**W 36x194**	**12.8**	**19.4**	**1330**
663	W 30x211	15.9	29.7	1330
623	**W 36x182**	**12.7**	**18.2**	**1250**
598	W 30x191	15.9	26.9	1200
580	**W 36x170**	**12.7**	**17.0**	**1160**
542	**W 36x160**	**12.7**	**15.7**	**1080**
539	W 30x173	15.8	24.2	1080
504	**W 36x150**	**12.6**	**14.6**	**1010**
502	W 27x178	14.9	27.9	1000
487	W 33x152	12.2	16.9	974
455	W 27x161	14.8	25.4	910
448	**W 33x141**	**12.2**	**15.4**	**896**
439	**W 36x135**	**12.3**	**13.0**	**878**
414	W 24x162	13.7	29.3	828
411	W 27x146	14.7	23.0	822
406	**W 33x130**	**12.1**	**13.8**	**812**
380	W 30x132	11.1	16.1	760
371	W 24x146	13.6	26.3	742
359	**W 33x118**	**12.0**	**12.6**	**718**
355	W 30x124	11.1	15.0	710
329	**W 30x116**	**11.1**	**13.8**	**658**
329	W 24x131	13.6	23.4	658
329	W 21x147	13.2	30.3	658
299	**W 30x108**	**11.1**	**12.3**	**598**
299	W 27x114	10.6	15.9	598
295	W 21x132	13.1	27.2	590
291	W 24x117	13.5	20.8	582
273	W 21x122	13.1	25.4	546

S_x	Shape	F_y = 36 ksi		
		L_c	L_u	M_R
In.^{3}		Ft.	Ft.	Kip-ft.
269	**W 30x 99**	**10.9**	**11.4**	**538**
267	W 27x102	10.6	14.2	534
258	W 24x104	13.5	18.4	516
249	W 21x111	13.0	23.3	498
243	**W 27x 94**	**10.5**	**12.8**	**486**
231	W 18x119	11.9	29.1	462
227	W 21x101	13.0	21.3	454
222	**W 24x 94**	**9.6**	**15.1**	**444**
213	**W 27x 84**	**10.5**	**11.0**	**426**
204	W 18x106	11.8	26.0	408
196	**W 24x 84**	**9.5**	**13.3**	**392**
192	W 21x 93	8.9	16.8	384
190	W 14x120	15.5	44.1	380
188	W 18x 97	11.8	24.1	376
176	**W 24x 76**	**9.5**	**11.8**	**352**
175	W 16x100	11.0	28.1	350
173	W 14x109	15.4	40.6	346
171	W 21x 83	8.8	15.1	342
166	W 18x 86	11.7	21.5	332
157	W14x 99	15.4	37.0	314
155	W 16x 89	10.9	25.0	310
154	**W 24x 68**	**9.5**	**10.2**	**308**
151	W 21x 73	8.8	13.4	302
146	W 18x 76	11.6	19.1	292
143	W 14x 90	15.3	34.0	286
140	**W 21x 68**	**8.7**	**12.4**	**280**
134	W 16x 77	10.9	21.9	268
131	**W 24x 62**	**7.4**	**8.1**	**262**
127	**W 21x 62**	**8.7**	**11.2**	**254**
127	W 18x 71	8.1	15.5	254
123	W 14x 82	10.7	28.1	246
118	W 12x 87	12.8	36.2	236
117	W 18x 65	8.0	14.4	234
117	W 16x 67	10.8	19.3	234
114	**W 24x 55**	**7.0**	**7.5**	**228**
112	W 14x 74	10.6	25.9	224
111	W 21x 57	6.9	9.4	222
108	W 18x 60	8.0	13.3	216
107	W 12x 79	12.8	33.3	214
103	W 14x 68	10.6	23.9	206
98.3	**W 18x 55**	**7.9**	**12.1**	**197**
97.4	W 12x 72	12.7	30.5	195

TABLE A.1. Section Modulus and Moment of Resistance for Selected Rolled Structural Shapes

S_x	Shape	F_y = 36 ksi		
		L_c	L_u	M_R
In.^{3}		Ft.	Ft.	Kip-ft.
94.5	W 21x50	6.9	7.8	189
92.2	W 16x57	7.5	14.3	184
92.2	W 14x61	10.6	21.5	184
88.9	W 18x50	7.9	11.0	178
87.9	W 12x65	12.7	27.7	176
81.6	W 21x44	6.6	7.0	163
81.0	W 16x50	7.5	12.7	162
78.8	W 18x46	6.4	9.4	158
78.0	W 12x58	10.6	24.4	156
77.8	W 14x53	8.5	17.7	156
72.7	W 16x45	7.4	11.4	145
70.6	W 12x53	10.6	22.0	141
70.3	W 14x48	8.5	16.0	141
68.4	W 18x40	6.3	8.2	137
66.7	W 10x60	10.6	31.1	133
64.7	W 16x40	7.4	10.2	129
64.7	W 12x50	8.5	19.6	129
62.7	W 14x43	8.4	14.4	125
60.0	W 10x54	10.6	28.2	120
58.1	W 12x45	8.5	17.7	116
57.6	W 18x35	6.3	6.7	115
56.5	W 16x36	7.4	8.8	113
54.6	W 14x38	7.1	11.5	109
54.6	W 10x49	10.6	26.0	109
51.9	W 12x40	8.4	16.0	104
49.1	W 10x45	8.5	22.8	98
48.6	W 14x34	7.1	10.2	97
47.2	W 16x31	5.8	7.1	94
45.6	W 12x35	6.9	12.6	91
42.1	W 10x39	8.4	19.8	84
42.0	W 14x30	7.1	8.7	84
38.6	W 12x30	6.9	10.8	77
38.4	W 16x26	5.6	6.0	77
35.3	W 14x26	5.3	7.0	71
35.0	W 10x33	8.4	16.5	70
33.4	W 12x26	6.9	9.4	67
32.4	W 10x30	6.1	13.1	65
31.2	W 8x35	8.5	22.6	62
29.0	W 14x22	5.3	5.6	58
27.9	W 10x26	6.1	11.4	56
27.5	W 8x31	8.4	20.1	55
25.4	W 12x22	4.3	6.4	51
24.3	W 8x28	6.9	17.5	49
23.2	W 10x22	6.1	9.4	46
21.3	W 12x19	4.2	5.3	43
21.1	M 14x18	3.6	4.0	42
20.9	W 8x24	6.9	15.2	42
18.8	W 10x19	4.2	7.2	38
18.2	W 8x21	5.6	11.8	36
17.1	W 12x16	4.1	4.3	34
16.7	W 6x25	6.4	20.0	33
16.2	W 10x17	4.2	6.1	32
15.2	W 8x18	5.5	9.9	30
14.9	W 12x14	3.5	4.2	30
13.8	W 10x15	4.2	5.0	28
13.4	W 6x20	6.4	16.4	27
13.0	M 6x20	6.3	17.4	26
12.0	M 12x11.8	2.7	3.0	24
11.8	W 8x15	4.2	7.2	24
10.9	W 10x12	3.9	4.3	22
10.2	W 6x16	4.3	12.0	20
10.2	W 5x19	5.3	19.5	20
9.91	W 8x13	4.2	5.9	20
9.72	W 6x15	6.3	12.0	19
9.63	M 5x18.9	5.3	19.3	19
8.51	W 5x16	5.3	16.7	17
7.81	W 8x10	4.2	4.7	16
7.76	M 10x 9	2.6	2.7	16
7.31	W 6x12	4.2	8.6	15
5.56	W 6x 9	4.2	6.7	11
5.46	W 4x13	4.3	15.6	11
5.24	M 4x13	4.2	16.9	10
4.62	M 8x 6.5	2.4	2.5	9
2.40	M 6x 4.4	1.9	2.4	5

TABLE A.1. (*Continued*)

TABLE A.2. Allowable Column Loads for Selected W Shapes[a]

	Effective Length (KL) in Feet										Bending Factor	
Shape	8	9	10	11	12	14	16	18	20	22	B_x	B_y
M 4 × 13	48	42	35	29	24	18					0.727	2.228
W 4 × 13	52	46	39	33	28	20	16				0.701	2.016
W 5 × 16	74	69	64	58	52	40	31	24	20		0.550	1.560
M 5 × 18.9	85	78	71	64	56	42	32	25			0.576	1.768
W 5 × 19	88	82	76	70	63	48	37	29	24		0.543	1.526
W 6 × 9	33	28	23	19	16	12					0.482	2.414
W 6 × 12	44	38	31	26	22	16					0.486	2.367
W 6 × 16	62	54	46	38	32	23	18				0.465	2.155
W 6 × 15	75	71	67	62	58	48	38	30	24	20	0.456	1.424
M 6 × 20	98	92	87	81	74	61	47	37	30	25	0.453	1.510
W 6 × 20	100	95	90	85	79	67	54	42	34	28	0.438	1.331
W 6 × 25	126	120	114	107	100	85	69	54	44	36	0.440	1.308
W 8 × 24	124	118	113	107	101	88	74	59	48	39	0.339	1.258
W 8 × 28	144	138	132	125	118	103	87	69	56	46	0.340	1.244
W 8 × 31	170	165	160	154	149	137	124	110	95	80	0.332	0.985
W 8 × 35	191	186	180	174	168	155	141	125	109	91	0.330	0.972
W 8 × 40	218	212	205	199	192	127	160	143	124	104	0.330	0.959
W 8 × 48	263	256	249	241	233	215	196	176	154	131	0.326	0.940
W 8 × 58	320	312	303	293	283	263	240	216	190	162	0.329	0.934
W 8 × 67	370	360	350	339	328	304	279	251	221	190	0.326	0.921
W 10 × 33	179	173	167	161	155	142	127	112	95	78	0.277	1.055
W 10 × 39	213	206	200	193	186	170	154	136	116	97	0.273	1.018
W 10 × 45	247	240	232	224	216	199	180	160	138	115	0.271	1.000
W 10 × 49	279	273	268	262	256	242	228	213	197	180	0.264	0.770
W 10 × 54	306	300	294	288	281	267	251	235	217	199	0.263	0.767
W 10 × 60	341	335	328	321	313	297	280	262	243	222	0.264	0.765
W 10 × 68	388	381	373	365	357	339	320	299	278	255	0.264	0.758
W 10 × 77	439	431	422	413	404	384	362	339	315	289	0.263	0.751
W 10 × 88	504	495	485	475	464	442	417	392	364	335	0.263	0.744
W 10 × 100	573	562	551	540	428	503	476	446	416	383	0.263	0.735
W 10 × 112	642	631	619	606	593	565	535	503	469	433	0.261	0.726
W 12 × 40	217	210	203	196	188	172	154	135	114	94	0.227	1.073
W 12 × 45	243	235	228	220	211	193	173	152	129	106	0.227	1.065
W 12 × 50	271	263	254	246	236	216	195	171	146	121	0.227	1.058
W 12 × 53	301	295	288	282	275	260	244	227	209	189	0.221	0.813
W 12 × 58	329	322	315	308	301	285	268	249	230	209	0.218	0.794
W 12 × 65	378	373	367	361	354	341	326	311	294	277	0.217	0.656
W 12 × 72	418	412	406	399	392	377	361	344	326	308	0.217	0.651
W 12 × 79	460	453	446	439	431	415	398	379	360	339	0.217	0.648
W 12 × 87	508	501	493	485	477	459	440	420	398	376	0.217	0.645
W 12 × 96	560	552	544	535	526	506	486	464	440	416	0.215	0.635
W 12 × 106	620	611	602	593	583	561	539	514	489	462	0.215	0.633
W 12 × 120	702	692	660	636	611	584	555	525	493	460	0.217	0.630
W 12 × 136	795	772	747	721	693	662	630	597	561	524	0.215	0.621
W 12 × 152	891	866	839	810	778	745	710	673	633	592	0.214	0.614
W 12 × 170	998	970	940	908	873	837	798	757	714	668	0.213	0.608

TABLE A.2. (*Continued*)

	Effective Length (KL) in Feet										Bending Factor	
Shape	8	10	12	14	16	18	20	22	24	26	B_x	B_y
W 12 × 190	1115	1084	1051	1016	978	937	894	849	802	752	0.212	0.600
W 12 × 210	1236	1202	1166	1127	1086	1042	995	946	894	840	0.212	0.594
W 12 × 230	1355	1319	1280	1238	1193	1145	1095	1041	985	927	0.211	0.589
W 12 × 252	1484	1445	1403	1358	1309	1258	1203	1146	1085	1022	0.210	0.583
W 12 × 279	1642	1600	1554	1505	1452	1396	1337	1275	1209	1141	0.208	0.573
W 12 × 305	1799	1753	1704	1651	1594	1534	1471	1404	1333	1260	0.206	0.564
W 12 × 336	1986	1937	1884	1827	1766	1701	1632	1560	1484	1404	0.205	0.558
W 14 × 43	230	215	199	181	161	140	117	96	81	69	0.201	1.115
W 14 × 48	258	242	224	204	182	159	133	110	93	79	0.201	1.102
W 14 × 53	286	268	248	226	202	177	149	123	104	88	0.201	1.091
W 14 × 61	345	330	314	297	278	258	237	214	190	165	0.194	0.833
W 14 × 68	385	369	351	332	311	289	266	241	214	186	0.194	0.826
W 14 × 74	421	403	384	363	341	317	292	265	236	206	0.195	0.820
W 14 × 82	465	446	425	402	377	351	323	293	261	227	0.196	0.823
W 14 × 90	536	524	511	497	482	466	449	432	413	394	0.185	0.531
W 14 × 99	589	575	561	546	529	512	494	475	454	433	0.185	0.527
W 14 × 109	647	633	618	601	583	564	544	523	501	478	0.185	0.523
W 14 × 120	714	699	682	663	644	623	601	578	554	528	0.186	0.523
W 14 × 132	786	768	750	730	708	686	662	637	610	583	0.186	0.521
W 14 × 145	869	851	832	812	790	767	743	718	691	663	0.184	0.489
W 14 × 159	950	931	911	889	865	840	814	786	758	727	0.184	0.485
W 14 × 176	1054	1034	1011	987	961	933	904	874	842	809	0.184	0.484
W 14 × 193	1157	1134	1110	1083	1055	1025	994	961	927	891	0.183	0.477
W 14 × 211	1263	1239	1212	1183	1153	1121	1087	1051	1014	975	0.183	0.477
W 14 × 233	1396	1370	1340	1309	1276	1241	1204	1165	1124	1081	0.183	0.472
W 14 × 257	1542	1513	1481	1447	1410	1372	1331	1289	1244	1198	0.182	0.470
W 14 × 283	1700	1668	1634	1597	1557	1515	1471	1425	1377	1326	0.181	0.465
W 14 × 311	1867	1832	1794	1754	1711	1666	1618	1568	1515	1460	0.181	0.459
W 14 × 342		2022	1985	1941	1894	1845	1793	1738	1681	1621	0.181	0.457
W 14 × 370		2181	2144	2097	2047	1995	1939	1881	1820	1756	0.180	0.452
W 14 × 398		2356	2304	2255	2202	2146	2087	2025	1961	1893	0.178	0.447
W 14 × 426		2515	2464	2411	2356	2296	2234	2169	2100	2029	0.177	0.442
W 14 × 455		2694	2644	2589	2430	2467	2401	2332	2260	2184	0.177	0.441
W 14 × 500		2952	2905	2845	2781	2714	2642	2568	2490	2409	0.175	0.434
W 14 × 550		3272	3206	3142	3073	3000	2923	2842	2758	2670	0.174	0.429
W 14 × 605		3591	3529	3459	3384	3306	3223	3136	3045	2951	0.171	0.421
W 14 × 665		3974	3892	3817	3737	3652	3563	3469	3372	3270	0.170	0.415
W 14 × 730		4355	4277	4196	4100	4019	3923	3823	3718	3609	0.168	0.408

[a] Loads in kips for shapes of steel with yield stress of 36 ksi [250 MPa]. Adapted from data in the *Manual of Steel Construction,* 8th ed. (Ref. 10), with permission of the publisher, American Institute of Steel Construction.

Nominal Dia.		12	10	8	6	5	4	$3\frac{1}{2}$	3
Wall Thickness		0.375	0.365	0.322	0.280	0.258	0.237	0.226	0.216
Weight per Foot		49.56	40.48	28.55	18.97	14.62	10.79	9.11	7.58
F_y		36 ksi							
Effective length in feet KL with respect to radius of gyration	0	315	257	181	121	93	68	58	48
	6	303	246	171	110	83	59	48	38
	7	301	243	168	108	81	57	46	36
	8	299	241	166	106	78	54	44	34
	9	296	238	163	103	76	52	41	31
	10	293	235	161	101	73	49	38	28
	11	291	232	158	98	71	46	35	25
	12	288	229	155	95	68	43	32	22
	13	285	226	152	92	65	40	29	19
	14	282	223	149	89	61	36	25	16
	15	278	220	145	86	58	33	22	14
	16	275	216	142	82	55	29	19	12
	17	272	213	138	79	51	26	17	11
	18	268	209	135	75	47	23	15	10
	19	265	205	131	71	43	21	14	9
	20	261	201	127	67	39	19	12	
	22	254	193	119	59	32	15	10	
	24	246	185	111	51	27	13		
	25	242	180	106	47	25	12		
	26	238	176	102	43	23			
	28	229	167	93	37	20			
	30	220	158	83	32	17			
	31	216	152	78	30	16			
	32	211	148	73	29				
	34	201	137	65	25				
	36	192	127	58	23				
	37	186	120	55	21				
	38	181	115	52					
	40	171	104	47					
Properties									
Area A (in.^{2})		14.6	11.9	8.40	5.58	4.30	3.17	2.68	2.23
I (in.^{4})		279	161	72.5	28.1	15.2	7.23	4.79	3.02
r (in.)		4.38	3.67	2.94	2.25	1.88	1.51	1.34	1.16
B Bending factor		0.333	0.398	0.500	0.657	0.789	0.987	1.12	1.29
* a		41.7	23.9	10.8	4.21	2.26	1.08	0.717	0.447

* Tabulated values of a must be multiplied by 10^6.
Note: Heavy line indicates Kl/r of 200.

TABLE A.3. Allowable Column Loads for Standard Steel Pipe of A36 Steel

TABLE A.4. Load–Span Values for Steel Beams[a]

Shape	Span (ft)	8	10	12	14	16	18	20	22	24	26	28	30
	Deflection Factor[b]	1.59	2.48	3.58	4.87	6.36	8.05	9.93	12.0	14.3	16.8	19.5	22.3
	L_c[c] (ft)												
M 8 × 6.5	2.4	9.24	7.39	6.16	5.28	4.62	4.11						
M 10 × 9	2.6	15.5	12.4	10.3	8.87	7.76	6.90	6.21	5.64				
W 8 × 10	4.2	15.6	12.5	10.4	8.92	7.81	6.94						
W 8 × 13	4.2	19.8	15.9	13.2	11.3	9.91	8.81						
W 10 × 12	3.9	21.8	17.4	14.5	12.5	10.9	9.69	8.72	7.93				
W 8 × 15	4.2	23.6	18.9	15.7	13.5	11.8	10.5						
M 12 × 11.8	2.7	24.0	19.2	16.0	13.7	12.0	10.7	9.60	8.73	8.00	7.38	6.86	
W 10 × 15	4.2	27.6	22.1	18.4	15.8	13.8	12.3	11.0	10.0				
W 12 × 14	3.5	29.8	23.8	19.9	17.0	14.9	13.2	11.9	10.8	9.93	9.17	8.51	
W 8 × 18	5.5	30.4	24.3	20.3	17.4	15.2	13.5						
W 10 × 17	4.2	32.4	25.9	21.6	18.5	16.2	14.4	13.0	11.8				
W 12 × 16	4.1	34.2	27.4	22.8	19.5	17.1	15.2	13.7	12.4	11.4	10.5	9.77	
W 8 × 21	5.6	36.4	29.1	24.3	20.8	18.2	16.2						
W 10 × 19	4.2	37.6	30.1	25.1	21.5	18.8	16.7	15.0	13.7				
W 8 × 24	6.9	41.8	33.4	27.9	23.9	20.9	18.6						
M 14 × 18	3.6	42.2	33.8	28.1	24.1	21.1	18.7	16.9	15.3	14.1	13.0	12.0	11.2
W 12 × 19	4.2	42.6	34.1	28.4	24.3	21.3	18.9	17.0	15.5	14.2	13.1	12.2	
W 10 × 22	6.1	46.4	37.1	30.9	26.5	23.2	20.6	18.5	16.9				
W 8 × 28	6.9	48.6	38.9	32.4	27.8	24.3	21.6						

TABLE A.4. (*Continued*)

Shape	Span (ft)	12	14	16	18	20	22	24	26	28	30	32	34
	Deflection Factor[b]	3.58	4.87	6.36	8.05	9.93	12.0	14.3	16.8	19.5	22.3	25.4	28.7
	L_c^c (ft)												
W 12 × 22	4.3	33.9	29.0	25.4	▮ 22.6	20.3	18.5	16.9	15.6	14.5			
W 10 × 26	6.1	37.2	▮ 31.9	27.9	24.8	22.3	20.3						
W 14 × 22	5.3	38.7	33.1	29.0	25.8	▮ 23.2	21.1	19.3	17.8	16.6	15.5	14.5	
W 10 × 30	6.1	43.2	▮ 37.0	32.4	28.8	25.9	23.6						
W 12 × 26	6.9	44.5	38.2	33.4	▮ 29.7	26.7	24.3	22.3	20.5	19.1			
W 10 × 33	8.4	46.7	▮ 40.0	35.0	31.0	28.0	25.4						
W 14 × 26	5.3	47.1	40.3	35.3	31.4	▮ 28.2	25.7	23.5	21.7	20.2	18.8	17.6	
W 16 × 26	5.6	51.2	43.9	38.4	34.1	30.7	▮ 27.9	25.6	23.6	21.9	20.5	19.2	18.1
W 12 × 30	6.9	51.5	44.1	38.6	▮ 34.3	30.9	28.1	25.7	23.8	22.0			
W 14 × 30	7.1	56.0	48.0	42.0	37.3	▮ 33.6	30.5	28.0	25.8	24.0	22.4	21.0	
W 10 × 39	8.4	56.1	▮ 48.1	42.1	37.4	33.7	30.6						
W 12 × 35	6.9	60.8	52.1	45.6	▮ 40.5	36.5	33.2	30.4	28.1	26.0			
W 16 × 31	5.8	62.9	53.9	47.2	41.9	37.8	▮ 34.3	31.5	29.0	27.0	25.2	23.6	22.2
W 14 × 34	7.1	64.8	55.5	48.6	43.2	▮ 38.9	35.3	32.4	29.9	27.8	25.9	24.3	
W 10 × 45	8.5	65.5	▮ 56.1	49.1	43.6	39.3	35.7						

TABLE A.4. (*Continued*)

Shape	Span (ft) / Deflection Factor[b] / L_e^c (ft)	16	18	20	22	24	26	28	30	32	34	36	38
		6.36	8.05	9.93	12.0	14.3	16.8	19.5	22.3	25.4	28.7	32.2	35.9
W 12 × 40	8.4	51.9	46.1	41.5	37.7	34.6	31.9	29.6					
W 14 × 38	7.1	54.6	48.5	43.7	39.7	36.4	33.6	31.2	29.1	27.3			
W 16 × 36	7.4	56.5	50.2	45.2	41.1	37.7	34.8	32.3	30.1	28.2	26.6	25.1	
W 18 × 35	6.3	57.8	51.4	46.2	42.0	38.5	35.6	33.0	30.8	28.9	27.2	25.7	24.3
W 12 × 45	8.5	58.1	51.6	46.5	42.2	38.7	35.7	33.2					
W 14 × 43	8.4	62.7	55.7	50.1	45.6	41.8	38.6	35.8	33.4	31.3			
W 12 × 50	8.5	64.7	57.5	51.7	47.0	43.1	39.8	37.0					
W 16 × 40	7.4	64.7	57.5	51.7	47.0	43.1	39.8	37.0	34.5	32.3	30.4	28.7	
W 18 × 40	6.3	68.4	60.8	54.7	49.7	45.6	42.1	39.1	36.5	34.2	32.2	30.4	28.8
W 14 × 48	8.5	70.3	62.5	56.2	51.1	46.9	43.3	40.2	37.5	35.1			
W 12 × 53	10.6	70.6	62.7	56.5	51.3	47.1	43.4	40.3					
W 16 × 45	7.4	72.7	64.6	58.2	52.9	48.5	44.7	41.5	38.8	36.3	34.2	32.3	
W 14 × 53	8.5	77.8	69.1	62.2	56.6	51.9	47.9	44.4	41.5	38.9			
W 18 × 46	6.4	78.8	70.0	63.0	57.3	52.5	48.5	45.0	42.0	39.4	37.1	35.0	33.2
W 16 × 50	7.5	81.0	72.0	64.8	58.9	54.0	49.8	46.3	43.2	40.5	38.1	36.0	

TABLE A.4. (*Continued*)

Shape	Span (ft) / Deflection Factor[b] / $L_c{}^c$ (ft)	16	18	20	22	24	27	30	33	36	39	42	45
		6.36	8.05	9.93	12.0	14.3	18.1	22.3	27.0	32.2	37.8	43.8	50.3
W 21 × 44	6.6	81.6	72.5	65.3	59.3	54.4	48.3	▮ 43.5	39.6	36.3	33.5	31.1	29.0
W 18 × 50	7.9	88.9	79.0	71.1	64.6	59.3	▮ 52.7	47.4	43.1	39.5	36.5		
W 14 × 61	10.6	92.2	81.9	▮ 73.8	▮ 67.0	61.5	54.6	49.2	44.7				
W 16 × 57	7.5	92.2	81.9	73.8	67.0	61.5	54.6	49.2	44.7	41.0			
W 21 × 50	6.9	94.5	84.0	75.6	68.7	63.0	56.0	▮ 50.4	45.8	42.0	38.8	36.0	33.6
W 18 × 55	7.9	98.3	87.4	78.6	71.5	65.5	▮ 58.2	52.4	47.7	43.7	40.3		
W 18 × 60	8.0	108	96.0	86.4	78.5	72.0	▮ 64.0	57.6	52.4	48.0	44.3		
W 21 × 57	6.9	111	98.7	88.6	80.7	74.0	65.8	▮ 59.2	53.8	49.3	45.5	42.3	39.5
W 24 × 55	7.0	114	101	91.2	82.9	76.0	67.5	60.8	▮ 55.3	50.7	46.8	43.4	40.5
W 16 × 67	10.8	117	104	93.6	▮ 85.1	78.0	69.3	62.4	56.7	52.0			
W 18 × 65	8.0	117	104	93.6	85.1	78.0	▮ 69.3	62.4	56.7	52.0	48.0		
W 18 × 71	8.1	127	113	102	92.4	84.7	▮ 72.2	67.7	61.5	56.4	52.1		
W 21 × 62	8.7	127	113	102	92.4	84.7	72.2	▮ 67.7	61.5	56.4	52.1	48.4	45.1
W 24 × 62	7.4	131	116	105	95.3	87.3	77.6	69.9	▮ 63.5	58.2	53.7	49.9	46.6
W 16 × 77	10.9	134	119	107	▮ 97.4	89.3	79.4	71.5	65.0	59.5			
W 21 × 68	8.7	140	124	112	102	93.3	83.0	▮ 74.7	67.9	62.2	57.4	53.3	49.8
W 18 × 76	11.6	146	130	117	106	97.3	▮ 86.5	77.9	70.8	64.9	59.9		
W 21 × 73	8.8	151	134	121	110	101	89.5	▮ 80.5	73.2	67.1	61.9	57.5	53.7
W 24 × 68	9.5	154	137	123	112	103	91.2	82.1	▮ 74.7	68.4	63.2	58.7	54.7
W 18 × 86	11.7	166	147	133	121	111	▮ 98.4	88.5	80.5	73.8	68.1		
W 21 × 83	8.8	171	152	137	124	114	101	▮ 91.2	82.9	76.0	70.1	65.1	60.8

TABLE A.4. *(Continued)*

Shape	Span (ft)	24	27	30	33	36	39	42	45	48	52	56	60
	Deflection Factor[b]	14.3	18.1	22.3	27.0	32.2	37.8	43.8	50.3	57.2	67.1	77.9	89.4
	L_c[c] (ft)												
W 24 × 76	9.5	117	104	93.9	▮ 85.3	78.2	72.2	67.0	62.6	58.7			
W 21 × 93	8.9	128	114	▮ 102	93.1	85.3	78.8	73.1	68.3				
W 24 × 84	9.5	131	116	104	▮ 95.0	87.1	80.4	74.7	69.7	65.3			
W 27 × 84	10.5	142	126	114	103	94.7	87.4	81.1	75.7	71.0	65.5	60.8	
W 24 × 94	9.6	148	131	118	▮ 108	98.7	91.1	84.6	78.9	74.0			
W 21 × 101	13.0	151	134	▮ 121	110	101	93.1	86.5	80.7				
W 27 × 94	10.5	162	144	130	113	108	▮ 99.7	92.6	86.4	81.0	74.8	69.4	
W 24 × 104	13.5	172	153	138	▮ 125	115	106	98.3	91.7	86.0			
W 27 × 102	10.6	178	158	142	129	119	▮ 109	102	94.9	89.0	82.1	76.3	
W 30 × 99	10.9	179	159	143	130	120	110	▮ 102	95.6	89.7	82.8	76.9	71.7
W 24 × 117	13.5	194	172	155	▮ 141	129	119	111	103	97.0			
W 27 × 114	10.6	199	177	159	145	133	▮ 123	114	106	99.7	92.0	85.4	
W 30 × 108	11.1	199	177	159	145	133	123	▮ 114	106	99.7	92.0	85.4	79.7
W 30 × 116	11.1	219	195	175	159	146	135	▮ 125	117	110	101	94.0	87.7
W 30 × 124	11.1	237	210	189	172	158	146	▮ 135	126	118	109	101	94.7

TABLE A.4. *(Continued)*

Shape	Span (ft)	30	33	36	39	42	45	48	52	56	60	65	70
	Deflection Factor[b] / L_c[c] (ft)	22.3	27.0	32.2	37.8	43.8	50.3	57.2	67.1	77.9	89.4	105	122
W 33 × 118	12.0	191	174	159	147	137	▌128	120	110	103	95.7	88.4	
W 30 × 132	11.1	203	184	169	156	▌145	135	127	117	109	101		
W 33 × 130	12.1	216	197	180	166	155	▌144	135	125	116	108	99.9	
W 27 × 146	14.7	219	199	183	▌169	156	146	137	126	117			
W 36 × 135	12.3	234	213	195	180	167	156	146	▌135	125	117	108	100
W 33 × 141	12.2	239	217	199	184	171	▌159	149	138	128	119	110	
W 33 × 152	12.2	260	236	216	200	185	▌173	162	150	139	130	120	
W 36 × 150	12.6	269	244	224	207	192	179	168	155	144	134	124	115
W 30 × 173	15.8	287	261	239	221	▌205	192	180	166	154	144		
W 36 × 160	12.7	289	263	241	222	206	193	181	▌167	155	144	133	124
W 36 × 170	12.7	309	281	258	238	221	206	193	▌178	166	155	143	132
W 30 × 191	15.9	319	290	268	245	▌228	213	199	184	171	159		
W 36 × 182	12.7	332	302	277	256	237	221	208	▌192	178	166	153	142
W 36 × 194	12.8	354	322	295	272	253	236	221	▌204	190	177	163	152
W 33 × 201	16.6	365	332	304	281	260	▌243	228	210	195	182	168	
W 36 × 210	12.9	383	349	319	295	274	256	240	▌221	205	192	177	164
W 33 × 221	16.7	404	367	336	310	288	▌269	252	233	216	202	186	
W 33 × 241	16.7	442	402	368	340	316	▌295	276	255	237	221	204	
W 36 × 230	17.4	446	406	372	343	319	298	279	▌257	239	223	206	191
W 36 × 245	17.4	477	434	398	367	341	318	298	▌275	256	239	220	204
W 36 × 260	17.5	508	462	423	391	363	339	318	▌293	272	254	234	218
W 36 × 280	17.5	549	499	458	422	392	366	343	▌317	294	275	253	235
W 36 × 300	17.6	592	538	493	455	423	395	370	▌341	317	296	273	254

[a] Total allowable uniformly distributed load in kips for simple span beams with yield stress of 36 ksi [250 MPa]. Loads to the right of the heavy vertical lines will cause deflections in excess of $\frac{1}{360}$ of the span.
[b] Maximum deflection in inches at the center of the span may be obtained by dividing the factor given by the depth of the beam in inches.
[c] Maximum permitted distance between points of lateral support. For greater distances use the charts in the AISC Manual (Ref. 10) to obtain the required beam size.

h/d	Rf	h/d	Rf	h/d	Rf	h/d	Rf	h/d	Rf	h/d	Rf
9.90	.0025	5.20	.0160	1.85	.2104	1.38	.3694	0.91	.7177	0.45	1.736
9.80	.0026	5.10	.0169	1.84	.2128	1.37	.3742	0.90	.7291	0.44	1.779
9.70	.0027	5.00	.0179	1.83	.2152	1.36	.3790	0.89	.7407	0.43	1.825
9.60	.0027	4.90	.0189	1.82	.2176	1.35	.3840	0.88	.7527	0.42	1.874
9.50	.0028	4.80	.0200	1.81	.2201	1.34	.3890	0.87	.7649	0.41	1.924
9.40	.0029	4.70	.0212	1.80	.2226	1.33	.3942	0.86	.7773	0.40	1.978
9.30	.0030	4.60	.0225	1.79	.2251	1.32	.3994	0.85	.7901	0.39	2.034
9.20	.0031	4.50	.0239	1.78	.2277	1.31	.4047	0.84	.8031	0.38	2.092
9.10	.0032	4.40	.0254	1.77	.2303	1.30	.4100	0.83	.8165	0.37	2.154
9.00	.0033	4.30	.0271	1.76	.2330	1.29	.4155	0.82	.8302	0.36	2.219
8.90	.0034	4.20	.0288	1.75	.2356	1.28	.4211	0.81	.8442	0.35	2.287
8.80	.0035	4.10	.0308	1.74	.2384	1.27	.4267	0.80	.8585	0.34	2.360
8.70	.0037	4.00	.0329	1.73	.2411	1.26	.4324	0.79	0.873	0.33	2.437
8.60	.0038	3.90	.0352	1.72	.2439	1.25	.4384	0.78	0.888	0.32	2.518
8.50	.0039	3.80	.0377	1.71	.2468	1.24	.4443	0.77	0.904	0.31	2.605
8.40	.0040	3.70	.0405	1.70	.2497	1.23	.4504	0.76	0.920	0.30	2.697
8.30	.0042	3.60	.0435	1.69	.2526	1.22	.4566	0.75	0.936	0.29	2.795
8.20	.0043	3.50	.0468	1.68	.2556	1.21	.4628	0.74	0.952	0.28	2.900
8.10	.0045	3.40	.0505	1.67	.2586	1.20	.4692	0.73	0.969	0.27	3.013
8.00	.0047	3.30	.0545	1.66	.2617	1.19	.4757	0.72	0.987	0.26	3.135
7.90	.0048	3.20	.0590	1.65	.2648	1.18	.4823	0.71	1.005	0.25	3.265
7.80	.0050	3.10	.0640	1.64	.2679	1.17	.4891	0.70	1.023	0.24	3.407
7.70	.0052	3.00	.0694	1.63	.2711	1.16	.4959	0.69	1.042	0.23	3.560
7.60	.0054	2.90	.0756	1.62	.2744	1.15	.5029	0.68	1.062	0.22	3.728
7.50	.0056	2.80	.0824	1.61	.2777	1.14	.5100	0.67	1.082	0.21	3.911
7.40	.0058	2.70	.0900	1.60	.2811	1.13	.5173	0.66	1.103	0.20	4.112
7.30	.0061	2.60	.0985	1.59	.2844	1.12	.5247	0.65	1.124	.195	4.220
7.20	.0063	2.50	.1081	1.58	.2879	1.11	.5322	0.64	1.146	.190	4.334
7.10	.0065	2.40	.1189	1.57	.2914	1.10	.5398	0.63	1.168	.185	4.454
7.00	.0069	2.30	.1311	1.56	.2949	1.09	.5476	0.62	1.191	.180	4.580
6.90	.0072	2.20	.1449	1.55	.2985	1.08	.5556	0.61	1.216	.175	4.714
6.80	.0075	2.10	.1607	1.54	.3022	1.07	.5637	0.60	1.240	.170	4.855
6.70	.0078	2.00	.1786	1.53	.3059	1.06	.5719	0.59	1.266	.165	5.005
6.60	.0081	1.99	.1805	1.52	.3097	1.05	.5804	0.58	1.292	.160	5.164
6.50	.0085	1.98	.1824	1.51	.3136	1.04	.5889	0.57	1.319	.155	5.334
6.40	.0089	1.97	.1844	1.50	.3175	1.03	.5977	0.56	1.347	.150	5.514
6.30	.0093	1.96	.1864	1.49	.3214	1.02	.6066	0.55	1.376	.145	5.707
6.20	.0097	1.95	.1885	1.48	.3245	1.01	.6157	0.54	1.407	.140	5.914
6.10	.0102	1.94	.1905	1.47	.3295	1.00	.6250	0.53	1.438	.135	6.136
6.00	.0107	1.93	.1926	1.46	.3337	0.99	.6344	0.52	1.470	.130	6.374
5.90	.0112	1.92	.1947	1.45	.3379	0.98	.6441	0.51	1.504	.125	6.632
5.80	.0118	1.91	.1969	1.44	.3422	0.97	.6540	0.50	1.539	.120	6.911
5.70	.0124	1.90	.1991	1.43	.3465	0.96	.6641	0.49	1.575	.115	7.215
5.60	.0130	1.89	.2013	1.42	.3510	0.95	.6743	0.48	1.612	.110	7.545
5.50	.0137	1.88	.2035	1.41	.3555	0.94	.6848	0.47	1.651	.105	7.908
5.40	.0144	1.87	.2058	1.40	.3600	0.93	.6955	0.46	1.692	.100	8.306
5.30	.0152	1.86	.2081	1.39	.3647	0.92	.7065				

TABLE A.5. Rigidity Coefficients for Cantilevered Masonry Walls

h/d	Rc	h/d	Rc	h/d	Rc	h/d	Rc	h/d	Rc	h/d	Rc
9.90	.0006	5.20	.0043	1.85	.0810	1.38	.1706	0.91	.4352	0.45	1.4582
9.80	.0007	5.10	.0046	1.84	.0821	1.37	.1737	0.90	.4452	0.44	1.5054
9.70	.0007	5.00	.0049	1.83	.0833	1.36	.1768	0.89	.4554	0.43	1.5547
9.60	.0007	4.90	.0052	1.82	.0845	1.35	.1800	0.88	.4659	0.42	1.6063
9.50	.0007	4.80	.0055	1.81	.0858	1.34	.1832	0.87	.4767	0.41	1.6604
9.40	.0007	4.70	.0058	1.80	.0870	1.33	.1866	0.86	.4899	0.40	1.7170
9.30	.0008	4.60	.0062	1.79	.0883	1.32	.1900	0.85	.4994	0.39	1.7765
9.20	.0008	4.50	.0066	1.78	.0896	1.31	.1935	0.84	.5112	0.38	1.8380
9.10	.0008	4.40	.0071	1.77	.0909	1.30	.1970	0.83	.5233	0.37	1.9098
9.00	.0008	4.30	.0076	1.76	.0923	1.29	.2007	0.82	.5359	0.36	1.9738
8.90	.0009	4.20	.0081	1.75	.0937	1.28	.2044	0.81	.5488	0.35	2.0467
8.80	.0009	4.10	.0087	1.74	.0951	1.27	.2083	0.80	.5621	0.34	2.1237
8.70	.0009	4.00	.0093	1.73	.0965	1.26	.2122	0.79	.5758	0.33	2.2051
8.60	.0010	3.90	.0100	1.72	.0980	1.25	.2162	0.78	.5899	0.32	2.2913
8.50	.0010	3.80	.0108	1.71	.0995	1.24	.2203	0.77	.6044	0.31	2.3828
8.40	.0010	3.70	.0117	1.70	.1010	1.23	.2245	0.76	.6194	0.30	2.4802
8.30	.0011	3.60	.0127	1.69	.1026	1.22	.2289	0.75	.6349	0.29	2.5838
8.20	.0012	3.50	.0137	1.68	.1041	1.21	.2333	0.74	.6509	0.28	2.6945
8.10	.0012	3.40	.0149	1.67	.1058	1.20	.2378	0.73	.6674	0.27	2.8130
8.00	.0012	3.30	.0163	1.66	.1074	1.19	.2425	0.72	.6844	0.26	2.9401
7.90	.0013	3.20	.0178	1.65	.1091	1.18	.2472	0.71	.7019	0.25	3.0769
7.80	.0013	3.10	.0195	1.64	.1108	1.17	.2521	0.70	.7200	0.24	3.2246
7.70	.0014	3.00	.0214	1.63	.1125	1.16	.2571	0.69	.7388	0.23	3.3845
7.60	.0014	2.90	.0235	1.62	.1143	1.15	.2622	0.68	.7581	0.22	3.5583
7.50	.0015	2.80	.0260	1.61	.1162	1.14	.2675	0.67	.7781	0.21	3.7479
7.40	.0015	2.70	.0288	1.60	.1180	1.13	.2729	0.66	.7987	0.20	3.9557
7.30	.0016	2.60	.0320	1.59	.1199	1.12	.2784	0.65	.8201	.195	4.0673
7.20	.0017	2.50	.0357	1.58	.1218	1.11	.2841	0.64	.8422	.190	4.1845
7.10	.0017	2.40	.0400	1.57	.1238	1.10	.2899	0.63	.8650	.185	4.3079
7.00	.0018	2.30	.0450	1.56	.1258	1.09	.2959	0.62	.8886	.180	4.4379
6.90	.0019	2.20	.0508	1.55	.1279	1.08	.3020	0.61	.9131	.175	4.5751
6.80	.0020	2.10	.0577	1.54	.1300	1.07	.3083	0.60	.9384	.170	4.7201
6.70	.0020	2.00	.0658	1.53	.1322	1.06	.3147	0.59	.9647	.165	4.8736
6.60	.0021	1.99	.0667	1.52	.1344	1.05	.3213	0.58	.9919	.160	5.0364
6.50	.0022	1.98	.0676	1.51	.1366	1.04	.3281	0.57	1.0201	.155	5.2095
6.40	.0023	1.97	.0685	1.50	.1389	1.03	.3351	0.56	1.0493	.150	5.3937
6.30	.0025	1.96	.0694	1.49	.1412	1.02	.3422	0.55	1.0797	.145	5.5904
6.20	.0026	1.95	.0704	1.48	.1436	1.01	.3496	0.54	1.1112	.140	5.8008
6.10	.0027	1.94	.0714	1.47	.1461	1.00	.3571	0.53	1.1439	.135	6.0261
6.00	.0028	1.93	.0724	1.46	.1486	0.99	.3649	0.52	1.1779	.130	6.2696
5.90	.0030	1.92	.0734	1.45	.1511	0.98	.3729	0.51	1.2132	.125	6.5306
5.80	.0031	1.91	.0744	1.44	.1537	0.97	.3811	0.50	1.2500	.120	6.8136
5.70	.0033	1.90	.0754	1.43	.1564	0.96	.3895	0.49	1.2883	.115	7.1208
5.60	.0035	1.89	.0765	1.42	.1591	0.95	.3981	0.48	1.3281	.110	7.4555
5.50	.0037	1.88	.0776	1.41	.1619	0.94	.4070	0.47	1.3696	.105	7.8215
5.40	.0039	1.87	.0787	1.40	.1647	0.93	.4162	0.46	1.4130	.100	8.2237
5.30	.0041	1.86	.0798	1.39	.1676	0.92	.4255				

TABLE A.6. Rigidity Coefficients for Fixed Masonry Walls

BENDING IN MASONRY WALLS

The following example illustrates the use of Figure A.1. Other uses are illustrated in Section 9.2.

Investigate the wall and the required reinforcing for the following data:

Wall height = 16.7 ft.

8-in. block, t = 7.625 in.

Use single row of reinforcing in center, d = 3.813 in. n = 40, allowable f_m = 250 × 1.33 = 333 psi.

Grade 40 bars, f_s = 1.33 × 20,000 = 26,667 psi.

Wind pressure is 20 psf.

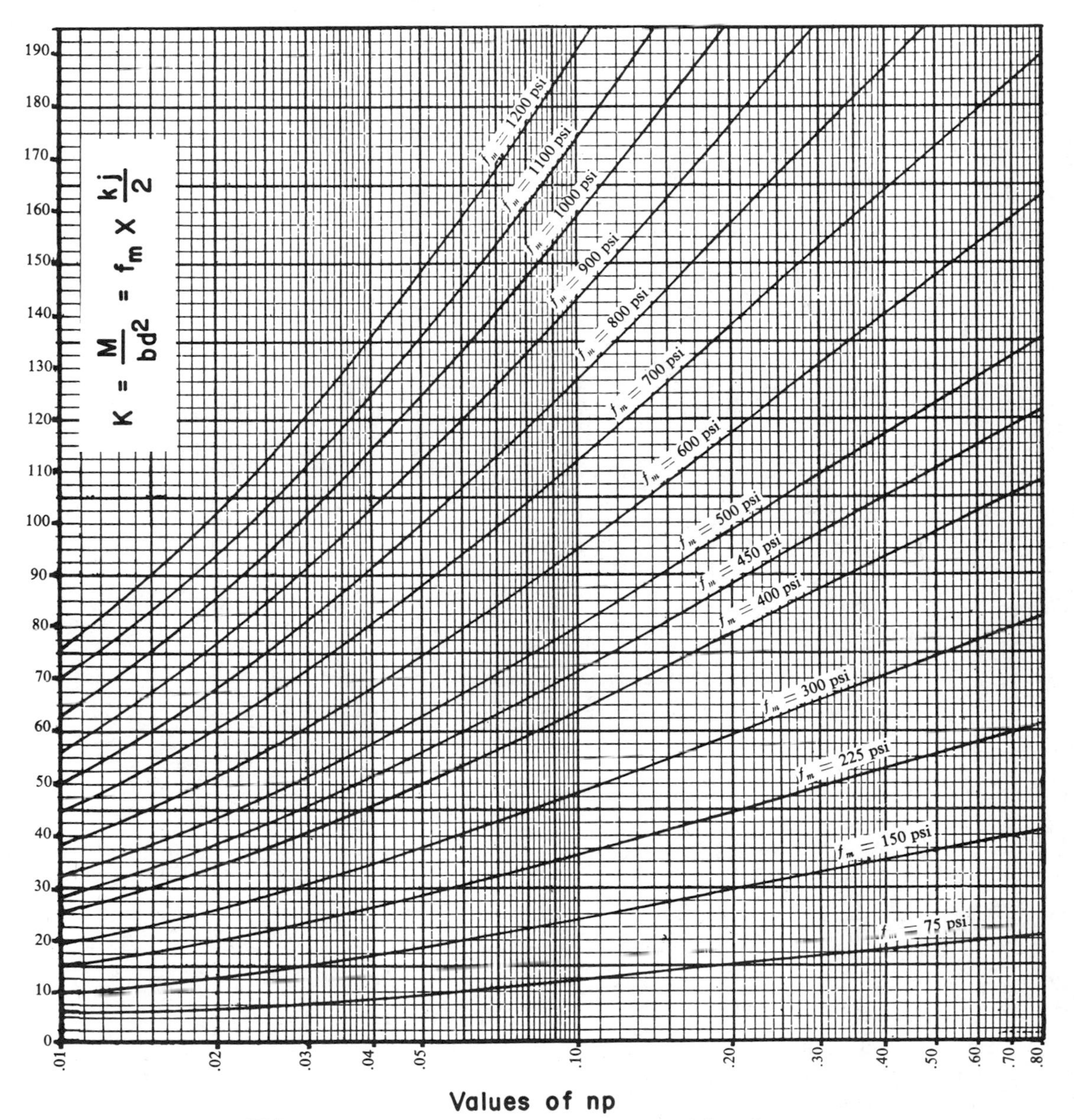

FIGURE A.1. Flexural coefficient k-chart for reinforced masonry.

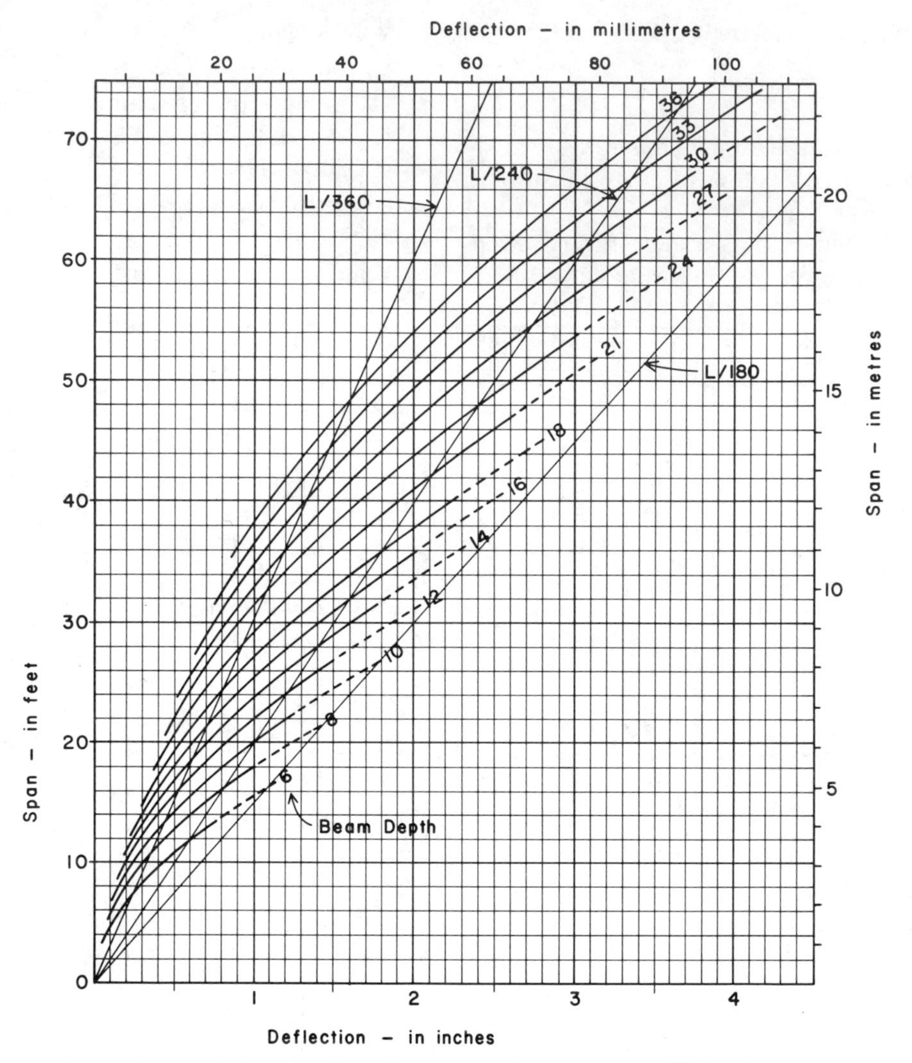

FIGURE A.2. Deflection of steel beams with bending stress of 24 ksi [165 MPa].

Find

$$M = \frac{wL^2}{8} = \frac{20(16.7)^2}{8} \times 12$$

$$= 8367 \text{ in.-lb}$$

$$K = \frac{M}{bd^2} = \frac{8367}{12(3.813)^2} = 48$$

Enter the diagram (Fig. A.1) at the left with $K = 48$, proceed to the right to intersect $f_m = 333$ psi, read at the bottom $np = 0.073$.

Then

$$p = \frac{0.073}{n} = \frac{0.073}{40} = 0.001825$$

$$A_s = pbd = 0.001825 \times 12 \times 3.813$$

$$= 0.0835 \text{ in.}^2/\text{ft}$$

Try No. 5 at 40 in.:

$$A_s = \frac{12}{40} \times 0.31 = 0.093 \text{ in.}^2/\text{ft}$$

Check f_s using approximate $j = 0.90$:

$$f_s = \frac{M}{f_s jd} = \frac{8367}{(0.093)(0.9)(3.813)}$$

$$= 26{,}217 \text{ psi}$$

The reinforcing is adequate unless a combined stress must be investigated.

DEFLECTION OF STEEL BEAMS

Figure A.2 may be used for the investigation of steel beams for deflection. The diagram is based on the use of A36 steel, for which the maximum allowable bending stress is 24 ksi. The following examples illustrate possible uses of the diagram.

Example 1

A beam spans 40 ft and has a critical deflection limit of $L/240$ under total load. Enter the diagram (Fig. A.2) at the left with the span of 40 ft and proceed to the right to intersect the line for $L/240$. Any beam with the depth curve above this point (21 in. or greater in depth) will have a deflection of less than $L/240$ if stress is limited to 24 ksi. This is a quickly found bit of information in preliminary selection of a beam size for design.

Example 2

An 18-in. deep beam is used on a span of 32 ft to carry a uniformly distributed load. The load is 19 k live load and 20 k dead load. Deflection is limited to $L/360$ under live load only and $L/240$ under total load. The maximum permitted total load from Table A.2 is 39.4 k. Is deflection critical?

It may be noted that the total load condition will be more severe. However, we will compute both deflections to illustrate the process.

Enter the diagram at the left with the span of 32 ft. Proceed to the right to intersect the curve for the beam depth of 18 in. Read at the bottom: Deflection is approximately 1.40 in. This is the deflection that corresponds to the total load of 39.4 k, with both relating to a maximum

TABLE A.7. Balanced Section Properties for Rectangular Concrete Sections with Tension Reinforcing Only

f_s		f'_c						R	
ksi	MPa	ksi	MPa	n	k	j	p	k-in.	kN-m
16	110	2.0	13.79	11.3	0.389	0.870	0.0109	0.152	1045
		2.5	17.24	10.1	0.415	0.862	0.0146	0.201	1382
		3.0	20.68	9.2	0.437	0.854	0.0184	0.252	1733
		4.0	27.58	8.0	0.474	0.842	0.0266	0.359	2468
20	138	2.0	13.79	11.3	0.337	0.888	0.0076	0.135	928
		2.5	17.24	10.1	0.362	0.879	0.0102	0.179	1231
		3.0	20.68	9.2	0.383	0.872	0.0129	0.226	1554
		4.0	27.58	8.0	0.419	0.860	0.0188	0.324	2228
24	165	2.0	13.79	11.3	0.298	0.901	0.0056	0.121	832
		2.5	17.24	10.1	0.321	0.893	0.0075	0.161	1107
		3.0	20.68	9.2	0.341	0.886	0.0096	0.204	1403
		4.0	27.58	8.0	0.375	0.875	0.0141	0.295	2028

bending stress of 24 ksi. For the actual deflections we compute the following: For live load only:

$$\Delta = \frac{\text{live load}}{\text{table load}} \times \text{diagram deflection}$$

$$= \frac{19}{39.4} \times 1.40 = 0.675 \text{ in.}$$

Compare to limit of:

$$\Delta = \frac{L}{360} = \frac{32 \times 12}{360} = 1.07 \text{ in.}$$

For total load:

$$\Delta = \frac{\text{total load}}{\text{table load}} \times \text{diagram deflection}$$

$$= \frac{39}{39.4} \times 1.40 = 1.39 \text{ in.}$$

Compare to limit of:

$$\Delta = \frac{L}{240} = \frac{32 \times 12}{240} = 1.60 \text{ in.}$$

Deflection is not critical.

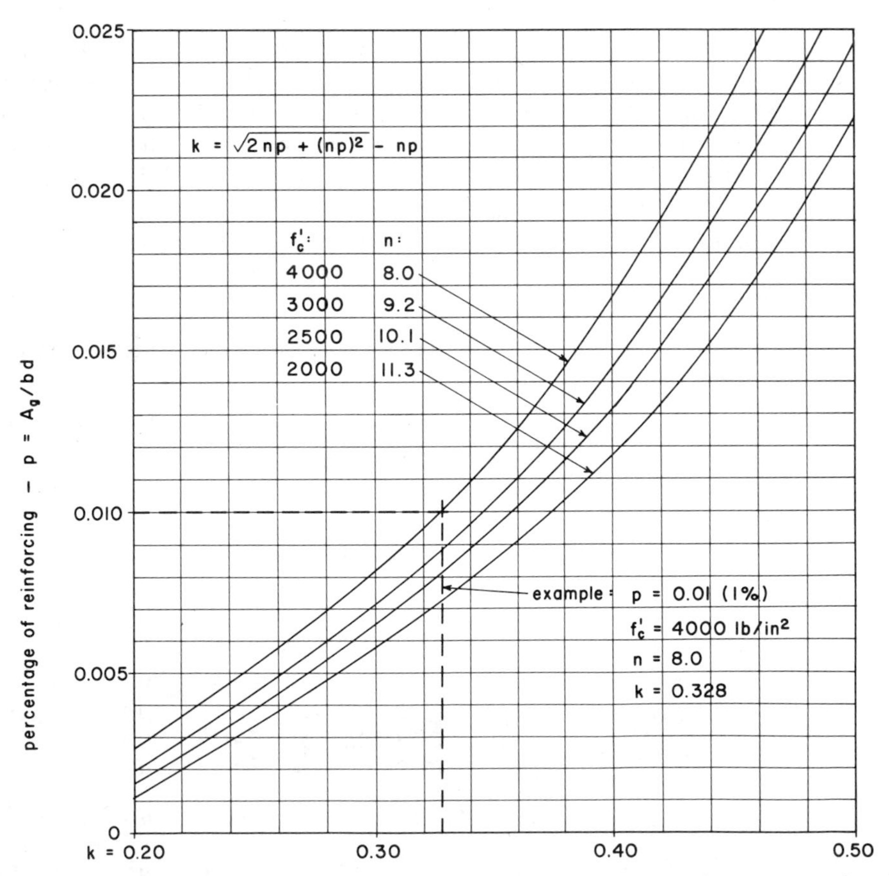

FIGURE A.3. *k* factors for rectangular concrete beams with tension reinforcing only, as a function of *n* and *p*.

TABLE A.8. Weights of Building Construction

	lb/ft²	kN/m²
Roofs		
3-ply ready roofing (roll, composition)	1	0.05
3-ply felt and gravel	5.5	0.26
5-ply felt and gravel	6.5	0.31
Shingles		
Wood	2	0.10
Asphalt	2–3	0.10–0.15
Clay tile	9–12	0.43–0.58
Concrete tile	8–12	0.38–0.58
Slate, $\frac{1}{4}$ in.	10	0.48
Fiber glass	2–3	0.10–0.15
Aluminum	1	0.05
Steel	2	0.10
Insulation		
Fiber glass batts	0.5	0.025
Rigid foam plastic	1.5	0.075
Foamed concrete, mineral aggregate	2.5/in.	0.0047/mm
Wood rafters		
2 × 6 at 24 in.	1.0	0.05
2 × 8 at 24 in.	1.4	0.07
2 × 10 at 24 in.	1.7	0.08
2 × 12 at 24 in.	2.1	0.10
Steel deck, painted		
22 ga	1.6	0.08
20 ga	2.0	0.10
18 ga	2.6	0.13
Skylight		
Glass with steel frame	6–10	0.29–0.48
Plastic with aluminum frame	3–6	0.15–0.29
Plywood or softwood board sheathing	3.0/in.	0.0057/mm
Ceilings		
Suspended steel channels	1	0.05
Lath		
Steel mesh	0.5	0.025
Gypsum board, $\frac{1}{2}$ in.	2	0.10
Fiber tile	1	0.05
Drywall, gypsum board, $\frac{1}{2}$ in.	2.5	0.12
Plaster		
Gypsum, acoustic	5	0.24
Cement	8.5	0.41
Suspended lighting and air distribution Systems, average	3	0.15

TABLE A.8. (*Continued*)

	lb/ft²	kN/m²
Floors		
Hardwood, ½ in.	2.5	0.12
Vinyl tile, ⅛ in.	1.5	0.07
Asphalt mastic	12/in.	0.023/mm
Ceramic tile		
¾ in.	10	0.48
Thin set	5	0.24
Fiberboard underlay, ⅝ in.	3	0.15
Carpet and pad, average	3	0.15
Timber deck	2.5/in.	0.0047/mm
Steel deck, stone concrete fill, average	35–40	1.68–1.92
Concrete deck, stone aggregate	12.5/in.	0.024/mm
Wood joists		
2 × 8 at 16 in.	2.1	0.10
2 × 10 at 16 in.	2.6	0.13
2 × 12 at 16 in.	3.2	0.16
Lightweight concrete fill	8.0/in.	0.015/mm
Walls		
2 × 4 studs at 16 in., average	2	0.10
Steel studs at 16 in., average	4	0.20
Lath, plaster; see Ceilings		
Gypsum drywall, ⅝ in. single	2.5	0.12
Stucco, ⅞ in., on wire and paper or felt	10	0.48
Windows, average, glazing + frame		
Small pane, single glazing, wood or metal frame	5	0.24
Large pane, single glazing, wood or metal frame	8	0.38
Increase for double glazing	2–3	0.10–0.15
Curtain walls, manufactured units	10–15	0.48–0.72
Brick veneer		
4-in., mortar joints	40	1.92
½-in., mastic	10	0.48
Concrete block		
Lightweight, unreinforced—4 in.	20	0.96
6 in.	25	1.20
8 in.	30	1.44
Heavy, reinforced, grouted—6 in.	45	2.15
8 in.	60	2.87
12 in.	85	4.07

APPENDIX B

Requirements of the Uniform Building Code

This appendix contains materials reprinted directly from the pages of the *Uniform Building Code,* 1985 edition (Ref. 7), with permission of the publishers. Although these materials are provided for the convenience of the reader, they should not be used for any actual design work unless this code corresponds with local jurisdiction.

TABLE NO. 25-A-1—ALLOWABLE UNIT STRESSES—STRUCTURAL LUMBER—(Continued)
Allowable Unit Stresses for Structural Lumber—VISUAL GRADING
(Normal loading. See also Section 2504)

SPECIES AND COMMERCIAL GRADE	SIZE CLASSIFICATION	ALLOWABLE UNIT STRESSES IN POUNDS PER SQUARE INCH							U.B.C. STDS UNDER WHICH GRADED
		Extreme Fiber in Bending F_b: Single-member Uses	Extreme Fiber in Bending F_b: Repetitive-member Uses	Tension Parallel to Grain F_t	Horizontal Shear F_v	Compression perpendicular to Grain $F_{c\perp}$ [21]	Compression Parallel to Grain F_c	MODULUS OF ELASTICITY E [21]	
DOUGLAS FIR - LARCH (Surfaced dry or surfaced green. Used at 19% max. m.c.)									
DOUGLAS FIR - LARCH (North)									
Dense Select Structural	2" to 4" thick 2" to 4" wide	2450	2800	1400	95	730	1850	1,900,000	25-2 25-3 and 25-4 (See footnotes 2 through 9, 11, 13, 15 and 16)
Select Structural		2100	2400	1200	95	625	1600	1,800,000	
Dense No. 1		2050	2400	1200	95	730	1450	1,900,000	
No. 1		1750	2050	1050	95	625	1250	1,800,000	
Dense No. 2		1700	1950	1000	95	730	1150	1,700,000	
No. 2		1450	1650	850	95	625	1000	1,700,000	
No. 3		800	925	475	95	625	600	1,500,000	
Appearance		1750	2050	1050	95	625	1500	1,800,000	
Stud		800	925	475	95	625	600	1,500,000	
Construction	2" to 4" thick 4" wide	1050	1200	625	95	625	1150	1,500,000	
Standard		600	675	350	95	625	925	1,500,000	
Utility		275	325	175	95	625	600	1,500,000	
Dense Select Structural	2" to 4" thick 5" and wider	2100	2400	1400	95	730	1650	1,900,000	
Select Structural		1800	2050	1200	95	625	1400	1,800,000	
Dense No. 1		1800	2050	1200	95	730	1450	1,900,000	
No. 1		1500	1750	1000	95	625	1250	1,800,000	
Dense No. 2		1450	1700	775	95	730	1250	1,700,000	
No. 2		1250	1450	650	95	625	1050	1,700,000	
No. 3 and Stud		725	850	375	95	625	675	1,500,000	
Appearance		1500	1750	1000	95	625	1500	1,800,000	
Dense Select Structural	Beams and Stringers [12]	1900	—	1100	85	730	1300	1,700,000	25-3 (See footnotes 2 through 9)
Select Structural		1600	—	950	85	625	1100	1,600,000	
Dense No. 1		1550	—	775	85	730	1100	1,700,000	
No. 1		1300	—	675	85	625	925	1,600,000	
Dense Select Structural	Posts and Timbers [12]	1750	—	1150	85	730	1350	1,700,000	
Select Structural		1500	—	1000	85	625	1150	1,600,000	
Dense No. 1		1400	—	950	85	730	1200	1,700,000	
No. 1		1200	—	825	85	625	1000	1,600,000	
Select Dex	Decking	1750	2000	—	—	625	—	1,800,000	
Commercial Dex		1450	1650	—	—	625	—	1,700,000	

TABLE NO. 25-J-1—ALLOWABLE SHEAR IN POUNDS PER FOOT FOR HORIZONTAL PLYWOOD DIAPHRAGMS WITH FRAMING OF DOUGLAS FIR-LARCH OR SOUTHERN PINE[1]

PLYWOOD GRADE	Common Nail Size	Minimum Nominal Penetration in Framing (In Inches)	Minimum Nominal Plywood Thickness (In Inches)	Minimum Nominal Width of Framing Member (In Inches)	BLOCKED DIAPHRAGMS — Nail spacing at diaphragm boundaries (all cases), at continuous panel edges parallel to load (Cases 3 and 4) and at all panel edges (Cases 5 and 6)				UNBLOCKED DIAPHRAGM — Nails spaced 6" max. at supported end	
					6	4	2½[2]	2[2]	Load perpendicular to unblocked edges and continuous panel joints (Case 1)	Other configurations (Cases 2, 3 & 4)
					Nail spacing at other plywood panel edges					
					6	6	4	3		
STRUCTURAL I	6d	1¼	5/16	2	185	250	375	420	165	125
				3	210	280	420	475	185	140
	8d	1½	3/8	2	270	360	530	600	240	180
				3	300	400	600	675	265	200
	10d	1⅝	15/32	2	320	425	640	730[2]	285	215
				3	360	480	720	820	320	240
C-D, C-C, STRUCTURAL II and other grades covered in U.B.C. Standard No. 25-9	6d	1¼	5/16	2	170	225	335	380	150	110
				3	190	250	380	430	170	125
			3/8	2	185	250	375	420	165	125
				3	210	280	420	475	185	140
	8d	1½	3/8	2	240	320	480	545	215	160
				3	270	360	540	610	240	180
			15/32	2	270	360	530	600	240	180
				3	300	400	600	675	265	200
	10d	1⅝	15/32	2	290	385	575	655[2]	255	190
				3	325	430	650	735	290	215
			19/32	2	320	425	640	730[2]	285	215
				3	360	480	720	820	320	240

[1]These values are for short-time loads due to wind or earthquake and must be reduced 25 percent for normal loading. Space nails 10 inches on center for floors and 12 inches on center for roofs along intermediate framing members.

Allowable shear values for nails in framing members of other species set forth in Table No. 25-17-J of U.B.C. Standards shall be calculated for all grades by multiplying the values for nails in STRUCTURAL I by the following factors: Group III, 0.82 and Group IV, 0.65.

[2]Framing shall be 3-inch nominal or wider and nails shall be staggered where nails are spaced 2 inches or 2½ inches on center, and where 10d nails having penetration into framing of more than 1⅝ inches are spaced 3 inches on center.

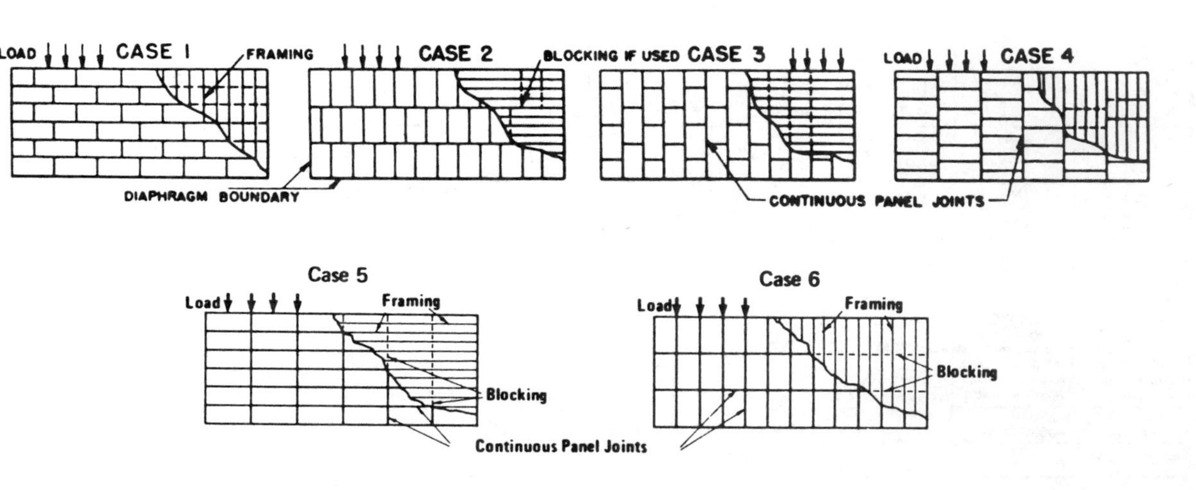

NOTE: Framing may be located in either direction for blocked diaphragms.

TABLE NO. 25-K-1—ALLOWABLE SHEAR FOR WIND OR SEISMIC FORCES IN POUNDS PER FOOT FOR PLYWOOD SHEAR WALLS WITH FRAMING OF DOUGLAS FIR-LARCH OR SOUTHERN PINE[1 4]

PLYWOOD GRADE	MINIMUM NOMINAL PLYWOOD THICKNESS (Inches)	MINIMUM NAIL PENETRATION IN FRAMING (Inches)	NAIL SIZE (Common or Galvanized Box)	PLYWOOD APPLIED DIRECT TO FRAMING — Nail Spacing at Plywood Panel Edges				NAIL SIZE (Common or Galvanized Box)	PLYWOOD APPLIED OVER ½-INCH GYPSUM SHEATHING — Nail Spacing at Plywood Panel Edges			
				6	4	3	2[2]		6	4	3	2[2]
STRUCTURAL I	5/16	1¼	6d	200	300	390	510	8d	200	300	390	510
	3/8	1½	8d	230[3]	360[3]	460[3]	610[3]	10d	280	430	550[2]	730[2]
	15/32	1½	8d	280	430	550	730	10d	280	430	550[2]	730
	15/32	1⅝	10d	340	510	665[2]	870	—	—	—	—	—
C-D, C-C STRUCTURAL II and other grades covered in U.B.C. Standard No. 25-9.	5/16	1¼	6d	180	270	350	450	8d	180	270	350	450
	3/8	1¼	6d	200	300	390	510	8d	200	300	390	510
	3/8	1½	8d	220[3]	320[3]	410[3]	530[3]	10d	260	380	490[2]	640
	15/32	1½	8d	260	380	490	640	10d	260	380	490[2]	640
	15/32	1⅝	10d	310	460	600[2]	770	—	—	—	—	—
	19/32	1⅝	10d	340	510	665[2]	870	—	—	—	—	—
			NAIL SIZE (Galvanized Casing)					NAIL SIZE (Galvanized Casing)				
Plywood panel siding in grades covered in U.B.C. Standard No. 25-9.	5/16	1¼	6d	140	210	275	360	8d	140	210	275	360
	3/8	1½	8d	130[3]	200[3]	260[3]	340[3]	10d	160	240	310[2]	410

[1]All panel edges backed with 2-inch nominal or wider framing. Plywood installed either horizontally or vertically. Space nails at 6 inches on center along intermediate framing members for 3/8-inch plywood installed with face grain parallel to studs spaced 24 inches on center and 12 inches on center for other conditions and plywood thicknesses. These values are for short-time loads due to wind or earthquake and must be reduced 25 percent for normal loading.

Allowable shear values for nails in framing members of other species set forth in Table No. 25-17-J of U.B.C. Standards shall be calculated for all grades by multiplying the values for common and galvanized box nails in STRUCTURAL I and galvanized casing nails in other grades by the following factors: Group III, 0.82 and Group IV, 0.65.

[2]Framing shall be 3-inch nominal or wider and nails shall be staggered where nails are spaced 2 inches on center, and where 10d nails having penetration into framing of more than 1 5/8 inches are spaced 3 inches on center.

[3]The values for 3/8-inch-thick plywood applied direct to framing may be increased 20 percent, provided studs are spaced a maximum of 16 inches on center or plywood is applied with face grain across studs.

[4]Where plywood is applied on both faces of a wall and nail spacing is less than 6 inches on center on either side, panel joints shall be offset to fall on different framing members or framing shall be 3-inch nominal or thicker and nails on each side shall be staggered.

TABLE NO. 25-S-1—ALLOWABLE SPANS FOR PLYWOOD SUBFLOOR AND ROOF SHEATHING CONTINUOUS OVER TWO OR MORE SPANS AND FACE GRAIN PERPENDICULAR TO SUPPORTS[1] [9]

PANEL SPAN RATING[3]	PLYWOOD THICKNESS (Inch)	ROOF[2] Maximum Span (In Inches)		Load (In Pounds per Square Foot)		FLOOR MAXIMUM SPAN[4] (In Inches)
		Edges Blocked	Edges Unblocked	Total Load	Live Load	
1. 12/0	15/16	12		135	130	0
2. 16/0	5/16, 3/8	16		80	65	0
3. 20/0	5/16, 3/8	20		70	55	0
4. 24/0	3/8	24	16	60	45	0
5. 24/0	15/32, 1/2	24	24	60	45	0
6. 32/16	15/32, 1/2, 19/32, 5/8	32	28	55	35[5]	16[7]
7. 40/20	19/32, 5/8, 23/32, 3/4, 7/8	40	32	40[5]	35[5]	20[7] [8]
8. 48/24	23/32, 3/4, 7/8	48	36	40[5]	35[5]	24

[1]These values apply for C-C, C-D, Structural I and II grades only. Spans shall be limited to values shown because of possible effect of concentrated loads.

[2]Uniform load deflection limitations 1/180 of the span under live load plus dead load, 1/240 under live load only. Edges may be blocked with lumber or other approved type of edge support.

[3]Span rating appears on all panels in the construction grades listed in Footnote No. 1.

[4]Plywood edges shall have approved tongue-and-groove joints or shall be supported with blocking unless 1/4-inch minimum thickness underlayment, or 1 1/2 inches of approved cellular or lightweight concrete is placed over the subfloor, or finish floor is 25/32-inch wood strip. Allowable uniform load based on deflection of 1/360 of span is 165 pounds per square foot.

[5]For roof live load of 40 pounds per square foot or total load of 55 pounds per square foot, decrease spans by 13 percent or use panel with next greater span rating.

[6]May be 24 inches if 25/32-inch wood strip flooring is installed at right angles to joists.

[7]May be 24 inches where a minimum of 1 1/2 inches of approved cellular or lightweight concrete is placed over the subfloor and the plywood sheathing is manufactured with exterior glue.

[8]Floor or roof sheathing conforming with this table shall be deemed to meet the design criteria of Section 2516.

TABLE NO. 25-S-2—ALLOWABLE LOADS FOR PLYWOOD ROOF SHEATHING CONTINUOUS OVER TWO OR MORE SPANS AND FACE GRAIN PARALLEL TO SUPPORTS[1 2]

	THICKNESS	NO. OF PLIES	SPAN	TOTAL LOAD	LIVE LOAD
STRUCTURAL I	15/32	4	24	30	20
		5	24	45	35
	1/2	4	24	35	25
		5	24	55	40
Other grades covered in U.B.C. Standard No. 25-9	15/32	5	24	25	20
	1/2	5	24	30	25
	19/32	4	24	35	25
		5	24	50	40
	5/8	4	24	40	30
		5	24	55	45

[1]Uniform load deflection limitations: 1/180 of span under live load plus dead load, 1/240 under live load only. Edges shall be blocked with lumber or other approved type of edge supports.

[2]Roof sheathing conforming with this table shall be deemed to meet the design criteria of Section 2516.

TABLE NO. 25-U-J-1—ALLOWABLE SPANS FOR FLOOR JOISTS—40 LBS. PER SQ. FT. LIVE LOAD

DESIGN CRITERIA: Deflection—For 40 lbs. per sq. ft. live load. Limited to span in inches divided by 360. Strength—Live load of 40 lbs. per sq. ft. plus dead load of 10 lbs. per sq. ft. determines the required fiber stress value.

JOIST SIZE (IN)	SPACING (IN)	Modulus of Elasticity, *E*, in 1,000,000 psi													
		0.8	0.9	1.0	1.1	1.2	1.3	1.4	1.5	1.6	1.7	1.8	1.9	2.0	2.2
2x6	12.0	8-6 720	8-10 780	9-2 830	9-6 890	9-9 940	10-0 990	10-3 1040	10-6 1090	10-9 1140	10-11 1190	11-2 1230	11-4 1280	11-7 1320	11-11 1410
2x6	16.0	7-9 790	8-0 860	8-4 920	8-7 980	8-10 1040	9-1 1090	9-4 1150	9-6 1200	9-9 1250	9-11 1310	10-2 1360	10-4 1410	10-6 1460	10-10 1550
2x6	24.0	6-9 900	7-0 980	7-3 1050	7-6 1120	7-9 1190	7-11 1250	8-2 1310	8-4 1380	8-6 1440	8-8 1500	8-10 1550	9-0 1610	9-2 1670	9-6 1780
2x8	12.0	11-3 720	11-8 780	12-1 830	12-6 890	12-10 940	13-2 990	13-6 1040	13-10 1090	14-2 1140	14-5 1190	14-8 1230	15-0 1280	15-3 1320	15-9 1410
2x8	16.0	10-2 790	10-7 850	11-0 920	11-4 980	11-8 1040	12-0 1090	12-3 1150	12-7 1200	12-10 1250	13-1 1310	13-4 1360	13-7 1410	13-10 1460	14-3 1550
2x8	24.0	8-11 900	9-3 980	9-7 1050	9-11 1120	10-2 1190	10-6 1250	10-9 1310	11-0 1380	11-3 1440	11-5 1500	11-8 1550	11-11 1610	12-1 1670	12-6 1780
2x10	12.0	14-4 720	14-11 780	15-5 830	15-11 890	16-5 940	16-10 990	17-3 1040	17-8 1090	18-0 1140	18-5 1190	18-9 1230	19-1 1280	19-5 1320	20-1 1410
2x10	16.0	13-0 790	13-6 850	14-0 920	14-6 980	14-11 1040	15-3 1090	15-8 1150	16-0 1200	16-5 1250	16-9 1310	17-0 1360	17-4 1410	17-8 1460	18-3 1550
2x10	24.0	11-4 900	11-10 980	12-3 1050	12-8 1120	13-0 1190	13-4 1250	13-8 1310	14-0 1380	14-4 1440	14-7 1500	14-11 1550	15-2 1610	15-5 1670	15-11 1780
2x12	12.0	17-5 720	18-1 780	18-9 830	19-4 890	19-11 940	20-6 990	21-0 1040	21-6 1090	21-11 1140	22-5 1190	22-10 1230	23-3 1280	23-7 1320	24-5 1410
2x12	16.0	15-10 790	16-5 860	17-0 920	17-7 980	18-1 1040	18-7 1090	19-1 1150	19-6 1200	19-11 1250	20-4 1310	20-9 1360	21-1 1410	21-6 1460	22-2 1550
2x12	24.0	13-10 900	14-4 980	14-11 1050	15-4 1120	15-10 1190	16-3 1250	16-8 1310	17-0 1380	17-5 1440	17-9 1500	18-1 1550	18-5 1610	18-9 1670	19-4 1780

NOTES:

(1) The required extreme fiber stress in bending (F_b) in pounds per square inch is shown below each span.

(2) Use single or repetitive member bending stress values (F_b) and modulus of elasticity values (E) from Tables Nos. 25-A-1 and 25-A-2.

(3) For more comprehensive tables covering a broader range of bending stress values (F_b) and modulus of elasticity values (E), other spacing of members and other conditions of loading, see U.B.C. Standard No. 25-21.

(4) The spans in these tables are intended for use in covered structures or where moisture content in use does not exceed 19 percent.

TABLE NO. 25-U-J-6—ALLOWABLE SPANS FOR CEILING JOISTS—10 LBS. PER SQ. FT. LIVE LOAD (Drywall Ceiling)

DESIGN CRITERIA: Deflection—For 10 lbs. per sq. ft. live load. Limited to span in inches divided by 240. Strength—Live load of 10 lbs. per sq. ft. plus dead load of 5 lbs. per sq. ft. determines the required fiber stress value.

JOIST SIZE (IN)	SPACING (IN)	Modulus of Elasticity, E, in 1,000,000 psi													
		0.8	0.9	1.0	1.1	1.2	1.3	1.4	1.5	1.6	1.7	1.8	1.9	2.0	2.2
2x4	12.0	9-10 710	10-3 770	10-7 830	10-11 880	11-3 930	11-7 980	11-10 1030	12-2 1080	12-5 1130	12.8 1180	12-11 1220	13-2 1270	13-4 1310	13-9 1400
	16.0	8-11 780	9-4 850	9-8 910	9-11 970	10-3 1030	10-6 1080	10-9 1140	11-0 1190	11-3 1240	11-6 1290	11-9 1340	11-11 1390	12-2 1440	12-6 1540
	24.0	7-10 900	8-1 970	8-5 1040	8-8 1110	8-11 1170	9-2 1240	9-5 1300	9-8 1360	9-10 1420	10-0 1480	10-3 1540	10-5 1600	10-7 1650	10-11 1760
2x6	12.0	15-6 710	16-1 770	16-8 830	17-2 880	17-8 930	18-2 980	18-8 1030	19-1 1080	19-6 1130	19-11 1180	20-3 1220	20-8 1270	21-0 1310	21-8 1400
	16.0	14-1 780	14-7 850	15-2 910	15-7 970	16-1 1030	16-6 1080	16-11 1140	17-4 1190	17-8 1240	18-1 1290	18-5 1340	18-9 1390	19-1 1440	19-8 1540
	24.0	12-3 900	12-9 970	13-3 1040	13-8 1110	14-1 1170	14-5 1240	14-9 1300	15-2 1360	15-6 1420	15-9 1480	16-1 1540	16-4 1600	16-8 1650	17-2 1760
2x8	12.0	20-5 710	21-2 770	21-11 830	22-8 880	23-4 930	24-0 980	24-7 1030	25-2 1080	25-8 1130	26-2 1180	26-9 1220	27-2 1270	27-8 1310	28-7 1400
	16.0	18-6 780	19-3 850	19-11 910	20-7 970	21-2 1030	21-9 1080	22-4 1140	22-10 1190	23-4 1240	23-10 1290	24-3 1340	24-8 1390	25-2 1440	25-11 1540
	24.0	16-2 900	16-10 970	17-5 1040	18-0 1110	18-6 1170	19-0 1240	19-6 1300	19-11 1360	20-5 1420	20-10 1480	21-2 1540	21-7 1600	21-11 1650	22-8 1760
2x10	12.0	26-0 710	27-1 770	28-0 830	28-11 880	29-9 930	30-7 980	31-4 1030	32-1 1080	32-9 1130	33-5 1180	34-1 1220	34-8 1270	35-4 1310	36-5 1400
	16.0	23-8 780	24-7 850	25-5 910	26-3 970	27-1 1030	27-9 1080	28-6 1140	29-2 1190	29-9 1240	30-5 1290	31-0 1340	31-6 1390	32-1 1440	33-1 1540
	24.0	20-8 900	21-6 970	22-3 1040	22-11 1110	23-8 1170	24-3 1240	24-10 1300	25-5 1360	26-0 1420	26-6 1480	27-1 1540	27-6 1600	28-0 1650	28-11 1760

NOTES:

(1)The required extreme fiber stress in bending (F_b) in pounds per square inch is shown below each span.

(2)Use single or repetitive member bending stress values (F_b) and modulus of elasticity values (E) from Tables Nos. 25-A-1 and 25-A-2.

(3)For more comprehensive tables covering a broader range of bending stress values (F_b) and modulus of elasticity values (E), other spacing of members and other conditions of loading, see U.B.C. Standard No. 25-21.

(4)The spans in these tables are intended for use in covered structures or where moisture content in use does not exceed 19 percent.

TABLE NO. 25-U-R-1—ALLOWABLE SPANS FOR LOW- OR HIGH-SLOPE RAFTERS 20 LBS. PER SQ. FT. LIVE LOAD (Supporting Drywall Ceiling)

DESIGN CRITERIA: Strength—15 lbs. per sq. ft. dead load plus 20 lbs. per sq. ft. live load determines required fiber stress. Deflection—For 20 lbs. per sq. ft. live load. Limited to span in inches divided by 240. RAFTERS: Spans are measured along the horizontal projection and loads are considered as applied on the horizontal projection.

RAFTER SIZE (IN)	SPACING (IN)	Allowable Extreme Fiber Stress in Bending F_b (psi).														
		500	600	700	800	900	1000	1100	1200	1300	1400	1500	1600	1700	1800	1900
2x6	12.0	8-6 0.26	9-4 0.35	10-0 0.44	10-9 0.54	11-5 0.64	12-0 0.75	12-7 0.86	13-2 0.98	13-8 1.11	14-2 1.24	14-8 1.37	15-2 1.51	15-8 1.66	16-1 1.81	16-7 1.96
	16.0	7-4 0.23	8-1 0.30	8-8 0.38	9-4 0.46	9-10 0.55	10-5 0.65	10-11 0.75	11-5 0.85	11-10 0.97	12-4 1.07	12-9 1.19	13-2 1.31	13-7 1.44	13-11 1.56	14-4 1.70
	24.0	6-0 0.19	6-7 0.25	7-1 0.31	7-7 0.38	8-1 0.45	8-6 0.53	8-11 0.61	9-4 0.70	9-8 0.78	10-0 0.88	10-5 0.97	10-9 1.07	11-1 1.17	11-5 1.28	11-8 1.39
2x8	12.0	11-2 0.26	12-3 0.35	13-3 0.44	14-2 0.54	15-0 0.64	15-10 0.75	16-7 0.86	17-4 0.98	18-0 1.11	18-9 1.24	19-5 1.37	20-0 1.51	20-8 1.66	21-3 1.81	21-10 1.96
	16.0	9-8 0.23	10-7 0.30	11-6 0.38	12-3 0.46	13-0 0.55	13-8 0.65	14-4 0.75	15-0 0.85	15-7 0.96	16-3 1.07	16-9 1.19	17-4 1.31	17-10 1.44	18-5 1.56	18-11 1.70
	24.0	7-11 0.19	8-8 0.25	9-4 0.31	10-0 0.38	10-7 0.45	11-2 0.53	11-9 0.61	12-3 0.70	12-9 0.78	13-3 0.88	13-8 0.97	14-2 1.07	14-7 1.17	15-0 1.28	15-5 1.39
2x10	12.0	14-3 0.26	15-8 0.35	16-11 0.44	18-1 0.54	19-2 0.64	20-2 0.75	21-2 0.86	22-1 0.98	23-0 1.11	23-11 1.24	24-9 1.37	25-6 1.51	26-4 1.66	27-1 1.81	27-10 1.96
	16.0	12-4 0.23	13-6 0.30	14-8 0.38	15-8 0.46	16-7 0.55	17-6 0.65	18-4 0.75	19-2 0.85	19-11 0.96	20-8 1.07	21-5 1.19	22-1 1.31	22-10 1.44	23-5 1.56	24-1 1.70
	24.0	10-1 0.19	11-1 0.25	11-11 0.31	12-9 0.38	13-6 0.45	14-3 0.53	15-0 0.61	15-8 0.70	16-3 0.78	16-11 0.88	17-6 0.97	18-1 1.07	18-7 1.17	19-2 1.28	19-8 1.39
2x12	12.0	17-4 0.26	19-0 0.35	20-6 0.44	21-11 0.54	23-3 0.64	24-7 0.75	25-9 0.86	26-11 0.98	28-0 1.11	29-1 1.24	30-1 1.37	31-1 1.51	32-0 1.66	32-11 1.81	33-10 1.96
	16.0	15-0 0.23	16-6 0.30	17-9 0.38	19-0 0.46	20-2 0.55	21-3 0.65	22-4 0.75	23-3 0.85	24-3 0.97	25-2 1.07	26-0 1.19	26-11 1.31	27-9 1.44	28-6 1.56	29-4 1.70
	24.0	12-3 0.19	13-5 0.25	14-6 0.31	15-6 0.38	16-6 0.45	17-4 0.53	18-2 0.61	19-0 0.70	19-10 0.78	20-6 0.88	21-3 0.97	21-11 1.07	22-8 1.17	23-3 1.28	23-11 1.39

NOTES:

(1)The required modulus of elasticity (E) in 1,000,000 pounds per square inch is shown below each span.

(2)Use single or repetitive member bending stress values (F_b) and modulus of elasticity values (E) from Tables Nos. 25-A-1 and 25-A-2. For duration of load stress increases, see Section 2504 (c) 4.

(3)For more comprehensive tables covering a broader range of bending stress values (F_b) and modulus of elasticity values (E), other spacing of members and other conditions of loading, see U.B.C. Standard No. 25-21.

(4)The spans in these tables are intended for use in covered structures or where moisture content in use does not exceed 19 percent.

TABLE NO. 25-U-R-14—ALLOWABLE SPANS FOR HIGH-SLOPE RAFTERS, SLOPE OVER 3 IN 12
30 LBS. PER SQ. FT. LIVE LOAD (Light Roof Covering)

DESIGN CRITERIA: Strength—7 lbs. per sq. ft. dead load plus 30 lbs. per sq. ft. live load determines required fiber stress. Deflection—For 30 lbs. per sq. ft. live load. Limited to span in inches divided by 180. RAFTERS: Spans are measured along the horizontal projection and loads are considered as applied on the horizontal projection.

RAFTER SIZE (IN)	SPACING (IN)	Allowable Extreme Fiber Stress in Bending F_b (psi).														
		500	600	700	800	900	1000	1100	1200	1300	1400	1500	1600	1700	1800	1900
2x4	12.0	5-3 0.27	5-9 0.36	6-3 0.45	6-8 0.55	7-1 0.66	7-5 0.77	7-9 0.89	8-2 1.02	8-6 1.15	8-9 1.28	9-1 1.42	9-5 1.57	9-8 1.72	10-0 1.87	10-3 2.03
	16.0	4-7 0.24	5-0 0.31	5-5 0.39	5-9 0.48	6-1 0.57	6-5 0.67	6-9 0.77	7-1 0.88	7-4 0.99	7-7 1.11	7-11 1.23	8-2 1.36	8-5 1.49	8-8 1.62	8-10 1.76
	24.0	3-9 0.19	4-1 0.25	4-5 0.32	4-8 0.39	5-0 0.47	5-3 0.55	5-6 0.63	5-9 0.72	6-0 0.81	6-3 0.91	6-5 1.01	6-8 1.11	6-10 1.21	7-1 1.32	7-3 1.43
2x6	12.0	8-3 0.27	9-1 0.36	9-9 0.45	10-5 0.55	11-1 0.66	11-8 0.77	12-3 0.89	12-9 1.02	13-4 1.15	13-10 1.28	14-4 1.42	14-9 1.57	15-3 1.72	15-8 1.87	16-1 2.03
	16.0	7-2 0.24	7-10 0.31	8-5 0.39	9-1 0.48	9-7 0.57	10-1 0.67	10-7 0.77	11-1 0.88	11-6 0.99	12-0 1.11	12-5 1.23	12-9 1.36	13-2 1.49	13-7 1.62	13-11 1.76
	24.0	5-10 0.19	6-5 0.25	6-11 0.32	7-5 0.39	7-10 0.47	8-3 0.55	8-8 0.63	9-1 0.72	9-5 0.81	9-9 0.91	10-1 1.01	10-5 1.11	10-9 1.21	11-1 1.32	11-5 1.43
2x8	12.0	10-11 0.27	11-11 0.36	12-10 0.45	13-9 0.55	14-7 0.66	15-5 0.77	16-2 0.89	16-10 1.02	17-7 1.15	18-2 1.28	18-10 1.42	19-6 1.57	20-1 1.72	20-8 1.87	21-3 2.03
	16.0	9-5 0.24	10-4 0.31	11-2 0.39	11-11 0.48	12-8 0.57	13-4 0.67	14-0 0.77	14-7 0.88	15-2 0.99	15-9 1.11	16-4 1.23	16-10 1.36	17-4 1.49	17-11 1.62	18-4 1.76
	24.0	7-8 0.19	8-5 0.25	9-1 0.32	9-9 0.39	10-4 0.47	10-11 0.55	11-5 0.63	11-11 0.72	12-5 0.81	12-10 0.91	13-4 1.01	13-9 1.11	14-2 1.21	14-7 1.32	15-0 1.43
2x10	12.0	13-11 0.27	15-2 0.36	16-5 0.45	17-7 0.55	18-7 0.66	19-8 0.77	20-7 0.89	21-6 1.02	22-5 1.15	23-3 1.28	24-1 1.42	24-10 1.57	25-7 1.72	26-4 1.87	27-1 2.03
	16.0	12-0 0.26	13-2 0.34	14-3 0.43	15-2 0.53	16-2 0.63	17-0 0.74	17-10 0.85	18-7 0.97	19-5 1.09	20-1 1.22	20-10 1.35	21-6 1.49	22-2 1.63	22-10 1.78	23-5 1.93
	24.0	9-10 0.19	10-9 0.25	11-7 0.32	12-5 0.39	13-2 0.47	13-11 0.55	14-7 0.63	15-2 0.72	15-10 0.81	16-5 0.91	17-0 1.01	17-7 1.11	18-1 1.21	18-7 1.32	19-2 1.43

NOTES: (1)The required modulus of elasticity (E) in 1,000,000 pounds per square inch is shown below each span.

(2)Use single or repetitive member bending stress values (F_b) and modulus of elasticity values (E) from Tables Nos. 25-A-1 and 25-A-2. For duration of load stress increases, see Section 2504 (c) 4.

(3)For more comprehensive tables covering a broader range of bending stress values (F_b) and modulus of elasticity values (E), other spacing of members and other conditions of loading, see U.B.C. Standard No. 25-21.

(4)The spans in these tables are intended for use in covered structures or where moisture content in use does not exceed 19 percent.

TABLE NO. 29-A—FOUNDATIONS FOR STUD BEARING WALLS—MINIMUM REQUIREMENTS[1] [2]

NUMBER OF FLOORS SUPPORTED BY THE FOUNDATION[3]	THICKNESS OF FOUNDATION WALL (Inches)		WIDTH OF FOOTING (Inches)	THICKNESS OF FOOTING (Inches)	DEPTH BELOW UNDISTURBED GROUND SURFACE (Inches)
	CONCRETE	UNIT MASONRY			
1	6	6	12	6	12
2	8	8	15	7	18
3	10	10	18	8	24

[1]Where unusual conditions or frost conditions are found, footings and foundations shall be as required in Section 2907 (a).

[2]The ground under the floor may be excavated to the elevation of the top of the footing.

[3]Foundations may support a roof in addition to the stipulated number of floors. Foundations supporting roofs only shall be as required for supporting one floor.

TABLE NO. 47-I—ALLOWABLE SHEAR FOR WIND OR SEISMIC FORCES IN POUNDS PER FOOT FOR VERTICAL DIAPHRAGMS OF LATH AND PLASTER OR GYPSUM BOARD FRAME WALL ASSEMBLIES[1]

TYPE OF MATERIAL	THICKNESS OF MATERIAL	WALL CONSTRUCTION	NAIL SPACING[2] MAXIMUM (In Inches)	SHEAR VALUE	MINIMUM NAIL SIZE[3]
1. Expanded metal, or woven wire lath and portland cement plaster	7/8"	Unblocked	6	180	No. 11 gauge, 1 1/2" long, 7/16" head No. 16 gauge staple, 7/8" legs
2. Gypsum lath, plain or perforated	3/8" Lath and 1/2" Plaster	Unblocked	5	100	No. 13 gauge, 1 1/8" long, 19/64" head, plasterboard blued nail
3. Gypsum sheathing board	1/2" x 2' x 8'	Unblocked	4	75	No. 11 gauge, 1 3/4" long, 7/16" head, diamond-point, galvanized
	1/2" x 4' 1/2" x 4'	Blocked Unblocked	4 7	175 100	
4. Gypsum wallboard or veneer base	1/2"	Unblocked	7	100	5d cooler or parker nails
			4	125	
		Blocked	7	125	
			4	150	
	5/8"	Blocked	4	175	6d cooler or parker nails
		Blocked Two-ply	Base ply 9 Face ply 7	250	Base ply—6d cooler or parker nails Face ply—8d cooler or parker nails

[1]These vertical diaphragms shall not be used to resist loads imposed by masonry or concrete construction. See Section 4713 (b). Values are for short-time loading due to wind or earthquake and must be reduced 25 percent for normal loading.

[2]Applies to nailing at all studs, top and bottom plates and blocking.

[3]Alternate nails may be used if their dimensions are not less than the specified dimensions.

Index